C0-ATG-407

DEC 1 4 2006

Monographs on
Theoretical and Applied Genetics 4

Edited by
R. Frankel (Coordinating Editor), Bet-Dagan
G. A. E. Gall, Davis · M. Grossman, Urbana
H. F. Linskens, Nijmegen · D. de Zeeuw, Wageningen

Leonid I. Korochkin

Gene Interactions
in Development

Translated and Edited by
Abraham Grossman

With 109 Figures

Springer-Verlag
Berlin Heidelberg New York 1981

QH
453
.K6713

Professor L. I. Korochkin
Institute of Developmental Biology
Vavilov Str., 26
117334 Moscow/USSR

Dr. A. Grossman
Dept. of Biological Science
The Florida State University
Tallahassee, FL 32306/USA

Translation of: Vsaimodeistvije genov v rasvitii,
© Nauka, 1977

ISBN 3-540-10112-8 Springer-Verlag Berlin Heidelberg New York
ISBN 0-387-10112-8 Springer-Verlag New York Heidelberg Berlin

Library of Congress Cataloging in Publication Data. Korochkin, Leonid Ivanovich. Gene
interactions in development. (Monographs on theoretical and applied genetics; 4) Biblio-
graphy: p. Includes index. 1. Developmental genetics. 2. Cell interaction. I. Grossman, Abraham,
1942-. II. Title. III. Series. [DNLM: 1. Phenotype. 2. Ontogeny. 3. Cytogenetics. W1 M0573N
v. 4/QH431 K844v]. QH453.K6713 574.3 80-18254.

This work is subject to copyright. All rights are reserved, whether the whole or part of the
material is concerned, specifically those of translation, reprinting, re-use of illustrations, broad-
casting, reproduction by photocopying machine or similar means, and storage in data banks.

Under § 54 of the German Copyright Law where copies are made for other than private use,
a fee is payable to "Verwertungsgesellschaft Wort", Munich

© by Springer-Verlag Berlin · Heidelberg 1981
Printed in Germany

The use of registered names, trademarks, etc. in this publication does not imply, even in the
absence of a specific statement, that such names are exempt from the relevant protective laws
and regulations and therefore free for general use.

Typesetting, printing, and binding: Brühlsche Universitätsdruckerei, Giessen.
2131/3130-543210

Preface

Since the Russian edition of this book was published in 1975 many new research works have appeared which have made necessary some additions for the English edition, to reflect progress in molecular developmental genetics. Recent important findings in this field have brought about essential corrections to the concept of genetic regulation of the process of cell differentiation. The discovery of the mosaic structure of a gene prompted the re-evaluation of our considerations about the regulation of gene activity in eukaryotes, and the data about transcriptional events during ontogenesis are of great importance as well.

Formerly it was generally accepted that a derepression of genes was responsible for cell differentiation in the process of development. Recently three important conclusions have been derived (Davidson and Britten, 1979) which help to pose the problem in a new way: 1) Only a small part of single copy sequences of DNA is represented in nuclear RNA of a given type of cell or tissue: 10% to 20% in sea urchin embryos, 11% in rat liver, 4% to 6% in *Drosophila* cell culture, etc. Since only about 10% of single copy sequences represent the structural genes (Davidson and Britten, 1973), transcription of almost the whole set of structural genes occurs. 2) The complexity of the transcripts from single copy sequences is the same in nuclei of cells of various tissues and in the majority of tissues in various stages of development, but polysomal mRNA sets are different. 3) The nuclear RNA includes copy families of moderately repeated DNA specific for various tissues and different stages of development. Thus the differentiating cells are characterized by an accumulation in high concentration in their cytoplasm of a rather limited number of types of specific mRNA, along with the presence in nuclei in significantly lower concentrations of many other types of mRNA which represent copies of the major part of the structural genes. If this is the case the control of the specificity of mRNA sets in various cell and tissue types is carried out at the post-transcriptional level with the involvement of transcripts from copies of the moderately repeated sequences of DNA (Davidson and Britten, 1979). Therefore some authors have supposed that during early embryogenesis all structural genes are active, and later some of them are repressed (Caplan and Ordahl, 1978).

Another important aspect of molecular-genetic events in the process of cell maturation was recently revealed as a result of detailed analysis of the differentiation of antibody-producing cells. From this it was concluded that rearrangement of genetic material can occur in the process of cell differentiation, and that the allelic exclusion takes place at the transcriptional level. It appears that in the process of lymphocyte differentiation V and C genes coding for variable and constant chains of immunoglobulin molecules join together. These genes are located on one chromosome but are separated by a long DNA segment. The beginning of antibody synthesis is accompanied by the excision of this segment so that V and C genes form one transcriptional unit. For instance, Ig heavy chain genetic material consists of at least three non-contiguous germ-line DNA segments – V_H gene segment, J_H gene segment, and associated C_μ and C_α gene segments.

In the process of lymphocyte differentiation these segments are joined together by means of two different types of DNA rearrangement: V–T joining and C_H switching. When the J_H segment is associated with the C_μ segment the lymphocyte produces immunoglobulin M (IgM). C_H switching denotes DNA rearrangement which substitutes the C_α segment for the C_μ segment, and initiates synthesis of α-chain and IgA molecules by the now fully differentiated lymphocyte (Davis et al., 1980).

The evidence favoring this assumption was obtained in experiments with embryonic and differentiated cells. The transcripts of V and C genes were hybridized with various DNA fragments isolated from mice embryos. However, if DNA was isolated from differentiated myeloma tissue (i.e., tissue that produces immunoglobulins very intensely), V and C gene transcripts hybridized with the same clearly defined DNA fraction (Honjo and Kataoka, 1978; Davis et al., 1980, Rabbits et al., 1980). Apparently the mechanism of allelic exclusion involves no joining of V and C genes in one of the homologous chromosomes which henceforth cannot produce Ig mRNA (Honjo and Kataoka, 1978).

Finally numerous data, some of which are presented in this book, indicate that the regulation of gene activity is carried out with the involvement of specific regulatory proteins. However, the fine molecular mechanisms of the action of these proteins are not yet clear, and their elucidation is one of the major aims of developmental genetics. This field of biology, now utilizing methods of molecular biology and genetic engineering, advances so rapidly that each year brings about impressive new results both of theoretic significance and of practical application to medicine and agriculture.

Caplan A, Ordahl C (1978) Science 201:120–130
Davidson E, Britten R (1973) Quart Rev Biol 48:565–613
Davidson E, Britten R (1979) Science 204:1052–1059
Honjo T, Kataoka T (1978) Proc Nat Acad Sci USA 75:2140–2144
Rabbits T, Forster A, Dunnick W, Bentley D (1980) Nature 283:351–356
Davis M, Calame K, Earls P, Livant D, Joho R, Weissmann I, Hood L (1980) Nature 283:733–742

October, 1980   LEONID KOROCHKIN

Foreword for Russian Edition

At the present time, the thesis that the development of an organism is the realization of its hereditary information has begun to be evidently trivial. This realization consists in the fact that genetic information described in DNA is initially recorded into the structure of protein molecules and later, at a higher level, is transformed into the properties and behavior of the differentiating cells. At a yet higher level, the hereditary information expresses itself in tissue and organ formation and, as a result, leads to the appearance of the traits of the parental organism.

It seems that in this accelerated hierarchy of various levels of interaction gene interaction is only a partial aspect of the problem and is more a concern of genetics than of developmental biology. In reality, Professor Korochkin's book demonstrates that gene interaction is an essential aspect of the developmental process, or, in other words, the realization of the heredity information occurs only through gene interaction.

It is difficult to imagine constant, direct interaction of genes. It is possible that an effect of the heterochromatin region on closely located genes (position effect) or the regulation of promotor and operator sites during transcription could be attributed to this kind of interaction. In fact, direct gene interaction occurs through compounds, synthesized under the direct control of certain genes and acting on another gene's functions. As yet we know only little about this kind of interaction. Still in question is the nature of the compounds through which direct intergenic interactions could be realized. In the light of certain assumptions these compounds could be a special type of nuclear RNA, if viewed under a different set of assumptions these regulators would also have a protein nature. Thus, the nature of direct gene interaction is still a matter of doubt and a subject of research, although it is an important part of the whole problem.

The merit of this book lies in that the concept of gene interaction is considered in its widest aspect. It includes not only the direct influence of one gene on another, but also the interaction of genetic information at all succeeding levels of realization. For an understanding of the mechanisms involved in erythropoiesis or simply in the synthesis of hemoglobin it is necessary to know whether interaction occurs between the genes controlling the synthesis of the α and β chains, between the mRNA templates, or between the polypeptide chains themselves. It is also necessary to know the role played by heme and the genes controlling its synthesis. It is necessary to find the factors which initiate the globin genes and to describe the pathways of genetic control of these factors. However, in one of Wainwright's works it was shown that the beginning of hemoglobin translation in early chick embryos is not dependent upon the activation of hemoglobin genes, since globin mRNA is transcribed early. The synthesis of hemoglobin began after the formation of hemin from the precursor delta-amino levulinic acid, which was synthesized with the participation of a specific enzyme. The translation of this protein on subsequent templates proceeds only after the transcription of an alanine tRNA. This is the crucial point of initial synthesis. This example demonstrates how in the regulation of the synthesis of one protein, i.e., hemoglobin, at least three kinds of gene are working together, i.e., interacting. These three genes are: structural globin genes; genes for the synthesis of delta-amino levulinic acid; and genes for the synthesis of alanine tRNA. The complete picture of this regulation is, obviously, more complex and includes the interaction of other genes.

Interactions between genes are realized also at higher levels of organization – on the level of the differentiation of tissues and organs. Our information about the genetic control of these processes is still incomplete. Possible theories in this field are very schematic and, in many respects, speculative.

The formation of the nervous system during embryogenesis is under the control of specific genes since mutations are known which effect particular neural differentiations. The genes which determine the first steps of neurogenesis should begin their function during the induction of the neural plate as a result of contact with the chordomesoderm. It is reasonable to expect that the inducing properties of chordomesodermal cells are also predetermined by specific gene activities and take place during the late blastula – early gastrula stages. It is possible to conclude that the formation of the nervous system in the spinal part of the embryo is due to an interaction of genes by means of embryonic induction. The genes are active in both nerve and chordomesodermal cells. The directed differentiation of some cells into chordomesoderm and others into ectoderm is accomplished through properties of the egg cytoplasm which are also predetermined by the functions of specific gene systems during oogenesis.

The puff activation of polytene chromosomes in the salivary glands of *Diptera* larvas occurs under the control of the steroid hormone ecdysone. The genes which control the formation of steroid metabolic enzymes are supposedly active in the gland cells where this hormone is synthesized. We can consider the regulation of salivary gland functions as a result of the indirect interactions of specific genes. All the major events belonging to developmental biology could be expressed in terms of interactions between genes; to paraphrase molecular, cellular, tissue, and organ development in terms of the genetic language is reasonable and constructive. Only in this manner is it possible to gain a complete understanding of the pathways of realization of the genetic information in the development of the organism.

To turn from these trivial and obvious propositions to the concrete, well-known mechanisms which are involved in the genetic basis of development is a complicated problem which requires an erudite approach. This book is dedicated to the description of these mechanisms. The author discusses in detail and at a current level the known facts and arranges them in the order which allows for a systematic conceptualization of the problems discussed. This monograph is also unique in that such a summary has not yet been published, although there is a great deal of original data on interactions of genes during development. This is why the appearance of this book in the *Developmental Biology* series is not only useful but also necessary.

At the same time, this book is not merely a simple compilation, devoid of the author's personality. Professor Korochkin's works are well known. In his research he fruitfully combines good genetic analysis with modern biochemical and cytochemical methods of molecular biology. Professor Korochkin's works, with the aid of ultra microchemical methods, which allowed to formulate the original concept of multilevel gene regulation of individual development, are of great interest.

All of these make Professor Korochkin's book both interesting and informative and, in some places, worthy of discussion.

Prof. A. A. NEYFACH
Institute of Developmental
Biology, Moscow, U.S.S.R.

Contents

Part 2 Gene Interaction at the Tissue Level

Introduction

Two general principles in the relationship between genes and characters were already formulated in the 1930's. These principles are: (1) Every gene influences all of the traits of an organism, although its effect on some of them could be small to the point of extinction. (2) Any character of an organism depends on all of the genes in the genome in general, although this dependence of some of them is not noticeable (Astaurov, 1968). This means that the development of each trait is due to a number of successive gene interactions, acting in specific conditions (Rokitsky, 1929).

The dominant–recessive interactions of allelic genes, gene complementation, epistasis, and the modification an interactions of nonallelic genes have been the object of investigations in genetics for a long time.

Dominant–recessive interactions are not always strictly fixed. In certain cases the dominance of one trait over another could depend upon the environment. For instance, the dominant mutation *Abnormal abdomen* in *Drosophila melanogaster* could appear as a recessive as it did when flies were cultivated with an excess of fresh food and under heightened humidity (Morgan, 1924). Moreover, the dominant–recessive nature of two genes could be changed as a result of the position effect of one of them. In 1926, Chetverikov proposed the role of the genotypic environment in the development of traits. He assumed a dependence of gene expression upon their placement in the chromosome and, upon replacement of a gene, a resultant change in the gene expression. Later, the position effect of genes was demonstrated (Sturtevant, 1925; Dubinin and Sidorov, 1934). Obviously there is a biochemical interaction between genes closely associated in a chromosome or between their primary products.

As a result of complementary interactions between dominant genes, it is possible for a new character to develop. This new character differs from the traits determined by each gene separately. For instance, in the formation of cyanide in white clover a complementary interaction is necessary between two genes, which is realized on two levels; the formation of the linamarin substrate and the synthesis of the linamarase enzyme which converts linamarin into cyanide. Each of the processes is controlled monogenically by the nonallelic dominant genes Li and Ac (Atwood and Sullivan, 1943).

The term epistasis designates the suppression of one gene by another, nonallelic, gene (suppressor or inhibitor) which may be either dominant or recessive. Genes may also display a modifying effect in increasing or decreasing the expression of a trait which is under the control of a structural gene.

Finally, in the case of so–called quantitative characters (rate of growth, weight of an animal, and so on) it is convenient to refer to interactions of polygenes. It

is supposed that some traits are dependent upon a number of genes with synony-
mous effects.

   Until recently, the role of gene interactions in the phenotype expression of traits
was analyzed basically at the morphological level. Only in some cases, as in the example of complementary interaction cited above, were concrete biochemical mechanisms discovered (see Gvozdev, 1968; Ratner, 1972, 1975).

   The current state of developmental biology urgently needs exact descriptions
of molecular events which take place in the various types of gene interaction. It is
necessary to know how each of these events is involved in gene interactions which
produce their effects at different stages of ontogenesis. In other words, "we have
to understand how a gene works", how the genotype controls the processes of biosynthesis, cell physiology, metabolism, differentiation, morphogenesis, and the
whole complex interacted process of ontogenesis. On the other hand, how does
"feedback" work? How do the biosynthetic products, the peculiarities of cytoplasmic differentiation, and the whole developmental system conduct the instruments
of the gene orchestra? Or, it may be better stated, how, on the basis of the dynamic
interaction between the nucleus and the cytoplasm, does the cybernetic system of
regulatory interrelations of ontogenesis realize the hereditary program of development? (Astaurov, 1968, 1974).

   The immediate characterization of gene activity is possible only through determination of direct products – the molecules of ribonucleic acid (RNA). But, on the
basis of methods currently available, we cannot, with rare exceptions, control the
synthesis of individual RNA molecules. In connection with this, indirect methods
for exposing separate gene activities are widely used. These methods are based on
analysis of the final products controlled by these genes, especially proteins and
isozymes (Ursprung et al., 1968). In this case, as in any classical genetic work, trait
polymorphism is analyzed with the aid of biochemical methods (electrophoresis,
immunoelectrophoresis, etc.) and established (for instance, protein variability for
electrophoretic mobility). After this, the following investigations are performed:

1. Genetic determination of the trait – the number of genes controlling its formation.
2. Specificity of various gene actions and the character of interactions among
   them. In this case it is necessary to take into account the pleiotropic effect of
   genes, and also the presence of multiple alleles for some of these genes
3. Gene localization in linkage groups.
4. Allelism of homologous genes of different species in interspecies crosses and
   combinative analysis (Smirnov and Vatti, 1971).

   On the basis of observed data, subsequent experiments could be performed for
the investigation of additional questions, such as: (a) the time during which genes
controlling corresponding biochemical traits began their action; (b) their connection with the particular stages and processes of cell and tissue differentiation; (c)
the factors which inhibit and activate these genes, regulating their activity. At the
same time, the influence of other genes on the expression of a controlled trait in
the phenotype plays a very important role. Figure 1 demonstrates the attempt to
summarize the most common principles of gene interactions during enzyme synthesis. It is obvious that a mutation, disturbing the function of gene-controlled syn-

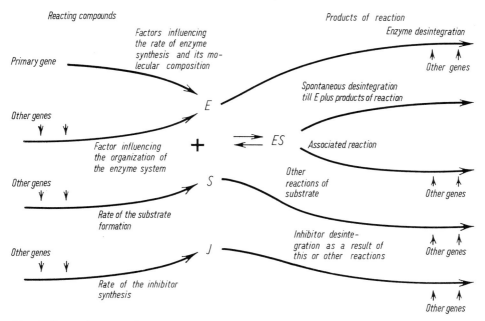

Fig. 1. Some factors of genetic origin influencing the rate of reactions (Wagner and Mitchell, 1958)

thesis of enzyme E, would prevent enzyme activity in the organism and corresponding reactions would not occur, if the organism did not possess another catalyzing agent of a different origin. On the other hand, numerous factors besides enzyme inactivation influence the rate of reaction, from decreasing it to zero to increasing it to a maximum permissible under conditions of the reaction.

Actually, the path from gene to trait is complicated enough, and the process of transcription of a particular gene does not mean that the trait controlled by this gene will be expressed in the phenotype. The events leading to its expression could be realized on the cellular as well as the tissue level, because, in reality, individual development is the processes occurring on these two levels (Olenov, 1972).

The cellular level implies the whole complex of processes which are determined by differential gene activity and consist of a system of organospecific syntheses. The latter determines the cell differentiation.

At the same time, it should not be considered, as it sometimes is, that the cellular level of regulation is based mainly upon transcriptional events. In reality, the genetic regulation of biochemical traits on the cellular level could be realized not only through differential transcription, but also through post-transcriptional, translational, and post–translational processes. The relative significance of genetic systems in the control of these processes could differ in each particular case for the total complex of differentiation.

The tissue and intercellular level of trait expression includes complicated intercellular interactions. These relationships are linked with the heterogeneity of differentiated systems and lead to the realization of the corresponding character through a complex of inductive interactions, intercellular competition, selective proliferation, or the elimination of corresponding cell clones.

The complicated mosaic of interactions between genes, realized at the cellular and tissue levels, finally leads to a specific differential distribution of gene activities in various parts of the developing embryo at different stages of its formation. This is the essence of the genetic control on ontogenesis.

The chapters of this book are dedicated to the analysis of the system of interactions determined by differential gene activity on different levels

Author gives deep acknowledgment to: A.A. Neyfach, T.A. Detlaf, I.I. Kiknadze, V.V. Chvostova, V.A. Gvozdev, M.D. Golubovsky, V.A. Ratner, S.N. Rodin, D.P. Furman, V.I. Mitashov, O.G. Stroeva, E.V. Savvateeva, M.B. Evgeniev, L.S. Korochkina, V.N. Pospelov, for their criticism and advice during the discussion of this book. Author also appreciates the declunical help of: N.M. Matveeva, M.O. Antonova, V.A. Prasolov, O.G. Prokofieva.

Part 1
Gene Interaction at the Cellular Level

1.1 Differential Activity of Genes as a Basis for Cell Differentiation

1.1.1 The Leading Role of the Nucleus in Ontogenic Regulation

The realization of hereditary information during ontogenesis is a multi-stage process and includes various levels of regulation – cell, tissue, and organism. A large number of genes act at each step of the development of an organism in the course of morphological, biochemical, and physiological differentiation of cell populations. Each of these genes controls one of many biochemical reactions and, through this, takes part in the realization of the formative processes (Wagner and Mitchell, 1958).

The localization of genes in the chromosomes of the nuclei determines the leading role of the nucleus in ontogenic regulation. This phenomenon was demonstrated in the classic experiments of Hammerling and Astaurov.

Hammerling found in the 1930's (Hammerling, 1953) that a hat or cap shape in the unicellular algae *Acetabularia* is dependent only on the nucleus. This reproductive organ developed on the top of the stem. If a rhizoid with nucleus is removed from the species of algae *A. mediterranea* and replaced with a rhizoid containing a nucleus from another species, *A. wettsteini*, the new cap will develop in a form peculiar to *A. wettsteini*.

Astaurov irradiated unfertilized silk worm eggs with high doses of X–rays; these eggs were subsequently fertilized by nonirradiated, normal spermatozoa. Several sperm commonly penetrated each silk worm egg. After this, the eggs were heat–treated for 135 min at 40 °C. As a result of heat treatment the egg nucleus remained at the periphery of the egg and did not take part in the processes of development. Those eggs which lacked a maternal nuclear apparatus developed through diploid androgenesis, and formed the division nucleus by the fusion of two sperm nuclei. The progeny were represented only by males and were easily recognizable with the aid of genetic markers. If, using this method, the egg cytoplasm of *Bombyx mandarina* is combined with the nucleus of *Bombyx mori*, two species which differ in many morphological, physiological, and behavioral traits, the developed progeny inherited exclusively paternal characteristics – the characteristics corresponding to the nuclear information (Astaurov, 1968). Fusion of *B. mandarina* nuclei with the cytoplasm of *B. mori* produced similar results (Fig. 2).

Galien (1971) investigated in detail the role of the nucleus in the development of amphibians. In his work two distinct species of the genus Pleurodeles, *P. waltlii* and *P. poireti*, were used. Through crosses the following hybrids were obtained:

n *P. waltlii*/n *P. poireti*   n *P. waltlii*/n *P. poireti*
cytoplasm *P. waltlii*   and   cytoplasm *P. poireti*

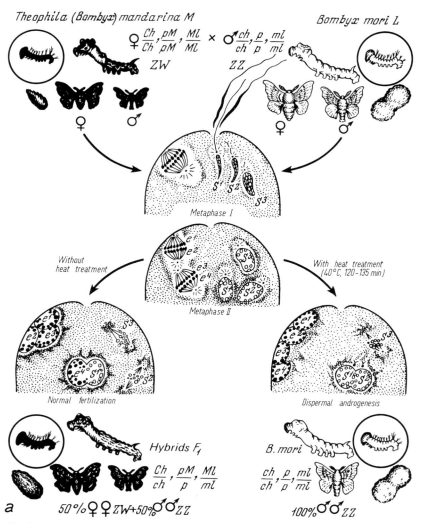

Fig. 2 a, b. Interspecific androgenesis. **a** The androgenic offspring (*Bombyx mandarina* cytoplasm + *B. mori* nucleus). **b** The reciprocal combination (*B. mori* cytoplasm + *B. mandarina* nucleus). Androgenic hybrids are phenotypically undistinguishable from regular hybrids. The method of the selective elimination of regular hybrids (the egg radiation treatment) was used for the obtaining of androgenic individuals. (After Astaurov, 1968)

The F_1 progeny were normally viable and fertile. From these data it can be determined that the presence of a haploid number of chromosomes from the same species as the cytoplasm determines normal embryonic development.

Then, through interspecific transplantation of nuclei, nuclei–cytoplasmic hybrids were established with the following constituents:

2n *P. waltlii*
cytoplasm *P. poireti*   and   2n *P. poireti*
cytoplasm *P. waltlii*

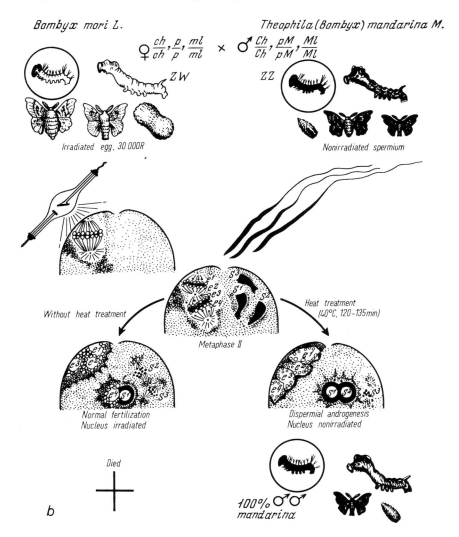

Bombyx mori L.   *Theophila (Bombyx) mandarina M.*

$\female \dfrac{ch}{ch}, \dfrac{p}{p}, \dfrac{ml}{ml}$ × $\male \dfrac{Ch}{Ch}, \dfrac{pM}{pM}, \dfrac{Ml}{Ml}$

ZW   ZZ

Irradiated egg, 30 000R   Nonirradiated spermium

Without heat treatment   Heat treatment
                                          (40°C, 120-135min)

Metaphase II

Normal fertilization   Dispermial androgenesis
Nucleus irradiated   Nucleus nonirradiated

Died   100% ♂♂
                        mandarina

b

Beginning in the early gastrula stage these hybrids demonstrated serious abnormalities in their development. However, a small number, about 2%, of them achieved the adult stage. All of these were morphologically, serologically, and karyotypically similar to the species from which the transplanted nucleus was obtained.

The nuclei controlled the various biochemical and physiological processes occurring in the cells which developed in different directions during morphogenesis, but, nevertheless, the nuclei did not undergo any irreversible changes, and retained the potential to control the normal development of the whole organism. This is clearly indicated by Gurdon's experiments on nuclei transplantation.

If a frog nucleus from an already differentiated cell of the intestine is transplanted into a frog egg whose nucleus was been destroyed by ultraviolet radiation,

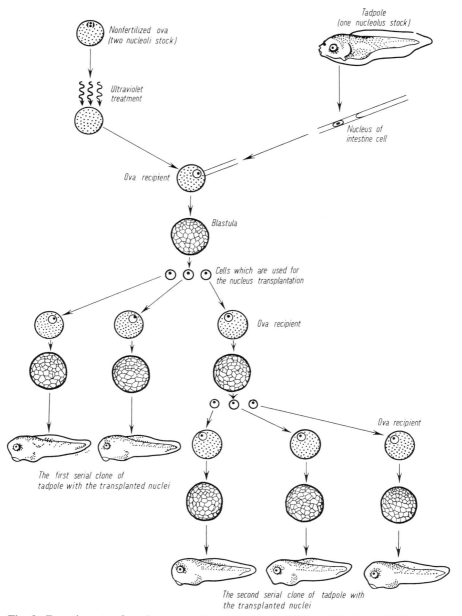

Fig. 3. Experiments of nucleus transplantation in amphibians (Gurdon, 1971)

about 1% of these eggs develop into adult frogs (Fig. 3). Lymphocyte nuclei of adult frogs, which were transplanted into *Xenopus* egg without nucleus, activated genes which control ontogenesis. In this case, the development continues till the stage of tadpole which stays alive for 12 days (Du Pasquier and Wabl, 1977). The nuclei of mosaic eggs, for instance *Drosophila* eggs, also retained the potential for normal control of developing embryos and larvae. This was demonstrated by Illmensee in experiments with the transplantation of nuclei from wild–type embryos into eggs of the y w sn₃ lz so mutant (Illmensee, 1972). Obviously, the nuclei of different specialized tissue systems are equipotential as to the information that they contain.

The basic problem of developmental genetics (or phenogenetics[1]) is the study of gene action during ontogenesis and the routes taken in the path from gene to character (Timofeev-Ressovsky and Ivanov, 1966). Particularly, how, with an identical gene constitution in all somatic cells, does cell diversity and the polyfunctional specialization of tissues and organs develop?

A general outline of the answer to this question was formulated by Morgan (1937). He wrote: "It is known that the protoplasm differs slightly in the various parts of the egg and a clear edge for these differences began to be more noticeable through the occurrence of displacement of compounds. It is possible to assume that an original difference in protoplasmic sections influenced gene activity. In turn, the genes will influence the protoplasm, whereupon a number of mutual reactions will begin. In this way, it is possible to imagine the increasing complexity and differentiation of different parts of the embryo."

Thus, cellular differentiation is considered as a consequence of differential gene activity, which, in turn, is determined by complicated interactions between the cytoplasm and nucleus specific for the various parts of the developing embryo. In this case the differentiated cell, for instance in the muscle, is characterized by a specific number of functioning genes, as is also the case for nerve cells. These differences determine the peculiarities in the spectrum of organospecific proteins provoked as a result of morphophysiological specification. At the present time convincing data supporting this point of view have been obtained.

1.1.2 The Expression of Differential Gene Activity in Transcriptional Level

With the aid of DNA–RNA hybridization methods, which are based on the ability of RNA molecules to join, under certain experimental conditions, with their complementary DNA sections, it was shown that mRNA's differ in various organs (Fig. 4) (McCarthy and Hoyer, 1964; Church and Brown, 1972). Later, the peculiarities of RNA synthesis in the ontogenesis of many organisms were investigated. Particularly in amphibian embryogenesis shortly before cell movements accompanying gastrulation, activation of nuclei was first obtained in endoderm and mesoderm. This pregastrular activation occurs in *Xenopus laevis* between stages 8

---

[1] The term phenogenetic was suggested in the 1920's by the German zoologist Valentine Hekker. He tried to determine the ontogenic phases in which the differences appeared between mutant and normal varieties. He named these phases phenocritical. At the present time, the term phenogenetic is rarely used, and usually the term developmental genetics is used instead

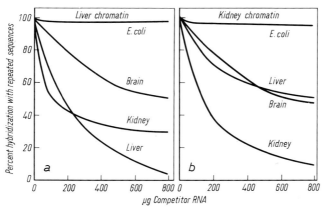

Fig. 4 a, b. Tissue specificity of RNA synthesized from chromatin originating from liver or kidney using mouse RNA-polymerase. **a** RNA was synthesized in vitro from a mouse liver chromatin template using mouse RNA-polymerase. The reaction was run in 5 ml buffer at 37 °C for 10 min and the RNA was extracted with hot phenol. RNA isolated from various tissues was used as competitor for the hybridization of the in vitro synthesized RNA to DNA. Hybridization with 12 µg of DNA in 0.2 ml of 2X SSC at 67 °C for 18 h; 384 cpm was hybridized in the absence of competitor. **b** RNA was synthesized in vitro from a mouse kidney chromatin template using mouse RNA-polymerase. The reaction was run in 5 ml buffer at 37 °C for 10 min and the RNA was extracted with hot phenol. RNA isolated from various tissues was used as competitor for the hybridization of the in vitro synthesized RNA to DNA. Hybridization was with 12 µg of DNA in 0.2 ml of 2X SSC at 67 °C for 18 h; 336 cpm was hybridized in the absence of competitor. (After Church and Brown, 1972)

and 9 (Davidson et al., 1965; Davidson and Mirsky, 1965). Autoradiography showed that within one hour the synthesis of RNA increased at least 30–40 times. Until this phase, the RNA synthesis activity per nucleus remains at a relatively constant, low level, and its increase per embryo is a result of the multiplication of cell nuclei. For a short time a large number of cells in the endoderm and mesoderm are involved in the process of activation (Bachvarova et al., 1966).

Synthesis of RNA in embryos of *Triturus alpestris* was active in the pregastrular and gastrular stages and followed by analysis of RNA sedimentation curves. During these periods, tRNA (4S) was synthesized as well as high molecular weight (32–38S) RNA. Both components of ribosomal RNA (18S and 28S) were synthesized during the gastrula stage in *Triturus alpestris* and *Xenopus laevis* (Tiedemann et al., 1965; Mariano, Schram-Doumont, 1965). Obviously, in the dorsal part of the amphibian embryo, including the nervous system, chord, and somites, the synthesis of RNA is active earlier than in other zones.

Gurdon (1969) analyzed nuclear activity during oogenesis and embryogenesis in *Xenopus laevis*; a summary of this data is given schematically in Fig. 5.

In mammals, for instance mice, unfertilized eggs and zygotes do not have RNA synthesis (Monesi and Salfi, 1967; Woodland and Graham, 1969). However, at the two blastomere stages, labeled nucleoli were found (Mintz, 1964a): Apparently the synthesis of tRNA, ribosomal RNA, and some of the fractions of mRNA become active during this period (Woodland and Graham, 1969; Church, 1970; Engel et

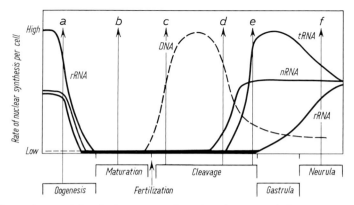

Fig. 5. Gross changes in nuclear activity during oogenesis and embryogenesis in *Xenopus laevis.* The evidence for these changes has been summarized by Brown (1968) and by Gurdon and Woodland (1969). Lines *a–f* represent stages at which the cytoplasmic control of nuclear activity was demonstrated by nuclear transfers (see text). rRNA, tRNA, nRNA: ribosomal, transfer, and nuclear RNA. (After Gurdon, 1969)

al., 1977). Recently, Clegg and Piko (1977) demonstrated that the rate of the RNA synthesis is twice as high in the morula–early blastocyst stage as in the 2–4 cells stage of mouse embryo development. However, it was characteristic that during the preimplantation period the concentration of RNA-polymerase per cell decreased. In connection with this, the suggestion was made that this polymerase does not play a direct regulatory role in RNA synthesis in early mice embryos (Siracusa, 1972).

The early synthesis of RNA in mammalian embryos was corroborated by analysis of lethal mice mutant t^{12}. The homozygous t^{12} mutation terminates development before the blastocyst stage (30 cells). The amount of RNA in homozygous mutant (t^{12}/t^{12}) blastomeres is lower than normal levels. This suggests that the genome of such mutants cannot provide a normal level of RNA synthesis during cleavage (Mintz, 1964b). If implantation of mouse embryos is artificially delayed, the synthesis of new RNA is necessary for its metabolic "activation" (Weitlauf, 1974). However, as a rule, the synthesis of ribosomal RNA takes precedence over mRNA synthesis before implantation. Only during the postimplantation period, accompanied by gastrulation and differentiation, is the synthesis of mRNA substantially increased (Ellem and Gwatkin, 1968; Gelderman et al., 1969). The successive increase in mRNA synthesis during ontogenesis was demonstrated in the sea urchin (Whiteley et al., 1966). The temporal and spatial heterogeneity of mRNA was discovered in amphibian embryogenesis. The temporal regulation of mRNA synthesis could be summarized as in Fig. 6.

1. Some genes are active in early embryogenesis, gastrula, and neurula stages, but their products (RNA) are absent in later periods of development. Consequently, these genes are active only during early phases of development. The RNA which was synthesized on these genes is designated as a separate rectangle on the left in Fig. 6. The existence of these genes is based on experiments with competitive hy-

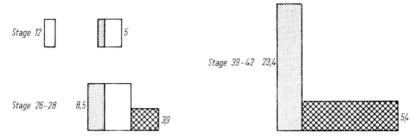

Fig. 6. Changes in gene activity occurring during embryonic development. Complementary RNA transcribed on the same DNA sites at various stages is represented by blocks lying below each other. The *hatched surfaces* correspond to stable RNA, i.e., RNA that is not labeled after 11 h of exposure to $^{14}CO_2$, but is labeled by ^{32}P, when the latter is present from the beginning of development. The *stippled surfaces* correspond to the mRNA which can be labeled by an 11-h-pulse at stage 42. The mRNA transcribed on the same DNA sites in earlier stages is marked in the same way. In early embryos, rapidly labeled RNA is represented by the sum of the hatched and open surfaces. The length of the block is proportional to the percentage of DNA which can be saturated by the RNA of each class. It is thus a measure of the number of DNA sites that are transcribed at one given stage. Messenger RNA present in gastrulae and absent in later embryos is represented by the separate block placed at the top left of the diagram. The area of the blocks is proportional to the amount of hybridizable RNA present at each stage. The height of the blocks is given in arbitrary units, obtained by dividing the amount (in mμg) of stable and unstable hybridizable RNA present in one embryo by the percentage of DNA to which each class of mRNA in complementary. Thus, the height of the blocks is a measure of the number of copies in which each class of mRNA is present. (After Denis, 1968)

bridization of DNA with labeled RNA extracted during the gastrula stage, and nonlabeled RNA from other stages of development.

2. Some genes are active during all postgastrula stages and possibly also in the adult organism (this RNA is designated by a hatched rectangle in Fig. 6). This RNA is quickly changed and is present in a higher number of copies in later embryos which could be caused by the increasing number of cells in the growing embryo. In amphibians this numerical increase is from 40,000 at the gastrula stage to 600,000 at the stage of differentiating tail-bud.

3. Some genes synthesize RNA only during a certain period after nerve tube formation and then they cease functioning. RNA synthesized by these genes is present in stable form in later stages and, apparently, in a small number of copies (Denis, 1968).

Greene and Flickinger (1970) distinguished the "early RNA" and the "later RNA". The former is represented by many copies and was possibly transcribed from polygenes. This RNA is synthesized mainly during gastrulation and neurulation periods. These authors suggested that the synthesis of this RNA has a connection with the determination process. The later RNA is apparently represented by a small number of stable copies.

The stable, long-lived molecules of mRNA appeared comparatively early (between the neurula and tail-bud stages) and slowly accumulated during development. Their appearance was accompanied by the process of cell differentiation.

This type of RNA is present in various kinds of differentiated cells: reticulocytes (Marx and Kovach, 1966); muscle cells (Fieldman et al., 1964); crystalline lens (Reeder and Bell, 1965); pancreas (Rutter et al., 1968); nerve tissue (Korochkin and Olenev, 1966; Korochkin and Korochkina, 1971); and so on. The long-lived mRNA was described in *Xenopus laevis* ovaries (Cabada et al., 1977; Ford et al., 1977). This type of RNA is characteristic of plant cells also, e.g., rye and pea embryos, and it demonstrated the distribution specificity among subcellular fractions (Payne et al., 1977). The early mouse embryos contained the deponated maternal mRNA's which are involved in protein synthesis during the early developmental stages (Young, 1977). A large amount of recent work has been done with the isolation of these stable mRNA templates and their injection into amphibian oocytes (generally *Xenopus laevis*) with a method devised by Gurdon (Gurdon, 1973, 1974a, b). RNA, when injected into oocytes, synthesized a corresponding specific product. As a result of the injection of individual mRNA molecules or of the total RNA, which contained RNA templates, the synthesis of globin (see review Gurdon, 1973), α-lactalbumin of lactating milk glands (Campbell et al., 1973), the heavy chain of immunoglobulin (Stevens and Williamson, 1973), tyroglobulin (DeNayar et al., 1974), the promelitin of honey bees (Kindas-Mügge et al., 1974), and the protein crystallin (Berns et al., 1972) was demonstrated.

In our laboratory the synthesis of rat albumin was discovered in oocytes of the frog *Rana ridibunda* after injection of polyribosomes isolated from rat liver (Borovkov et al., 1975).

Interesting data were recently gathered by Gurdon and his co-workers (Gurdon et al., 1974; Woodland et al., 1974). The authors injected mouse and rabbit globin mRNA into fertilized eggs of *Xenopus laevis* and found the synthesis of mouse and rabbit globin in the developed *Xenopus* embryo. It was found that mouse and rabbit globin was synthesized at a relatively constant rate, at least during 8 days of development, covering 20 cycles of cell division. Obviously, the globin mRNA remained stable during cell divisions. It is of interest that the injected mRNA was distributed more or less evenly in the developed embryos. The analysis of various zones of the embryo for intensity of mouse globin synthesis demonstrated that mRNA is similar in endoderm and axial regions, which includes functioning myotomes, chord, and nerve tube. These experiments demonstrated the high level of globin mRNA stability and, possibly, its ability to function in an experimentally created environment.

Some authors (e.g., Spirin et al., 1965) have supposed that the early-synthesized, stable mRNA apparently remained in the cytoplasm of differentiated cells in special structures. These structures, called informosomes, are a specific type of ribonucleoproteid particle and represent a complex of mRNA and protein. The authors suggest that before gastrulation there is a period when this RNA assembled with ribosomes. In this event the partial or total degradation of the old maternal mRNA is possible. Periodicity in ribosomal reprograming could guarantee replacement in the synthesis of specific proteins which then control the sequence of stages during ontogenesis.

According to Georgiev, the newly synthesized mRNA united in the nucleus with informomers, 30S particles made up of three major and several minor protein compounds. It is suggested that informophers are formed in the separation of nas-

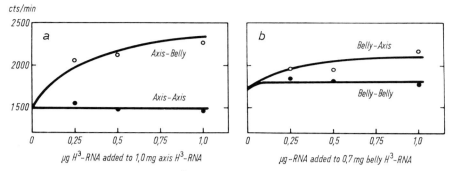

Fig. 7. a Further annealing of H³-RNA of tail-bud bellies to DNA previously saturated with H³-RNA of tail-bud axial regions *(Axis-Belly)*, and the lack of further annealing of H³-RNA of axial regions *(Axis-Axis)*. **b** Reciprocal experiment shows further annealing of H³-RNA of axial regions to DNA saturated with H³-RNA of bellies *(Belly-Axis)* and the absence of further annealing with H³-RNA of bellies *(Belly-Belly)*. (After Flickinger et al., 1966)

cent DNA-like RNA (dRNA) from chromosomal templates. The processing of RNA, i.e., the transformation of the primary synthesized, giant nuclear RNA into the relatively low molecular weight dRNA, also occurs in the informopheric complexes (Georgiev, 1973).

In this way the different phases of embryogenesis are characterized by the synthesis of different, stage-specific mRNA molecules, which reflect the differential gene activity at the time.

At the same time, Flickinger and his co-workers demonstrated that the different zones of amphibian embryos possessed different RNA constitutions. It was found that, during the gastrula stage, the earlier differentiated cells of the ectoderm and mesoderm produced a larger diversity of mRNA molecules, specifically the DNA-like RNA, than did the cells of the endoderm in which differentiation proceeded slowly. However, more DNA sequences are transcribed in cells of the gastral part of the embryo than in organs of the dorsal axial zone during the tailbud stage (Fig. 7; Flickinger et al., 1966; Flickinger, 1971). Therefore, the various parts and cell types of both embryo and adult organisms are characterized by a specific mRNA constitution. This phenomenon indicates spatial differential gene activities.

In all of the cases cited above, the DNA-RNA hybridization method made possible the detection of RNA transcribed from redundant DNA. From 2%–5% of oocyte RNA was transcribed from redundant DNA and remained in the developed egg at least until the late blastula stage (Hough and Davidson, 1972; Davidson E., 1972). However, similar tendencies apparently occurred in the transcription of unique DNA sequences (Flickinger, 1971; Gelderman et al., 1971).

Rachkus et al. (1969a, b) demonstrated that in early stages of loach embryogenesis mainly the repetitive sequences of DNA were transcribed, but in later stages loci with unique sequences began to function. The heterogenous nuclear RNA of the sea urchin could be separated into three fractions: (1) α-heterogenous RNA contains the internal oligo (A) segments with length of 12–125 mononucleotides;

(2) β-heterogenous RNA contains the internal oligo (A) and 3' terminal oligo (A) segments with length of 175 nucleotides; and (3) γ-heterogenous nuclear RNA which does not contain either oligo (A) or poly (A) sequences. The γRNA dominates during the early stages of development. The maximum amount of such RNA (75% of the total heterogenous nuclear RNA) was observed in the early blastula stage and associated with the maximum of histone synthesis.

The amount of βRNA began to increase in the early blastula stage which correlated with the accumulation of the cytoplasmic mRNA which contained poly (A). The maximum amount of αRNA (25% of the total heterogenous nuclear RNA) was obtained in the middle blastula stage (Dubroff and Nemer, 1975/76). The ratio of mRNA's with and without poly (A) sequences correlated with the increasing of size of free polysomes in the sea urchin embryo (Nemer and Surrey, 1975/76; Nemer et al., 1975). It is possible to translate different mRNA fractions extracted from sea urchin and amphibian embryos in the cell-free system (Ruderman and Pardu, 1977).

The stimulation of the synthesis of poly (A) containing mRNA was obtained when the *Artemia salina* embryos development was stimulated. It was demonstrated that the proportion of mRNA containing poly (A) is increased in the nuclear RNA and in the soluble fraction of the cytoplasmic RNA (Christy and Jayaraman, 1975). At the same time the increasing of the RNA-polymerase activity was registered (Renart and Sebastian, 1976; D'Alession and Bagshaw, 1977).

In embryos of *Drosophila melanogaster* the maximal percentage of unique DNA sequences hybridized with RNA is equal to 15%; with larval RNA, 14%; in pupae, 17%; and in adult flies, 10%. Because only one of the two complementary DNA chains is transcribed, it is possible to say that more than 30% of the unique sequences of *D. melanogaster* DNA was transcribed during the pupal stage. It is known that 80% of the *Drosophila* genome is represented by unique sequences and, therefore, 24% of this genome is transcribed during the pupal period (Turner and Laird, 1973). Church and Robertson (1966) found that the RNA/DNA ratio in *D. melanogaster* was equal to 10. Therefore, the 4C type cell in *Drosophila* containing 0.6 picograms (pg) of DNA could contain 6 pg of RNA, from which 0.5% or 0.03 pg should be complementary with unique DNA. If 15% of the unique DNA is transcribed in the cell, this is equal to about 0.07 pg (0.6 pg DNA per cell $\times 0.8$ unique DNA $\times 0.15$ hybridized DNA). Therefore, 0.03 pg hybridized, unique RNA corresponded in the mean to less than one ($0.03/0.07 = 0.43$) copy per each DNA sequence in the cell (Turner and Laird, 1973).

In mammals the hybridization of the total embryonic RNA with repeated sequences of DNA rapidly increased in the preimplantation period of development and remained at a constant level during the following stages of embryogeneses (Fig. 8). At the same time, approximately 1% of unique DNA sequences were transcribed before implantation. This percentage increased to 10% at birth (Brown and Church, 1972; Church and Brown, 1972; Church and Schultz, 1974). More than 2% from newly formed RNA in the preimplantation mouse embryo are the mRNA (Warner and Hearn, 1977). Rabbit blastocyst RNA was able to hybridize with 1.8% of the unique sequences of the genome. This corresponds to 60,000 unique sequences with 1,000 nucleotides in each. RNA, extracted from implanted 12-day-old embryos, hybridized with 2.5% of the unique DNA sequences. Approximately

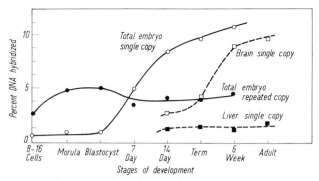

Fig. 8. The complexity of transcription of total nuclear RNA isolated from whole mouse embryos, liver and brain. The reaction of repeated DNA sequences with RNA isolated from all embryonic tissue was carried out by the filter technique. The reaction of nonrepeated H^3DNA (Cot 220) with unlabeled nuclear RNA isolated from embryos, liver, and brain was carried out according to the author's method. The time scale is not linear and is diagrammatic. (After Church and Brown, 1972)

70% of the newly synthesized heterogenous RNA in 4–6-day-old rabbit blastocysts hybridized with slowly renaturated or nonrepeating sites of the genome. Because mRNA comprises only 20% of the heterogenous RNA , it is supposed that the majority of copies transcribed from unique DNA sequences during the preimplantation period of rabbit development are not molecules of mRNA (Schultz et al., 1973). Specialized tissues, represented by many cell types, as in brain tissue, are characterized by an extraordinary complexity and heterogeneity of synthesized RNA. In mouse brain tissue there are transcribed approximately 300,000 various unique DNA sequences with 1,000 nucleotides each. These numbers obviously reflect the differential gene activity in the multitude of various cell types which are organized into brain tissue (Hahn and Laird, 1971; Soga and Takahashi, 1975). Some authors have suggested that transcribed copies from repeated sequences are necessary for realization of functions specific to all cell types, and transcribed copies from unique sequences determine the specificity of cell differentiation (Campo and Bishop, 1974).

1.1.3 Temporal Variability of Morphogenetic Nuclear Activity

The spatial-temporal pattern of transcription determines the specificity of ontogenic processes and is realized in the variability of morphophysiological nuclear activity. This was described in detail by Neyfach (1962). Nuclear activity is characterized by a relatively high sensitivity to irradiation with X-rays and to actinomycin treatment (Neyfach, 1963, 1964). In this case, dosages of X-rays which suppressed morphogenetic nuclear activity also inhibited the synthesis of mRNA. The synthesis of different RNA fractions was inhibited to an equal degree (Kafiani et al., 1966). Nevertheless, the rate of RNA synthesis in the whole embryo is apparently not correlated with morphogenetic periods. For instance, in the loach embryo RNA synthesis is maintained at a high, constant level beginning in the early blastula stage (Kafiani and Timofeeva, 1965; Timofeeva and Kafiani, 1964). However, as was already mentioned, the increasing RNA content could be explained through an increasing number of cells synthesizing RNA or through an increase in the rate of RNA synthesis per cell. Apparently, nuclear morphogenetic activities are accompanied by increasing RNA synthesis per nucleus (Bachvarova et al., 1966; Neyfach, pers. comm.). The possible presence of specific mRNA molecules of demonstrable morphogenetic significance is not excluded (Belicina et al., 1963). It was demonstrated that 1 h of morphogenetic nuclear activity made possible subsequent embryonic development for the next 3.5 h (Neyfach, 1962).

Corresponding variability in nuclear morphogenetic activity was demonstrated in later developmental stages and during postnatal ontogenesis, when the processes of cell and tissue specialization were complete. This was demonstrated for the differentiation of the crystalline lens (Reeder and Bell, 1965), muscle differentiation (Fieldman et al., 1964), hemopoiesis (Marx and Kovach, 1966), and early neuroembryogenesis (Akimova and Diban, 1967). The cortex of 1–5-day-old newborn rats is sensitive to the injection of the RNA synthesis inhibitor actinomycin D. Under these conditions cortex development is slowed. Injection of actinomycin D after the fifth day postpartum does not affect development of 5–6 layers of cortex (Korochkin and Olenev, 1966; Korochkin, 1970).

In Korolev and Neyfach's investigations of loach, embryos were irradiated during the blastula (5 h) and early gastrula (12 h) stages, to determine when the morphogenetics activities of nuclei play their role in the development of separate embryonic parts. It was shown that variability of morphogenetic activity exists in all of the separate embryonic parts. Active nuclear operation occurs in all zones of the embryo simultaneously from 6 to 8.5 h and after 14 h, whereas gastrulation proceeds without direct nuclear control, having been determined by its previous activity. At the same time, the levels of development of separate embryonic parts differ at various stages of ontogenesis. The inactivation of nuclei in the period 8.5–14 h of development induced the earliest termination of development in the ectoderm. Endodermic cells are located between the periblast and the chordomesoderm (this corresponds to 15 h of normal development) and thereafter their further development is terminated. Shortly after this they degenerated. Compared to ectodermal and endodermal cells, mesodermal elements are at a higher level of development. In this case, the stages of developmental termination differ in regard to two components of the mesodermic complex – the chord and the mass of nondetermined mesoderm. Subsequent chordal cells formed the solid chord typical of the end of the gastrula stage (18 h). The material of the neuroectoderm remains structurally the same during the period from 14–15 to 17–18 h of development and, therefore, it is only indirectly possible to determine its developmental stage after irradiation between 8.5 and 14 h of development. One criterion which could serve for developmental termination is a reduction in the increase of mass of neuroectoderm in the head part of the embryo. In regard to this criterion, the development of the neural part is inhibited at the end of gastrulation (17–18 h). Nuclear inactivation during the period from 22–24 h of development results in termination of development of the nervous system components at 26–28 h and of the somites from 30–35 h.

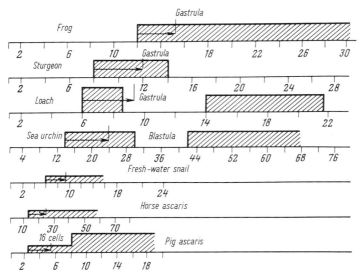

Fig. 9. Time of the expression of morphogenetic activity of nuclei in various species of embryos (Neyfach, 1962)

These differences lead us to suggest that, in this case, the peculiarities of nucleo-cytoplasmic interactions in the different parts of the embryo played an important role. Apparently, nuclear activities differ in the presumptive cells of any embryo component. Such differences determine the further development of the various presumptive layers in gastrulation (Korolev and Neyfach, 1965).

Developmental regulation due to morphogenetic nuclear function began in the embryo during the early, middle, and late blastula stage when the embryo contained from 100–200 (sea urchin) to many thousands of cells (amphibian). This is outwardly noticeable in sea urchins during the middle blastula stage, or in the stage of beginning gastrulation in fish and amphibians.

Morphogenetic nuclear function both began and became apparent earlier in mosaic eggs. It began at the 12 and 22–24 blastomere stages in Mollusca, and in ascaris at the two blastomere stages (Fig. 9; after Neyfach, 1962).

However, the development of mosaic eggs does not differ principally from the development of eggs of the regulatory type. The development of the mollusk *Ilyanassa* is also characterized by spatial and temporal differential gene activity and a well-expressed periodicity in morphogenetic nuclear activity. The sequence of transcription in the development of the different tissues of *Ilyanassa* is given in Table 1.

The analysis of actinomycin inhibition on the differentiation of a number of organs (Table 2) demonstrated the appearance of stable mRNA during certain periods of embryogenesis, differing for the various embryonic parts. This stable mRNA determined the resistance of developed organs to antibiotic action (Collier, 1966).

The discovery of temporal variability of morphogenetic nuclear activity makes possible the explanation of classical postulates on critical periods of ontogenesis

Table 1. Gene schedule for *Ilyanassa* embryogenesis. (After Collier, 1966)

Oogenesis	Transcription for stages of development				
	Day 1	Day 4	Day 5		Day 6
Cleavage		Foot	Eyes	Shell	Esophagus Digestive gland
Epiboly		Operculum	Otocysts		Intestine Heart
					Stomach

Table 2. The repression of morphogenesis is *Ilyanassa* by Actinomycin D[a]. (After Collier, 1966)

Time of treatment	Foot	Otocyst	Oper-culum	Shell	Velum	Eyes	Esoph-agus	Stomach	Digestive gland	Intestine	Heart
Day 4	1.8	0.3	1.8	0.2	1.0	0.2	0.0	0.0	0.0	0.1	0.0
Day 5	2.3	2.3	2.8	1.4	2.2	3.0	1.3	0.0	0.0	1.1	0.0

[a] Embryos were treated at either 4 or 5 days for 6 h with Actinomycin D, 25 µg/ml, and allowed to complete their differentiation in plain sea water. A normal organ was scored as 3.0. All values are the mean from the scores of at least 25 embryos

in terms of molecular biology. This hypothesis was created by Stockard (1921) and developed by Svetlov (1960), and may be so stated: during early embryogenesis the critical stages are those in which morphogenetic nuclear functions are manifested.

1.1.4 Puffing as a Cytological Indicator of Differential Gene Activity

It is supposed that, in certain cases, it is possible to detect differential gene activity cytologically. This question concerns the formation of swellings – puffs – in portions of the polytenic chromosomes of the salivary glands in *Diptera* which contain large amounts of RNA (Fig. 10; Beerman, 1964; Kröger, 1964; Clever, 1968; Serfling et al., 1969; Kiknadze, 1972; Berendes, 1973). Puffing is accompanied by the RNA synthesis in the corresponding region of chromosome. The results of the radioautographic experiments demonstrated the increasing of the labeled precursors incorporation into puff RNA (Fig. 10). The electrophoretic analysis of the RNA synthesis in the Balbiani's ring of *Chironomus* demonstrated that the chang-

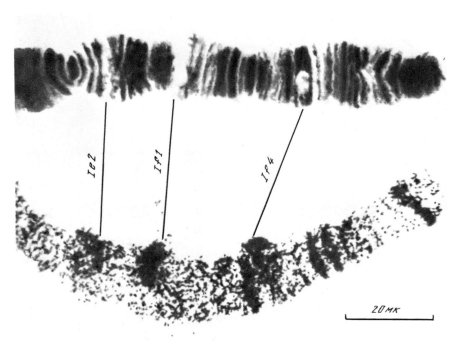

Fig. 10. Incorporation of H³-uridine in RNA of chromosome 1 of the salivary gland in *Chironomus thummi* with the induced heterozygous puffs. *Top,* the squash preparation, aceto-acid carmine staining. *Bottom,* the autography of the heterozygous puffs in the *Ie2, If1,* and *If4* regions. Only two puffs (*If1* and *If4*) are heterozygous on the squash preparation. (After Serfling et al., 1969)

ing of the size of Balbiani's ring correlated with the intensity of some RNA fractions, which confirmed the idea of the increasing of RNA synthesis during puffing (Fig. 11; Daneholt et al., 1969). Berendes found tissue-specific puffs in *Drosophila hydei*; three in malpighian tubule cells, four in the middle intestine, and seven in the salivary glands (Berendes, 1965, 1966a, b). The tissue specificity of puffing was also found in *Calliphora* and *Sarcophaga* (Ribbert, 1972). In Edström's laboratory it was demonstrated that in the polytenic chromosomes of the salivary gland of *Chironomus tentans*, the stable, high molecular weight, 75S RNA was synthesized in the region of a giant puff, named Balbiani's ring (Fig. 17). This RNA, absent in other organs, later concentrated in the cytoplasm and, presumably, was used for the synthesis of a specific secretory product of the salivary glands (Daneholt, 1972, 1974, 1976; Daneholt and Hosick, 1973). With the aid of cytological hybridization between this RNA and the DNA of the polytenic chromosome, it was demonstrated that this RNA hybridized only with that part of chromosome IV which formed Balbiani's ring (BR-2) (Lambert, 1972).

Paralleling the tissue specificity of puffing, peculiarities in its stage specificity have also been demonstrated. The analysis of puffing in the giant chromosomes of the salivary gland in *Chironomus dorsalis* during various periods of development makes possible the determination of tendencies in their dynamics. It was found that the distribution of puffing remained stable in the periods between moultings, but changed dramatically in the crucial phases of larval moulting and metamorphosis. For instance, in *Chironomus thummi*, at the end of the last phase of the fourth stage and directly at the moment of pupamation, 40% of "between line" puffs demonstrated a noticeable size increase, 11 new "metamorphosis" puffs appeared, and 4 disappeared or essentially decreased in size (Fig. 12). In the late pupal stage, before lysis of salivary gland cells, the main portion of puffs disappeared and only a small part of the active center remained. A large number of metamorphic puffs also appeared during larval moultings although they did not reach their maximal size during this period (Fig. 12). It is of interest that this altered picture of puffing remained for several hours after larval moulting (the so-called red head stage), after which it returned to the standard picture characteristic of periods between moultings. The periods of changes in puffing are actinomycin- and puromycin-sensitive in Chironomidae (Clever et al., 1969; Vlasova and Kiknadze, 1975). It is supposed that during this time a kind of reprograming of the salivary gland cells takes place. This could be caused by certain puffs of the complex acting, not constantly, but only at certain steps of development, dependent upon metamorphosis hormone concentrations (Vlasova and Kiknadze, 1975).

Stage-specific puffs were found in *D. melanogaster* (Becker, 1959; Aschburner, 1967). Particularly, puffing was found in 66 regions of the X chromosome. In 52 of these puffing is constant, and in 14 it is stage-specific. The maximum in puffing activities occurs during the transformation from larval to pupal stages (Belyaeva et al., 1974).

In some cases the genetic constitution of the organism can influence puffing. These influences can affect quantitative characteristics through timing in the appearance of activity and the maximization of individual puff sizes, and qualitative characteristics through a loss in the ability of individual discs to make a puff or a change in the puff activities of the whole group.

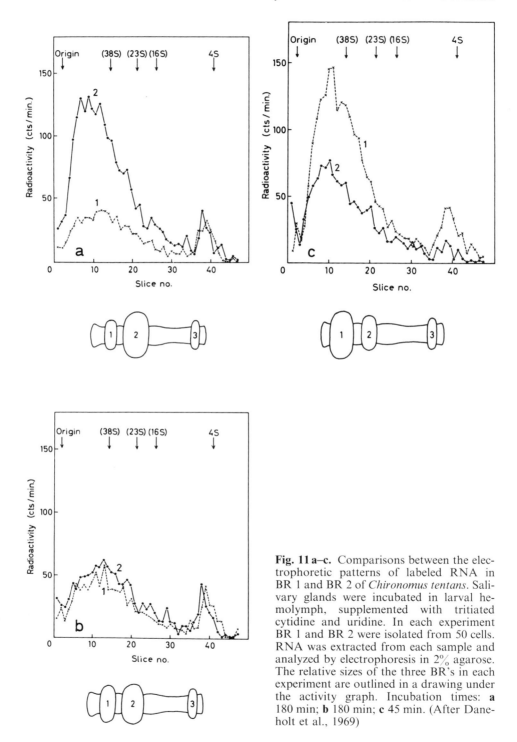

Fig. 11 a–c. Comparisons between the electrophoretic patterns of labeled RNA in BR 1 and BR 2 of *Chironomus tentans*. Salivary glands were incubated in larval hemolymph, supplemented with tritiated cytidine and uridine. In each experiment BR 1 and BR 2 were isolated from 50 cells. RNA was extracted from each sample and analyzed by electrophoresis in 2% agarose. The relative sizes of the three BR's in each experiment are outlined in a drawing under the activity graph. Incubation times: **a** 180 min; **b** 180 min; **c** 45 min. (After Daneholt et al., 1969)

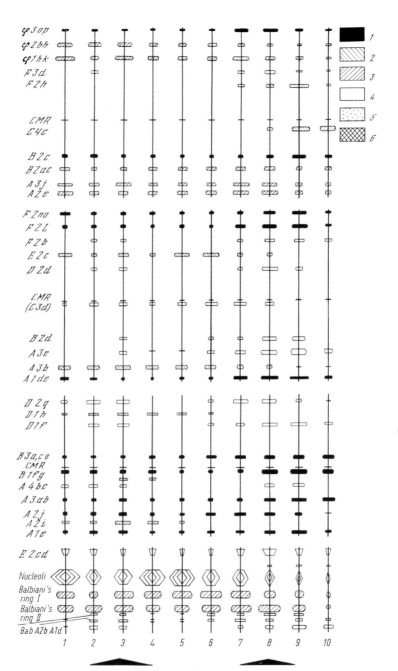

Fig. 12. Changing of activity of the main puffs in salivary glands chromosomes during *Chironomus thummi* development.
Stages of development: 1, before moulting from 3rd to 4th stage; *2*, larva moulting, the beginning of the red head stage; *3*, the red head stage; *4–5*, are corresponding periods of 4th stage; *6–8*, are the prepupae stage periods; *9*, the red pupae stage; *10*, the grey pupae stage.
Types of puffs: 1, puffs, which significantly enlarged during moulting; *2* and *6*, larvae puffs; *3*, puffs, which diminished during moulting; *4*, metamorphosis puffs, which appeared periodically during moulting; *5*, methamorphosis puffs, which appeared only during metamorphosis. (After Kiknadze, 1972)

Table 3 demonstrates the peculiarities of puffing dynamics in a control line of *D. melanogaster* and in two mutants of lethal giant larvae [1(2) gl]. The larvae in these strains died either before or shortly after pupation. Interstrain differences, concerning mainly the time of puffing activations, were noticeable. At the same time, Belyaeva and Zhimulev (1974) demonstrated high puffing identity in *D. melanogaster* during the prepupal stage. The investigation of salivary glands in three ordinary laboratory strains and two inbred strains, encompassing 100 generations of inbreeding, did not exhibit any differences in puffing characteristics. Genetic analysis of individuals which lose the ability to form puffs demonstrates that, in some cases, this is caused by mutation. Analysis of the distribution of the "puff–nonpuff condition" in F_2 progeny of the cross Oregon R × Pacific strains for 22B puff, and vg6 × Oregon R strains for 64C puff, demonstrates the typical Mendelian segregation ratio 3:1 (17:4 and 68:20 respectively). The backcross hybrid × Oregon R demonstrated a 1:1 segregation ratio (Aschburner, 1969a, b, 1970a).

The mutations causing the loss of the band's ability to form a puff represented a heterogeneous group. In regard to their functional character they could be joined into groups which differ in their expression in heterozygous combination with respective nonmutant strains.

Group One: These mutations demonstrated independent behavior. In the synapsis, i.e., in chromosome conjugation, a puff developed on both of the homologs. In the case of asynapsis, when conjugation is distorted, puffing showed up only in the nonmutant homolog.

Group Two: The mutation demonstrated autonomy in the heterozygotes, i.e., the mutant strain band remained inactive regardless of the conjugation of chromosomes in this region.

Group Three: The mutation demonstrated dependence upon its homolog. These mutant bands formed puffs in heterozygous combinations when synapsis or asynapsis took place (Aschburner, 1970a, b).

It should be noticed that the group of mutations characterized by the loss in ability for either band to form a puff is small.

From Kiknadze's data (Kiknadze, 1972) it follows that puffs 4AB and 17E which specifically appeared in *Drosophila* mutants, as well as a number of other puffs, are mainly dependent upon the genotype of the organisms.

The investigation of *D. melanogaster* mutant lethal giant larvae demonstrated the appearance of a new puff in the 63BC region which was unusual for metamorphosis (Becker, 1959). In the mutant fat, changes in the 24DE locus where the corresponding gene is located, as well as the appearance of several new puffs in some regions of the left arm of chromosome II were obtained (Slizynski, 1964). In larvae of the tumor-head (th-h) mutant a complex puff disappeared in the 74–75 region of the left arm of chromosome III (Rodman, 1964). However, there is evidence that the distribution of puffing is highly stable and does not demonstrate noticeable interstrain differences (Zhimulev and Belyaeva, 1975a, b).

Taken as a whole, this factual evidence makes the hypothesis of puffing as a cytological expression of gene activity very alluring. In some cases, however, it was demonstrated that the processes of transcription and translation in the salivary glands of *Drosophila* occurred before puff formation and before a change in the puffing distribution of polytenic chromosomes (Kress, 1972; Korochkin et al.,

Table 3. Percentages of puffs, observed during ontogenesis in mutants 1 (2) gl. *D. melanogaster*, which differ in the time of death. (After Korochkina, unpublished)

Puff	Control, Oregon R						Mutant			
	Age of larvae, h				Prepupae, 0 h	Prepupae, 1/2 h	With pupation		Without pupation	
	96	110	115	118			Larvae, 120–170 h	Prepupae	Larvae, 120–170 h	Larvae, 170–220 h
1E	100	100	100	0	0	–	100	89	94	100
3CD	100	100	95	80	0	25	100	65	100	100
$22B_{4-5}$	0	50	90	100	100	100	30	50	45	25
$29F_{1-2}$	0	0	40	90	100	100	100	82	100	100
45C	0	0	0	0	100	100	60	100	20	50
46A	0	0	0	0	100	100	27	45	0	0
46D	0	0	0	0	100	100	0	0	0	0
46F	0	0	0	100	100	100	25	95	0	0
$47D_3$	0	0	30	100	100	100	100	94	100	100
51F	0	0	0	0	100	50	100	60	100	100
52C	100	100	100	0	0	0	100	33	100	100
$55C_{1-5}$	100	100	33	50	100	100	100	100	100	100
57B	0	0	0	0	100	30	80	35	60	80
61F	0	0	0	0	100	100	0	70	0	0
$63E_g$	0	0	0	0	100	100	0	70	0	0
$66B_g$	0	0	0	100	100	100	55	75	0	0
$67F_{1-3}$	0	0	0	100	100	100	25	85	25	0
78D	0	0	100	100	100	100	13	27	0	0
$85D_{1-2}$	0	0	0	80	90	70	30	80	10	16
$97B_{1-2}$	0	0	14	50	30	100	0	20	0	0
97C	0	25	14	60	100	0	0	20	0	0

1974). In particular, it was demonstrated that the organospecific formation of secretory products in the salivary glands of *D. melanogaster* and *D. virilis* showed up very early in the second larval stage (Kolesnikov and Zhimulev, 1974). This was also the case for an increase in esterase activity, as well as some dehydrogenases and phosphatases in *D. melanogaster* and some species of the virilis group (Korochkin et al., 1972b, c). These data contradict the hypothesis of Edström (personal communication) which states that in large puffs Balbiani's ring-type, organospecific mRNA is synthesized and small puffs are the sites of synthesis of mRNA for various cellular enzymes. Thus, the question of the significance of puffing remains open (see also Harris, 1973).

All critical considerations were arranged by Thomson (1969) as follows:

1. The majority of hereditary material in polytenic chromosomes is inactive. No more than 10%–15% of the 5000 bands in *D. melanogaster* and 1900 bands in *Chironomus* formed puffs. The maximum proportion of bands which are simultaneously active is smaller and is approximately equal to 6% (Pelling, 1966).

2. The number of tissue-specific puffs is small and it is possible that they may be completely absent (Clever, 1966).

3. A correlation between puffing and metabolic processes has not been found (Clever, 1966).

4. What are the principles of regulation in nonpolytenic chromosomes, if the activation of genetic units forms puffs and the synthesis of RNA is regulated by the number of polytenic chromosomes?

5. There is only little evidence of correlation between individual bands and cystrons (Welshons, 1965). The middle-sized chromomere in the giant chromosomes of *Chironomus* salivary glands contains sufficient DNA for the coding of 100 intermediate-size protein molecules (Edström, 1965).

6. As a result of experiments on puffing inhibition in the polytenic chromosomes of *Sciaria corpophila* due to cortison injection, it was found that puffing is not necessary for normal development (Goodman et al., 1967).

Furthermore, no differences were found in the electrophoretic spectra of salivary gland proteins in two strains of *Chironomus thummi* with either two or three Balbiani's rings. In the strain with an additional ring of Balbiani, intensive RNA synthesis was observed (Wobus et al., 1971).

The strict differences observed in the antigenic spectrum of proteins in wild-type and mutant lethal giant larval of *D. melanogaster* do not correlate well with a relatively good similarity observed in the puffing activities of these strains. The protein differences were observed with the aid of two-dimensional immunoelectrophoresis (Karakin, personal communication).

Finally, in spite of functional differences in the proximal and distal parts of the salivary gland in *D. melanogaster*, the puffing picture in these locations is identical (Belyaeva and Zhimulev, 1974; Zhimulev, 1974).

In all fairness, it should be mentioned that not all of these objections are of equal strength. Goodman's data have been seriously criticized (Rasch and Lewis, 1968). Nevertheless, it is still in doubt whether puffing is really the expression of locus activity, or, conversely, if its function lies in the inhibition of a gene product or the promotion of its degradation. In other words, whether the puff is a peculiar "tombstone" of former activity. Observation of ribonuclease activity in the polytenic chromosomes of *Drosophila* and *Chironomus* is of interest in regard to the last suggestion (Korochkina et al., 1975). It is also possible that puffing could serve in both capacities, activation and inhibition of RNA production, depending upon the circumstances.

Recently, new data were obtained which demonstrated that, at least in some cases, the active gene function is accompanied by the phenotypic expression of the product coded for this gene and is reflected in puff formation. Firstly, a correlation was found between the puff sizes and the quantitative ratios of various protein fractions that were electrophoretically distinguished in the salivary glands of *D. melanogaster* (Tissieres et al., 1974). Then, it was found that the high-temperature induction of new puffs in *D. melanogaster* is accompanied by the regression of the existing puffs and, especially, one of them (McKenzie et al., 1975; Spradling et al., 1975, 1977). A similar change was obtained in *Drosophila* embryonic cells cultures after temperature treatment. The authors obtained polysomes from such cultures and extracted from these polysomes the labeled RNA which contained poly (A). This RNA was bonded mainly with the 87B region of III chromosome during in situ hybridization on the chromosomes preparates from salivary glands. It is of interest, that exactly in this area new puff is induced by temperature shock, and among newly synthesized proteins one fraction is intensively labeled.

In *D. melanogaster, D. simulans, D. virilis, D. funebris, D. buscii* and *D. ananassae*, the synthesis of six–seven new proteins was induced by heat treatment – 37 °C, duration 20 min (Lewis et al., 1975). There is good agreement between the number of induced proteins and the number of new puffs obtained in these experiments.

High transcriptional activity was demonstrated for the 87A and 87C puffs induced by temperature shock in *D. melanogaster* (Henikoff and Meselson, 1977). Scalenghe and Ritossa (1976) demonstrated that the temperature-induced 93D6–7 puff coded for the glutamine synthetase A subunits in *D. melanogaster*. Six new polypeptides and their corresponding new mRNA's were produced as a response to temperature treatment (Arrigo et al., 1977). It was demonstrated also that although the synthesis of the major part of the proteins obtained in normal (25 °C) conditions is inhibited after heat treatment, the mRNA's corresponding to these proteins are not degraded. Therefore, the presence of regulatory processes was suggested for the translational level.

At the same time, it was found that the formation of temperature-sensitive puffs is sometimes not coordinated. The puff patterns could be variable and were dependent on the constitution of the medium when the puffs were induced in vitro cultivated salivary gland cultures (Bonner and Pardue, 1976). The newly induced deletion in *D. melanogaster* (87C1, III chromosome) did not change the composition of the temperature-induced puffs (Ish-Horowitz et al., 1976). On the other hand, detailed autoradiographic study of RNA synthesis in *Drosophila* salivary glands, together with in situ hybridization demonstrated that there is no strong correlation between the active puff size and the amount of RNA in the cell. The amount of RNA that hybridized with a large puff is smaller than the amount of RNA that hybridized with a small puff, and at the same time is smaller than the amount of RNA that hybridized with nonpuffing regions of the chromosome. This observation yielded the conclusion that puffing cannot be a simple reflection of the rate of transcription of the corresponding genes (Bonner and Pardue, 1977).

Table 4. Localization of the gene of the "special cells" in the progeny of C. *tentans* × C. *pallidivittatus* crosses. (After Grossbach, 1969)

Indi-vidual	Arm of chromosome					Special cells	Indi-vidual	Arm of chromosome					Special cells
	2L	2R	3L	3R	4			2L	2R	3L	3R	4	
1	tt	tp	pp	pp	tp	+	14	tt	tp	tt	tt	tp	+
2	tt	pp	pp	pp	pp	+	15	tt	tp	pp	pp	tp	+
3	tp	tp	pp	pp	pp	+	16	tp	tp	tt	tt	tp	+
4	pp	tp?	tt	tt	tp	+	17	tt	tp	tp	tp	tp	+
5	tp	tp	tp	tp	pp	+	18	tt	tp	tp	tp	tt	−
6	tt	tt	pp	pp	pp	+	19	tp	tt	tt	tt	tp	+
7	tp	pp	pp	pp	tp	+	20	tp	tp	pp	pp	tp	+
8	tp	pp	tp	tp	pp	+	21	tt	tp	tp	tp	tp	+
9	tt	tp	pp	pp	pp	+	22	pp	pp	tt	tt	tp	+
10	tp	tt	tp	tp	pp	+	23	pp	pp	tp	tp	tp	+
11	tt	tp	tp	tp	tp	+	24	tp	tp	tp	tp	pp	+
12	tt	tp	tp	tp	tt	−	25	tp	pp	tp	pp	tt	−
13	tt	tt	pp	tp	tp	+							

The most convincing data that suggested the connection between puff formation and the synthesis of a certain product were obtained when cytogenetic and biochemical approaches were combined with the genetic approach (Korge, 1975, 1977a, b). It was demonstrated that saliva proteins are synthesized in the *Drosophila* salivary glands autonomously rather than transported into the glands with the hemolymph flow. The electrophoretic variants of fractions for three salivary proteins were described. With the aid of recombination analysis and cytogenetic methods, the gene localization of one of the genes that was responsible for one of these proteins (Sgs-4) was estimated (3C10-3D1, X chromosome). The puffs in these chromosomal regions of the X chromosome are active before and during secretion and terminated their activity with secretional termination. The absence of the Sgs-4 protein in the larvae from one of the investigated stocks of *D. melanogaster* was accompanied by puffing extinction in the corresponding area of the chromosome. The correlation for other puffs and the synthesis at specific secretory proteins was demonstrated as well (Korge, 1975). A similar correlation between amylase synthesis and the puffing in the chromosomal region, where the amylase gene is located, was found in *Drosophila* intestinal cells (Doane, 1971, 1975; Dickinson and Sullivan, 1975).

Genetic confirmations of a similar correlation were demonstrated in the papers dealing with *Chironomus* larvae. The locus that is responsible for the specific type of protein granules was described. It was demonstrated also that the salivary glands of two related species, *Chironomus tentans* and *Chironomus pallidivittatus*, are different in the external trait of so-called special cells surrounding the excretory duct of salivary glands. The genetic analysis shows that there is a locus which determines the formation of protein granules in these cells. Interspecies progeny analysis (*C. tentans* × *C. pallidivittatus*) demonstrated that this trait has monogenic heredity and normal Mendelian segregation. On the basis of cytogenetic data the suggestion was made in favor localization of this gene on the distal end of chromosome 4. *C. tentans* does not have a puff in this region, however *C. pallidivittatus* has two. F_1 hybrids from crosses of these two species, in which chromosome 4 from both species is present, formed only one puff (Beermann, 1961; Grossbach, 1969). Table 4 shows that protein granules are absent in the special cells when both chromosomes 4 are represented by chromosomes of *C. tentans*.

1.1.5 Gene Amplification

One peculiar expression of differential gene activity is termed gene amplification. This phenomenon has the effect of causing those parts of DNA which should synthesize a large amount of RNA to be repeatedly copied or amplified.

The most demonstrative and possibly only example of this phenomenon could be the synthesis of additional amounts of ribosomal DNA (rDNA)[1] in amphibian oocytes.

Long ago the suggestion was made that excess of DNA in *Bufo* oocytes could be explained by the formation of extra copies of the nucleolus organizer (Painter and Taylor, 1942). This hypothesis was further developed by Lima-de-Faria (Lima-de-Faria and Mozes, 1966; Lima-de-Faria et al., 1969).

Experiments with molecular hybridization showed that oocytes contained tremendously large amounts of rDNA (Brown and Dawid, 1968; Lima-de-Faria et al., 1969). This extra rDNA occurs as numerous nucleoli which are localized inside the oocyte nucleus. These additional copies exist mainly as rings (Miller and Beatty, 1969; Brown and Blackler, 1972). The synthesis of extra DNA in *Xenopus* and *Bufo* occurs during meiosis, at the pachytene stage (Coggins and Gall, 1972). Nuclei contained only one nucleolus at the leptotena stage, however at the pachytene stage additional nucleoli began to appear. The number of nucleoli increased very rapidly during the early diplotene stage and reached about 1000. As a result of the amplification process, the number of copies of the nucleolus organizer increased to 2500–5000 units (Gall, 1969).

With the aid of DNA–RNA hybridization of cytological preparations it was shown that 20–40 copies of amplified DNA were already present during the premeiotic interphase (Gall and Pardue, 1969). The cascade mechanism of amplification has been suggested (McGregor, 1968). In this case each copy of rDNA serves as a template for the next. It is possible to calculate that the first five cycles of synthesis ($2^5 = 32$) will complete development to the beginning of the pachytene. In general, there are 11 such cycles, each cycle continuing for 1.2 days (13 days/11 cycles) (Coggins and Gall, 1972).

Specific amplification of rDNA has been demonstrated not only in amphibians, but also in *Roccus* fish, *Colymbetes* beetle, cricket, and other animals (Brown and Dawid, 1968; Gall, 1969).

Recently, it was demonstrated that *Xenopus laevis* spermagonia also contain additional nucleoli and extra copies of rDNA, as do oogoniums of the same developmental stage (Pardue, 1969).

---

[1] rDNA is DNA which synthesizes ribosomal RNA

Is amplification usual for other types of DNA? Denis reported on the possible amplification of genes which synthesize tRNA, although these data were not corroborated (Denis, 1970). There is evidence that during differentiation of hen cartilage amplification of the moderately repeated DNA sequences could occur (Strom and Dorfman, 1976).

It is possible that the formation of DNA puffs in *Sciaria* is the same phenomenon as gene amplification. Their puffs appeared at the end of larval development in specific chromosomal loci (Crouse and Kryl, 1968). In this case conclusive proof of gene amplification was not established.

Is amplification characteristic of unique DNA sequences?

In the investigation of differentiated cells which synthesized a large amount of a protein, it was shown that the respective gene amplifications did not occur. This was demonstrated for instance in the kinetic analysis of mouse somatic DNA hybridized with globine DNA obtained from the system by reverse transcription on the globine RNA template (Bishop and Rosbach, 1973; Gambrini et al., 1974; Harrison et al., 1974).

In the silk-producing glands of *Bombyx mori* about 300 µg of the protein fibroin is synthesized during the last 4 days of the fifth larval stage. In each haploid genome there is only one copy of this gene (Suzuki and Brown, 1972; Suzuki et al., 1972; Lizardi and Brown, 1975).

Presumably, the same conclusions could be reached in regard to ovalbumin synthesis induced in the oviduct by estradiol (Harrison et al., 1974; Sullivan et al., (1973), and also in regard to the light chains of immunoglobulin (Faust et al., 1974). The results of these experiments, confirmed the idea that the amplification is characteristic only of rDNA.

What is the mechanism of amplification which leads to the differential increase in the content of ribosomal and, possibly, other types of RNA through the transcription of extra copies of corresponding DNA?

Two main hypotheses are discussed in regard to this question:
1. The extra copies are replicates of one (or more) of several hundred repeated sequences of rDNA of the nucleolus organizer. This is a mechanism of chromosomal amplification.
   In this case it is not important how successive copies are synthesized, from the original or from the replicate chromosomal rDNA of the nucleolus organizer.
2. Some of the extra rDNA copies remained in the nucleus of oocytes during mitosis and one or more copies remained in the embryonic cells and were incorporated into the germinal cells. In regard to this hypothesis all extra rDNA copies are originated from one (or some) stored copies. This is a mechanism of episomic heredity.

The main differences between these two hypotheses are that the former predicts the nuclear heredity of amplified rDNA, while the episomal mechanism requires maternal heredity. If the presence of paternal repeated rDNA were demonstrated in oocytes, the first hypothesis would be considered as more correct. A corresponding model experiment could be made using the fact that the rDNA of two interbreeding species – *Xenopus laevis* and *Xenopus mülleri* – differ in regard to their nucleotide sequences in "spacers" which divide genes for 28S and 18S RNA (Brown et al., 1972; Wellauer and Reeder, 1975). In this case, crosses between these two species and hybrid analysis result in a convenient test system for the assessment of these formulated hypotheses. Molecular hybridization of cytoplasmic RNA with DNA demonstrated the presence of hybrid rDNA of both species in the somatic cells, although the relative proportions of each were difficult to establish. At the same time, in such a hybrid, the amplification proceeds mainly, if not entirely, with rDNA of *X. laevis*. In hybrids from crosses ♀*X. mülleri* × ♂*X. laevis* only the paternal (*X. laevis*) rDNA was amplified.

The results exclude maternal heredity of amplified rDNA and make the hypothesis of "chromosomal amplification" preferable.

Some authors (Crippa et al., 1972; Tocchini-Valentini et al., 1974) have proposed that amplification of rDNA genes in amphibians occurs, not through gene amplification of the chromosomal matrix, but on the rRNA template in the cytoplasm. These authors demonstrated that amplification is inhibited by the same rifampycine derivatives which differentially inhibited the action of the reverse transcriptases. However, this interesting hypothesis contradicts available experimental material and, without convincing proof, cannot be accepted (Brown and Blackler, 1972).

Speculations on the more commonplace role of amplification in the genetic control of cell differentiation are also in doubt. In these speculations, the assumption is made that extra copies of rDNA could be a crucial factor in determination and control of differential gene activity during ontogenesis (Brown and Dawid, 1969). However, at this time there is no serious basis for this kind of wide generalization.

Apparently amplification is only particular and unusual mechanism for realization of differential gene activity and possibly occurs only for those genes represented in the genome as polycopies.

1.1.6 The Interactional Effect of Certain Genes on the Transcription of Ribosomal Genes

The synthesis of rRNA is a convenient model which allows study of the regulatory mechanisms of the differential transcription in the ontogenesis of eukaryotes. A large amount of this type of work was done on *Drosophila*, and especially on the bobbed mutant of *D. melanogaster*. This mutation is localized on the X chromosome (66.0) and on the short arm of the Y chromosome (Lindsley and Grell, 1967) and is stipulated by deletion in the nuclear organizer region. This mutation is phenotypically manifested as short and thin bristles, a decrease in the chitin cover thickness, and a reduction in the growth rate.

The locus bobbed contained genes which synthesized rRNA. It is supposed that the bobbed phenotype could be determined not only by deletion in the ribosomal RNA genes but, in some cases, by a mutation which does not appreciably change the number of these genes (Howells, 1972).

A connection between molecular events in the cells (transcription peculiarities) and morphogenetic effects was found in the case of the mutation bobbed (*bb*). Particularly, the rate of rRNA synthesis is proportional to the length of scutellar brist-les. In this case the development of the trichogenic cells proceeds normally and the size of scutellar bristles is dependent upon RNA synthesis (Weinman, 1972). How-ever, the synthesis of rRNA is not necessary for the normal differentiation of cer-tain *Drosophila* cells (neurons, myocytes) in in vitro cultivation (Donady et al., 1973).

The action of the mutation *bb* is supposedly of two types: in some tissues the delay in development continues until sufficient ribosomes have been collected. In other tissues development may continue, but it is characterized by an inhibition of protein synthesis. This, in turn, produces a noticeable phenotypic effect (Mohan and Ritossa, 1970).

Weinman (1972) found that the amount of rRNA is constant in mutant *bb* in-dividuals, but the rate of synthesis of ribosomal RNA is reduced to only 30% as compared to wild-type flies. The rate of synthesis of 5S RNA, the genes for which are located in the 56EF region of chromosome II of *D. melanogaster* (Tartof and Perry, 1970), does not differ from normal. Therefore, the rates of rRNA and 5S RNA synthesis are not coordinated although they are present in the ribosomes in equimolar ratios. However, the accumulation of 5S RNA is small which may, pos-sibly, be associated with its degradation. This degradation is apparently a direct result of a low synthesis rate of rRNA since, in this case, 5S RNA cannot be incor-porated into ribosomes. The impossibility for integration into ribosomes could ac-tivate a specific mechanism for degradation of 5S RNA (Mohan and Ritossa, 1970; Weinman, 1972).

An increase in rRNA synthesis was found in tissues of XO individuals with a mutant allele at the *bb* locus. Using genital tissues, it was demonstrated that the rate of synthesis of rRNA in XO tissues of bobbed flies is higher than in flies with a normal nuclear organizer. However, tissues of flies with the same XO constitu-tion, but carrying a wild-type allele at the *bb* locus, do not demonstrate an increase in the rate of synthesis of rRNA. (Ritossa et al., 1971; Krider and Plaut, 1972).

The amount of RNA is lower in bobbed mutant flies as compared to wild-type flies during all developmental stages, while the amount of DNA in both cases is similar. Therefore, the RNA/DNA ratio is lower during all developmental stages in mutant bobbed flies. The maximal rate of RNA synthesis per 12-h periods dur-ing the third larval stage equals 0.28 mg/h (wild-type), 0.33 mg/h (XXY control), and 0.19 mg/h (bobbed mutant).

After hatching of the flies this rate is equivalent to 50% in five mutants as com-pared to the wild type, and 32% in comparison to the XXY control. The results of this experiment demonstrate that bobbed mutants cannot synthesize rRNA at the same rate as wild-type flies (Howells, 1972).

Using the example of bobbed mutants it was also found that ribosomal genes of *D. melanogaster* are capable of undergoing disproportional replication (Tartof, 1971, 1973). This disproportionality could be obtained in somatic cells as well as in embryonic cells under certain genetic conditions. In the latter case, the amount of rDNA in mutant flies with a partial deficiency of rDNA could reach a level char-acteristic of wild-type flies. These hereditary peculiarities will transfer to sub-sequent generations.

It is known that there are about 250 copies of the rRNA gene in the region of the nuclear organizer (NO) in the wild-type flies of *D. melanogaster*. When NO is represented by one dose in flies of ♂ X/O or ♀ X/X NO constitutions, the number of rRNA gene copies increases to 400 in the X chromosome. This tremendously sharp increase in gene number occurred during ontogenesis in the somatic cells of the F_1 generation and could not be a result of recombination between homologous chromosomes since this phenomenon is observed in the male X/O where one of the homologs is absent (Atwood, 1969; Tartof, 1973).

Apparently, this disproportionality of rDNA replication determines the reversion of bb mutants to the wild-type regarding their phenotypes and their amount of rRNA genes. This reversion occurred when the X chromosome was paired for several generations with the Ybb^- chromosome[1] in males (bb/Ybb^-) and was termed magnification (Ritossa, 1968a, b; Malva et al., 1972). This process obviously could noticeably influence development during ontogenesis and cell differentiation. Ritossa suggested the following mechanism as an explanation for magnification (Fig. 13):

1. Magnification began in males (but not in females) of the bobbed phenotype.
2. The first stage of this process occurs in these bobbed mutants and entails synthesis of extra copies of rDNA. This does not manifest itself phenotypically in the first generation (premagnification). These extra copies have perhaps a circular shape, and the presence of the additional circular molecules of DNA, which are complementary to rRNA during magnification, was demonstrated (Gargano and Graziani, 1977).
3. It is in the progeny of these males that the described molecular events have their phenotypic expression. It is supposed that the extra copies of rDNA synthesized during the premagnification stage are integrated into chromosomes of cells in the germinal line.
4. The efficacy of this process varied from nearly 100% to very small amounts.
5. The magnification could continue in subsequent generations (Ritossa, 1972).

Recently Tartof (1974a, b) suggested the hypothesis that gene magnification could occur through unequal crossing over between sister mitotic chromatids. He proposed this on the basis that: (1) rDNA magnification occurred in the mitotically active embryonic cells; (2) it is possible with the aid of genetic methods to obtain the reduction of the extra copies of rDNA; (3) magnification and reduction are reversible and reciprocal events; (4) it is possible to obtain somatic mosaics with bb^+ and bb bristles; (5) magnification is reduced in the ring chromosomes. However, on the basis of the experimental data the suggestion was made that the increase of rRNA genes during magnification is not associated with the stable event which is similar to the uneven crossing over (La Mantia and Graziani, 1977).

Interesting results were observed in experiments with rRNA synthesis in *Xenopus laevis* with mutations p^{1-1} and p^{1-2}. These mutations caused an insufficiency of the nucleolus organizer (NO) and are similar in their expression to bobbed mutation in *Drosophila*. It was found that these mutants synthesized rRNA at a rate

---

[1] The nature of Ybb^- is not clear. It apparently contains an amount of rDNA close to that found in the wild type, however, it is unable to transcribe 28S and 18S RNA (Tartof, 1973)

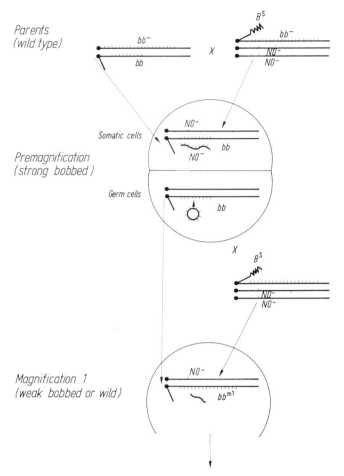

Fig. 13. A working hypothesis on the first steps of rDNA magnification. A bobbed male is here generated by crossing Xbb^+/Ybb males with X_{NO}-$/X_{NO}$-$/B^sY$ females. (The X_{NO}-chromosome carries no gene for rRNA while the B^sY has a wild bobbed locus). In such bobbed male combination (X_{NO}-$/Ybb$) magnification starts. It consists in the synthesis of rDNA not bound to the chromosome and unable to produce mature ribosomal RNA. Only in the germ line of such males can the extra copies of rDNA be anchored to the chromosome. The extra rDNA is here assumed to be in a circular form and anchorage is supposed to be an integration event. In the successive generation (magnification 1) in those zygotes which receive a Ybb chromosome where integration of extra rDNA has not occurred, the process will start all over again; while those zygotes which receive a Ybb chromosome where rDNA integration has occurred, will generate individuals (X_{NO}-$/Ybb^{m1}$) whose phenotype will be normal or less severely bobbed than that of their fathers. In appropriate conditions a further step of magnification, similar to that which occurred previously, can go on in this, magnification 1, generation. (After Ritossa, 1972)

Table 5. Relative amount of ribosomal and soluble RNA and relative amount of their radio-activities in embryos *Xenopus laevis* with different nucleolar genotypes. (After Miller and Knowland, 1972)

Nucleolar genotype	Percent RNA radioactivity in		Percent of total RNA in	
	28S + 18S	4S	28S + 18S	4S
$+/+$nu	67.3 ± 1.1	9.4 ± 0.4	35.3	6.7
$+/p^1$nu	58.1 ± 2.0	10.4 ± 0.2	30.1	7.1
p^{1-1}/p^{1-2}nu	33.1 ± 1.1	20.6 ± 1.3	21.5	6.1

Table 6. The relative rate of rRNA synthesis in embryos with different amounts of rDNA. (After Miller and Knowland, 1972)

	Nucleolar genotype				
	$+/+$	$+/p^{1-1}$	$+/0$	p^{1-1}/p^{1-2}	$p^{1-1}/0$
Relative amount of rDNA (%)	100	60	45	35	23
Relative rate of rRNA synthesis (%)	100	100	100	50	25

50% of normal and have about 70% of the haploid amount of rDNA distributed between two NO's (Table 5).

However, there is the possibility that the deletion overlapped not only the region of rDNA loci, but other loci as well. In connection with this the question then came into being: perhaps the observed changes in rRNA synthesis could be explained, not only through an inadequacy in the ribosomal gene itself, but also through a mutation at one of the *regulatory* sites. For instance, some bobbed mutants, in which the lethal effect was induced by ethylmethansulfonate, do not differ from the wild type in regard to their amount of rDNA (Marrakechi and Prud'homme, 1971).

It is known that the amount of rDNA does not vary dramatically with different synthesis rates in tissues which synthesize ribosomal RNA (Brown, 1966) (Table 6).

It was demonstrated that the reduction in rate of rRNA synthesis in the $p^{1-1}/0$ nu r^{1-1}/p^{1-2} nu embryos was not a result of blockage in the precursor rRNA to mature rRNA transformation. The data (Table 6) showed that the ratio of the synthesis rate to the number of corresponding genes is lower in the $p^{1-1}/0$ and p^{1-1}/p^{1-2} embryos than in embryos with a $+/0$ nu genotype. These data suggested that the rDNA is not transcribed with full effectiveness in mutants with a partial insufficiency in the NO. One possible explanation of this phenomenon is the assumption that the NO contains the genes controlling the transcription of big blocks of rRNA genes (Perry and Kelly, 1970). A mutation in one of these regions reduced the activity of a large number of corresponding genes. It was suggested that in the case of the bobbed mutation in *Drosophila* a similar mutation could have occurred (Miller and Knowland, 1972).

Chromosomal rearrangements, especially inversions, have a noticeable influence on ribosomal gene transcription. It was shown that the transcriptional character depends on the position of the ribosomal genes in the chromosome. Suppression at the rRNA cystrons was first demonstrated with the aid of genetic methods (Baker, 1971) by analysis of the reduced viability of X/O males carrying the inversions on the X chromosome. An extreme reduction in viability of these individuals was demonstrated when the one breakage point was located between the NO and the centromere. The activity of the genes controlling the individual's viability during ontogenesis changed as a result of certain inversions. At the same time, Baker suggested that two lethals, sc^{L8} and sc^{S1}, when included in the inversion, changed development through the suppression of rRNA gene activity. To investigate this suggestion, the author combined X chromosomes with inversions with Y chromosomes whose rDNA is not transcribed (Ybb$^-$), and checked the ability of these chromosomes for complete reparation of viability. The strains Ybb-A, Ybb and Ybb-SDf were used in these crosses. These mutations did not prevent the lethal effect of chromosomes carrying sc^{4L} and sc^{8R} and lacking an NO. It was found that these three Ybb chromosomes reinstated the viability in no more than 50% of sc^{L8}/0 and sc^{S1}/0 males. The author interpreted these data as evidence that these two inversions suppressed rRNA synthesis (Baker, 1971).

This hypothesis was soon corroborated biochemically. It was found that the amount of 28S and 18S ribosomal RNA in sc^{S1}/0 larvae is 15% lower than the control (w/o). With the aid of DNA/RNA hybridization it was demonstrated that the amount of rRNA in sc^{S1} chromosomes does not differ from normal and, therefore, reduction in the amount of rRNA cannot depend on deletion (Nix, 1973). The observation makes it possible to conclude that the disturbance in synthesis of 18S and 28S ribosomal RNA in sc^{S1}/0 individuals was induced by transcription suppression of the corresponding genes as a result of position effect. Different mechanisms for the reduction in amount of ribosomal RNA could be postulated: (1) transition of the 18S and 28S genes to the "masked" condition when transcription is not possible; (2) absence of the enzyme controlling the transition of the 38S ribosomal precursor into 18S and 28S mature RNA; (3) absence of the ribosomal proteins and, as a result of this, various defects in the ribosomal assembly; (4) absence of the specific enzyme RNA-polymerase for rRNA (Nix, 1973). It was demonstrated that in oocytes of the strain In(1)sc^{S1} + S y sc^{S1}B and In(1)sc^{L8}, sc^{L8}w^{a}mcar the amount of rRNA is reduced by 10%–12%. At the same time, the synthesis rate of 28S and 18SRNA during oocyte development is normal. This could be explained by defects in the ribosomal assembly. In any case, the position effect had an adverse influence on the late phase of rRNA synthesis during ontogenesis (Puckett and Snyder, 1974).

Therefore, the different types of chromosomal rearrangements (deletions, inversions) involving genes coding for rRNA led to a change in the transcriptional character of the corresponding DNA sites.

The position effect was discovered in *D. melanogaster* also using the example of transcriptional activity of structural genes. Particularly, transfer of the Pgd gene from the left chromosomal end, where this gene is located, to the heterochromatin region of the same chromosome as a result of the chromosomal rearrangement Dp(1:f)R, reduced the activity of the 6-PGD enzyme which is under control of this gene. The activity of the homologous locus in the X chromosome of the same individual was not changed (Gvozdev et al., 1973). A similar picture was demonstrated for the isoamylases of *D. melanogaster* (Bahn, 1971).

Some of the mutations which change the transcription of ribosomal genes in *D. melanogaster* are known. However, the mechanisms of these gene actions are not clear.

In the first place, the mutation *lethal-translucida* (ltr, 3-20 ± 0.8; Lindsley and Grell, 1967) should be mentioned. This mutation is characterized by the additional accumulation of hemolymph during development and by the reduction of the fat body. Pupation is delayed in such a strain and the imago does not emerge. The relative rates of synthesis of the various RNA molecules, particularly rRNA in the ltr/ltr larval genotypes are similar to the rate of synthesis in normal larvae, although this rate is delayed during ontogenesis. Shortly before pupation, the synthesis of all types of RNA is inhibited in the mutant flies while it continues in the wild-type flies. The 28S and 18S fractions of RNA demonstrated the lowest label incorporation. The synthesis inhibition was also demonstrated for RNA fractions which have a mobility between 5S and 18S on polyacrylamide gel. These RNA fractions, possibly mRNA, greatly stimulated the introduction of amino acids into proteins (Chen, 1971; Kubli et al., 1971; Weideli, 1971).

The mutation *lethal-meander* (lme, 2-71 and 73; Lindsley and Grell, 1967) has an early effect on larval development. This mutation inhibits larval growth and is characterized by a very low concentration of DNA and RNA in the salivary glands, digestive tube, and other organs. The RNA/DNA ratio (2, 2–3, 4) is also reduced as compared to the normal (8.8–10.8). However, 28S and 18S RNA are synthesized with delay. Therefore, this mutation affects only the time of transcription (Chen et al., 1963; Weideli et al., 1969; Kubli, 1970; Chen, 1971; Kubli et al., 1971).

It was suggested (Chen, 1971) that a similar mechanism of RNA synthesis distortion occurs in the mutant larva *lethal giant* (lgl, X chromosome, 2,0 ±; Lindsley and Grell, 1967).

The dose compensation is an example where the gene interaction in the regulation of the nonribosomal RNA synthesis was demonstrated for various animals and especially *Drosophila*.

1.1.7 Dose Compensation of Genes and Transcription

The genetic mechanism of dose compensation is such that one dose of the sex-linked genes of the heterogametic sex leads to the appearance of the same phenotype as does the homogametic sex possessing two doses of the genes. This phenomenon was described by Bridges (Bridges, 1922) and was assigned its terminology by Müller (Müller et al., 1931). The equalization of X chromosome-linked doses in mammals is provided by the total (or almost total) inactivation of one of the female X chromosomes through its heterochromatization ("Lyonization", Lyon, 1961). This case will be discussed in more detail in Chapter 1.3.3.2. Such an inactivation

Table 7. 6PGD relative activities for various gene dosages in *D. melanogaster*. (After Seecoff et al., 1969)

Genotype	Phenotype	Number of PGD gene	Enzyme activity[a]
g/Df(g)	Female with deletion	1	2.81 ± 0.04
g/g	Normal female	2	4.66 ± 0.03
g/Y	Normal male	1	4.65 ± 0.10
g/Dp(g)/Y	Male with duplication	2	6.26 ± 0.26

[a] Enzyme activities expressed as $mM \times 10^3 NAD/ml/min/mg$ protein

does not exist in *Drosophila*, and other specific regulatory mechanisms are in action. These compensate for the differences in gene number in the male and female (see review of Lucchesi, 1973).

As a biochemical example of dose compensation in *D. melanogaster* we may use the activity of 6-phosphogluconate dehydrogenase (6-PGD), controlled by gene Pgd which is located on the X chromosome (Table 7). The table clearly shows a direct correlation between enzyme activity levels and the structural gene dose in each sex. At the same time, one dose of the gene in the male determines the same 6-phosphogluconate dehydrogenase level as do two doses in the female. Similar data were also observed for other enzymes whose synthesis is controlled by sex-linked genes. Examples of this are glucoso-6-phosphate dehydrogenase (gene Zw; Gvozdev, 1968; Gvozdev et al., 1969, 1972; Seecoff et al., 1969; Bowman and Simmons, 1973; Lucchesi and Rawls, 1973) and tryptophan pyrrolase (Baillie and Chovnick, 1971; Tobler et al., 1971).

X-chromosomal segments controlling various traits demonstrated the ability of dose compensation regardless of their position in the genome (Gvozdev et al., 1972). It was demonstrated that dose compensation ability remained in the translocation mutants $T(1;3)ras^v$ and y^+Yv+ of *D. melanogaster*. In the former mutation, the translocation carried the segment of X chromosome with the vermilion locus, which controls the synthesis of tryptophan pyrrolase into the centromeric or heterochromatic region. The second mutation has a translocation which carries a shorter segment with the same vermilion locus to the short arm of the Y chromosome (Table 8; Tobler et al., 1971). At the same time, the autosomal gene coding

Table 8. Tryptophan pyrrolase activities in $T(1; 3) ras^v$ homozygotes, and heterozygotes[a]. (After Tobler et al., 1971)

Genotype	Sex	Doses of v^+	Activity	SE	No. of experiments
T/Y	M	1	9.9	0.7	6
T/T	F	2	8.6	0.6	5
T/v^{36f}	F	1	5.3	0.5	6
T/v^+	F	2	9.9	0.7	5

[a] Enzyme activities expressed as mm kynurenine/mg protein/2 h incubation. T presents v^+ located on the third chromosomes

of aldehyde oxidase (3 chromosome) in *D. melanogaster* does not demonstrate dose compensation when it is translocated into the X chromosome (Roehrdanz et al., 1977).

Concerning the example of phenotypic expression of traits, it was proven that, in *Drosophila*, dose compensation is the result of the decrease of activity in both of the female homologous X chromosomes. In fact, females that are heterozygous for electrophoretic variants of the Pdg gene possess both parental dimer fractions of 6-phosphoglucate dehydrogenase and a third, hybrid fraction, which is the result of parental monomer combinations. These data led to conclusions on the activity of both homologs (Kazazian et al., 1965). Characteristically, dose compensation occurred not only in adult flies, but also during larval stages of development (Young, 1966).

There are two hypotheses which have been proposed to explain the phenomenon of dose compensation. According to the first hypothesis, the X chromosome contains specific gene modifiers which reduce the effect of the compensated gene. A double dose of these gene modifiers eliminates effects of the extra dose of the compensated gene in the female (Stern, 1929; Müller et al., 1931; Müller, 1950). Müller observed that white-apricot (w^a) eyes in males which carried a duplication of this mutant allele were significantly darker than the eyes of w^a females, although both sexes had the same gene doses. This phenomenon was explained as follows: the X chromosome, where these genes are located, contains other genes as well. These genes are able to reduce the eye color from the level of two doses in the female to the single dose level in the male.

The second hypothesis on the phenomenon of dose compensation was suggested by Goldschmidt. This hypothesis rejected the presence of specific compensatory gene modifiers and explained dose compensation through physiological peculiarities in male development as compared with female development (Goldschmidt, 1961).

Goldschmidt based his hypothesis on the fact that the rates of male and female development differ, the growth phases are not synchronous, the determination time of various tissues varied in individuals of different sexes, and the rhythms of individual organ differentiations differed. It is precisely the specificity of sex physiology during ontogenesis which determines dose compensation, according to Goldschmidt.

A number of authors have tried to prove the former hypothesis (Müller, 1950) as well as the latter (Komma, 1966; Lee, 1968), but in both cases serious objections were found. For instance, it was demonstrated that an additional X chromosome with a deletion of the Pgd structural gene does not change the total 6-phosphogluconate dehydrogenase activity in the body of the fly, while a third intact X chromosome increased the total enzyme activity by 50% as compared to females containing two X chromosomes. This observation makes doubtful the hypothesis concerning the presence of specific gene compensators in *Drosophila* (Gvozdev et al., 1972; Khesin, 1972). Stewart and Merriam (1975) were unable to find specific dosage modifiers.

The mutations dsx and tra were used to investigate the Goldschmidt hypothesis. The mutation dsx (double sex) transformed XX females and XY males into 2X and 1X diploid intersexes and the mutation tra (transformer) induced the develop-

Table 9. Pigment determination [a]: X-linked females versus 2X intersexes. (After Smith and Lucchesi, 1969)

Strain [b]	Female	2X intersex	Ratio 2X intersex/female
Prune	151.43 ± 8.49	141.68 ± 8.45	0.94 ± 0.06
White-apricot	42.12 ± 3.22	39.49 ± 2.02	0.94 ± 0.05
White-eosin	20.75 ± 1.61	22.48 ± 1.75	1.09 ± 0.11

[a] Amount of pigment expressed as the coloration intensity determined spectrophotometrically/100 heads/ml alcohol/one unit of the eye area $\times 10^4$
[b] White-apricot comparisons could not be made due to the extreme inviability of 2X intersexes

ment of the XX potential female into individuals with completely male traits (pseudomales). If sex physiology is of crucial significance in dose compensation mechanisms, the functional activity of the X chromosome should correlate with the sex characteristics, independently of chromosomal constitution. This independent correlation was not observed.

Komma (1966) suggested that there is a gradient of the physiological surroundings toward male or female development and that the corresponding parameters of the medium reduced or increased gene activity. In this case XX intersexes and pseudomales should have a higher enzyme activity than their normal sisters since they are similar to males. At the same time, XY intersexes should have a lower gene activity than their normal brothers, because they are similar to females. On the basis of glucoso-6-phosphate dehydrogenase activity determinations in individuals carrying either the gene tra or dsx, the author concluded that the results observed in the XX intersex and pseudomale analyses corresponded to expectation. On the other hand, Komma did not find confirmation for his assumption on XY intersexes. In general, the author suggested that sex physiology plays an important role in the regulation of enzyme activity.

However, other authors' data are in disagreement with the hypothesis suggested by Komma (Smith and Lucchesi, 1969). Although XY intersexes demonstrated a pigmentation level significantly lower than their normal brothers in two out of three cases in mutations controlling eye pigmentation in *Drosophila* (prune and white-apricot[2]), a non significant increase in pigment was observed in the mutant white-apricot. Moreover, XX intersexes demonstrated a statistically significant decrease in dose compensation as expressed in a low amount of pigmentation as compared to their normal sisters. These data are in conflict with Goldschmidt's hypothesis (Tables 9 and 10).

Therefore, at the present time the actual data do not allow the unreserved acceptance of either the gene-compensator hypothesis or Goldschmidt's physiological hypothesis as a complete explanation for dose-compensation mechanisms.

Concomitantly, the possibility of casting some light on the nature of this phenomenon has appeared in connection with progress in the development of biochemical and cytological methods of investigation. First of all, at what level does dose compensation take place, transcriptional or post-transcriptional?

Table 10. Pigment determination[a]: X-linked males versus 1X intersexes (gene dsx). (After Smith and Lucchesi, 1969)

Strain	1X intersex	Male	Ratio 1X intersex/male
Prune	152.08 ± 11.36	179.52 ± 7.96	0.85 ± 0.05
White-apricot	11.05 ± 0.59	10.26 ± 1.01	1.10 ± 0.11 [b]
White-apricot^2	40.27 ± 3.53	44.53 ± 3.14	0.91 ± 0.10
White-eosin	7.71 ± 0.84	7.13 ± 0.81	1.09 ± 0.10

[a] See Table 9
[b] Statistically not significant

A large number of papers have testified that the transcriptional level plays the main role. Using autoradiography, several authors have demonstrated that the incorporation of H^3-uridine into RNA per unit length of the normal male X chromosome proceeds twice as fast as that found in the normal female X chromosome (Mukherjee, 1966; Kaplan and Plaut, 1968; Lakhotia and Mukherjee, 1969; Korge, 1970b; Chatterjee and Mukherjee, 1971a; Ananiev, 1974).

Korge (1970b) demonstrated that the dose compensation (the 3B puff of $Df(1)w^{258-13}$, $y/++$ strain D. *melanogaster* was chosen as an experimental model) or the dose effect (duplication in female and male) occurs on the transcriptional level (Fig. 14). Although direct comparison was not possible, there is good correlation between the RNA synthesis and the phenotypic expression in the case of the dose compensation as well as in the case of the dose effect, when the trait expression is associated with the gene number. Two results represented the special interest:

1. The same chromosomal segment -3B puff in some cases can demonstrate dose compensation [in the $Df(1)w^{258-11}$, $y/++$ flies] whereas in other cases it exhibits the dose effect [in ywbb; $Dp(1;3)$ w^{vco} flies].
2. The increase of the RNA synthesis in males is significantly greater than in females. However, the enzyme activity in $Df(1)w$ was twice as low as in normal males (Gvozdev et al., 1972, 1975). This observation contradicted the conclusions of Korge.

At the same time, it was demonstrated that the X chromosome reduplication in males proceeds more rapidly than in females, especially toward the very end of the DNA synthesis period. The major part of replicones in the male X chromosomes completed the DNA synthesis earlier than those in the female X chromosome. The character of reduplication of autosomes differs also for sexes; the sequence of the replication of various chromosomal segments is similar, but the H^3-timidine incorporation into DNA in male is higher than in female (Lakhotia and Mukherjee, 1970; Chatterjee and Mukherjee, 1971a; Ananiev, 1974).

The change in position of any part of the chromosome in the genome (for instance, the transfer of an autosomal segment from chromosome 3 to the X chromosome) does not change the dose compensation or the dose effect at the level of direct phenotypic expression or at the transcriptional level of RNA synthesis (Lakhotia, 1970).

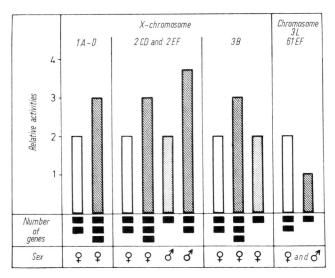

Fig. 14. Dose compensation and dose effect on the level of the RNA synthesis in small segment of X chromosome and in autosome 3, autoradiographic method. *Open bars,* normal activity; *striped bars,* dose effect; *stippled bars,* dose compensation. (After Korge, 1970b)

Therefore, dose compensation is determined by differences in the transcriptional activity of the X chromosome in the male and female and is an example of the well-proven regulation of gene activity in animals. This regulation is apparent in the synthesis on only a certain amount of RNA. Various parts of the chromosome have demonstrated a kind of autonomy in this respect.

What is the cause of these differences in the transcription rate which have been described – the hyperactivity of male X chromosomes or the reduction in activity of female X chromosomes?

According to previous (Offerman, 1936) and recent (Lakhotia and Mukherjee, 1969; Holmquist, 1972) cytological data, the first explanation seems to be the more probable one. For instance, larva gynandromorphs were obtained when the unstable ring X chromosome was used. These larvae are mosaics and contain the elements of both the female (ring X/regular X) and the male (regular X/O) cell constitutions. Morphologically (the width) and functionally (the RNA synthesis) the one X chromosome of "male" cells in this larva is equivalent to the double X chromosome of the "female" cells (Lakhotia and Mukherjee, 1969).

What are the concrete mechanisms of gene activity regulation in dosage compensation?

The role of the number of interactions among genes and, possibly, the complexity of interactions on dosage compensation is undoubted. Apparently, there is a system for the synthesis of regulated compounds which migrate into the nucleus and affect the functional condition of chromosomes. It was suggested that this system involved the presence of some activating and inhibiting factors. The ratio between these compounds is the regulatory factor for the X chromosome activity level (Seecoff et al., 1969). For example, there exists a factor in the extracts of male, pseudomale, and intersex which were obtained by sex transformation. The addition of this factor could change the electrophoretic mobility of glucoso-6-phosphate dehydrogenase. It is possible that this factor directly participates in dose compensation regulation by interaction with the G-6-PDH molecules (Gvozdev et al., 1969).

Investigation of RNA-synthesis and glucoso-6-phosphate dehydrogenase synthesis in triploid intersexes (2X3A) and triploid females (3X3A) demonstrated that there is a tendency toward equalization of the relative activities of genes located on the X chromosome in 2X3A and 3X3A individuals. The expression of dose compensation was not complete, and the level of RNA synthesis and enzyme activity depended on the ratio between the number of X chromosomes and autosomes (Maroni and Plaut, 1973 a, b).

It is possible that autosomal genes control the synthesis of regulatory molecules which are necessary for the activation of any X-linked gene activity. Therefore, the activity of this gene is directly proportional to the amount of the regulatory molecules (Lucchesi, 1973).

One of the working hypotheses formulated the following basic principles:
1. The formation of the factor which limits the transcription rate of X chromosome-linked genes is controlled by autosomes and this factor is necessary for RNA synthesis.
2. Most of the sex-linked genes reacted in the same manner to this regulatory factor. Therefore, all regions which belong to one functional group competed for this specific factor.
3. The binding constant of the regulatory molecules is high and their synthesis rate is much lower than the maximum potential rate of their usage by the X chromosome. Therefore, the pool of unbound factors is small and there is always a surplus of "free" genes (Maroni and Plaut, 1973 a, b). It has been suggested that DNA-dependent RNA polymerases could serve as these regulatory molecules (Schwartz, 1973 a).

Data obtained by Ananiev and coauthors supported the suggestion that autosomes played some role in the realization of dose compensation (Ananiev et al., 1974). They investigated the H^3-uridine incorporation into the polytene X chromosomes of males (1X2A), females (2X2A), superfemales (3X2A), intersexes (2X3A), supermales (1X3A), and triploid females(3X2A). The results of this experiment are presented in Table 11. The transcriptional activity of one X chromosome in the supermale (1X3A) is of the same level as in the normal male (1X2A), while the activity of one X chromosome in the female (2X2A and 3X3A) is twice as low. The introduction of an extra X chromosome into superfemales (3X2A) does not reduce the level of X-chromosomal transcription.

These data are in good agreement with data on the activity of 6-phosphogluconate dehydrogenase from the same individuals (Fig. 15). Therefore, in the case of dose compensation, the correlation between transcriptional activity and the biochemical expression of the trait was strictly demonstrated (Faizullin and Gvozdev, 1973).

These authors considered that the system control and the autonomous response of the genes of the X chromosome allow the supposition of the presence of similar

Table 11. Incorporation of H^3-uridine into polytenic chromosomes of the salivary gland of individuals with various X:A ratios. (After Ananiev et al., 1974b)

Geno-type	X:A	Phenotype	Number of nuclei	Number of the silver granules on		$\bar{X}/\bar{A} \pm SE\%$	q [a]
				X-chromo-some	Autosomes		
1X2A	0.50	Male	20	189 ± 13	786 ± 51	24.0 ± 0.75	0.48
2X2A	1.00	Female	15	345 ± 71	1429 ± 280	24.2 ± 0.64	0.25
3X2A	1.50	Superfemale	17	682 ± 78	1925 ± 225	35.6 ± 0.72	0.24
1X3A	0.33	Supermale	11	613 ± 95	3728 ± 566	16.5 ± 0.63	0.50
2X3A	0.66	Intersex	22	522 ± 63	2182 ± 228	24.6 ± 0.92	0.36
3X3A	1.00	Female	9	323 ± 32	10403 ± 121	23.0 ± 0.83	0.23

[a] The characteristic of the transcriptional activity on the $\bar{X}$ chromosome: $q-(\bar{X}/p):(\bar{A}/n)$, where p=number of X chromosomes in cell, n=number of autosomal sets. There are no significant differences between the mean ratio (X/A%) in diploid male, diploid female, triploid female, and intersexes. The mean X/A% for the all ratios in diploid and triploid females is 23.7 ± 0.51. There are no significant differences in the labeling of X chromosomes between two sexes

regulatory elements in all X-linked loci. These regulatory sites reacted to a common controlling factor. Because this factor affected all genes located on the X chromosome independently of their localization, it is necessary to suppose that it is diffusible.

There are differences between data observed by J. Lucchesi's group and those found in Gvozdev's laboratory. In the former, the smooth modulation of X-linked gene activities was dependent upon the X:A ratio (Lucchesi et al., 1974, 1977). In the latter, such a smooth dependence was not observed, as is shown in Fig. 15 (Ananiev et al., 1974a, b). Additional data are necessary in order to make a final determination.

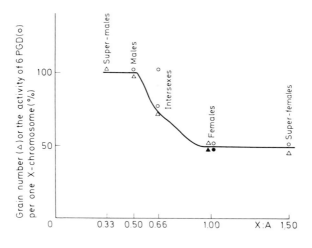

Fig. 15. Effect of X : A ratio on the activity of the X-linked genes. *Dark circle* and *triangle*, triploid females. (After Ananiev et al., 1974a).

An interesting consideration was expressed by Holmquist (1972). Holmquist determined that up to 100% of dosage compensation, on the phenotypic level of gene expression (e.g., enzyme activity) in *D. melanogaster*, corresponded to only 40% of the transcription rate increase in the male X chromosome as compared to the female. Thus, two possible mechanisms for dose compensation were suggested:

1. The selective stabilization and transport of chromosomal RNA in males. In this case, the high maintenance of proteins in the male X chromosome as compared to the female (the ratio of protein/DNA is 10% higher in the male; (Rudkin, 1964) could result in the 40% increase in transcriptional activity as well as in its product stabilization. As a result, a more effective mRNA transport into the cytoplasm is obtained. Final compensation is completed on the post-transcriptional level. Thereafter the "amounts" of synthesized mRNA began to equalize as a result of the heightened degradation of the female mRNA as compared to the males.

2. It is possible that the RNA fraction which is quickly degraded has a control function and the stable fraction plays the role of mRNA. The surplus protein in the male chromosomes (as well as in the puff regions) could selectively stimulate stable RNA (mRNA) synthesis. Hyperactivity in the male X chromosomes related especially to the synthesis of this fraction of RNA. At the same time the synthesis of the quickly degraded RNA remained at the same level. If the latter has a "repressor" function, its concentration in the female could be higher than in the male.

There are two possible types of response of X-linked genes to the action of their corresponding regulatory molecules: (1) all genes which are located on the X chromosome demonstrate a response to this type of regulator; (2) genes are grouped into blocks and there are block-specific regulating compounds (Maroni and Plaut, 1973 a, b). The situation in the second case conformed well with the gene material organization model in eukaryotes suggested by Britten and Davidson (1969) and Georgiev (1973). According to this hypothesis, each operon is made up of two major parts: the larger part is the acceptor or noninformational site; the smaller is the structural or informational site.

There have been some genetic data which have corroborated the presence of regulatory sites in X-chromosomal loci. The mutation apricot (w^a) of the white gene demonstrated the ability for dose compensation, hence the eye color of males X^{w^e} Y and females X^{w^e} X^{w^e} is the same. Another mutation, eosin (w^e), of the gene white does not demonstrate the dose compensation effect. Thus, the eye color of X^{w^e} X^{w^e} females is twice as dark as that found in Y^{w^e} Y males. These two mutations, w^a and w^e, are located in different parts, left and right respectively, of the white gene. These two parts of the white gene differ in their functional significance. The crimson (w^c) mutation often induced a transposition of the left part of the white gene into different zones of the third chromosome (Fig. 16) where it continued to function. It is of interest that when the mutation in the right part of the locus inactivated the entire white gene (no pigment synthesized), the transposition of the left part of the gene to the third chromosome may repair gene activity (pigment is synthesized). It was suggested that the white locus is composed of at least two parts: structural–between mutations w^{BWX} and w^c and regulatory–including mutations wch and w^{sp}–sites (Green, 1969 a, b).

Recently, Leibovitch and Khesin (1974) obtained single male X chromosomes which absorbed twice as much H^3-RNA polymerase from *E. coli* as did each female

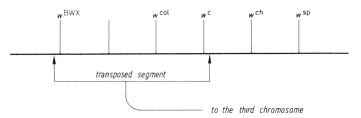

Fig. 16. A micromap of the w region indicating the postulated maximum genetic length of the transport segment. (After Green, 1969 a, b)

X chromosome in cytological preparations. The authors explained this phenomenon through the additional decondensation DNP in the male X chromosome which makes possible an easy access for exogenic proteins to the DNA. They suggested (Khesin and Leibovitch, 1976) that possibly a change in the structure of the chromosome is the cause for the additional transcription. The 38B-40 deletion in the second chromosome included the histone genes between 39 D2-3 and 39 E1-2, which induced X chromosome decondensation in intersexes and increased the transcription rate in vivo as well as in vitro. It is possible that histone incorporation into nuclei Df_{hist} slowed down slightly which, in turn, initiated the switch in the X chromosome to another activity level.

It was suggested that, in dose compensation, the regulation of transcription in chromosomes is realized on two levels: (1) the change in degree of DNA condensation which, in turn, determines the activation of corresponding loci, and (2) the activation of corresponding loci by the regulatory protein activators which are specific to particular genes. According to the first mechanism the decondensation of X-chromosomal DNP made male sex-linked genes twice as receptive to the activators as compared to female sex-linked genes (Leibovitch et al., 1974; Khesin, Leibovitch, 1974).

The X chromosome of *Drosophila* aroused specific interest because at least four levels of gene activity were obtained: The highest level of sex-linked gene activity was registered in males, the lowest in metafemales, and two intermediate levels were found in triploid intersexes and aneuploid females. The relative levels of gene activity are established comparatively early in development, before the last larval state. The elucidation of causal factors for the specific levels of X-linked gene activity in *Drosophila* will clarify an understanding of gene activity regulation in eukaryotes (Lucchesi et al., 1974).

It is possible that dose compensation is not an exceptional mechanism for gene activity regulation, but is a part of the total system of gene regulation in eukaryotes, characterized by groups of functionally united genes. In that case, this regulation is apparently realized through different chromosomal interactions, whereas the regulatory molecules, possibly proteins, are synthesized by the autosomes. These molecules reacted with the site regulators of gene complexes and, therefore, stipulated the maintenance of a specific level of gene activity.

Finally, some data (Holmquist, 1972) have shown that, on the phenotypic level of dose compensation, post-transcriptional processes are also important.

1.2 The Realization of Differential Gene Activity on the Post-Transcriptional, Translational, and Post-Translational Levels

1.2.1 The Stages of Gene Activity Realization in the Phenotype

Trait realization on the molecular level undergoes a number of stages, from transcription to RNA maturation and transport, then translation and, finally, the post-translational level.

At each stage in the realization of any particular character, the complex direct and indirect influences of other genes are possible. In this sense, differential gene transcription is only the first step in the pathway of biochemical specification of different cell and tissue types. In this case, differences in RNA-polymerase specificity and efficacy could also play an important role. After this stage followed the selective stabilization and maturation of the potential mRNA from the total pool of nonstable, heterogeneous nuclear RNA. The third stage comprised the regulation of transport of the "mature" potential mRNA from the nucleus into the cytoplasm. This stage also includes changes in the nonrandom transport specificity through physiological peculiarities and hormone interactions. The fourth stage is the differential stabilization of the mRNA in the cytoplasm. The fifth stage is concerned with the peculiarities in the translational effectiveness of the protein-synthesizing mRNA and ribosome complexes. The sixth and last stage involves the differential stabilization of the definite protein products (Church and Brown, 1972).

Indeed, the stability and half-life of the RNA itself is dependent upon the type of cell in which this mRNA acts. For instance, in reticulocytes the globin RNA has a half-life of 1 day, whereas, in oocytes, after injection, its half-life is no less than 2 weeks (Gurdon, 1973; Hunt, 1974).

Stabilization of molecules of mRNA occurred at the moment of joining with informophers. The mRNA products of different genes complexed with different proteins, therefore gene products, are characterized not only by the uniqueness of their nucleotide sequences, but also by the uniqueness of the proteins which will form complexes with them (Pederson, 1974). In this connection, it is possible that various nuclear RNA's require different times for intranuclear processing. Some authors have supposed that mRNA stabilization depends upon its polyadenylization. It was suggested that the presence and extent of polyadenylate sequences determine the longevity and translation period of mRNA (see review Gaizchoki and Kiselev, 1974; Greenberg, 1975).

Later on, a certain role was played by differences in the rate of transport mRNA's from the nucleus into the cytoplasm. In particular, the histone mRNA is obtained in the cytoplasm very soon as compared with other fractions (Schochterman and Perry, 1972). Edström and his co-workers were able to distinguish two types of heteromorphic RNA in the salivary glands of chironomides–an early and

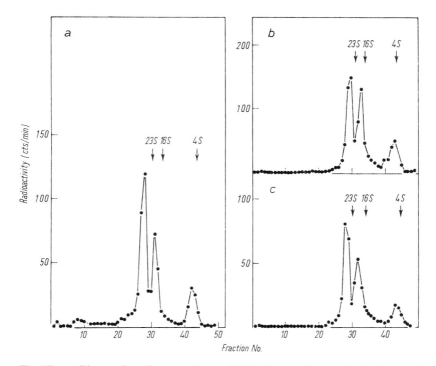

Fig. 17 a–c. Electrophoretic separations of RNA from salivary gland cell cytoplasm **a,** gastric caeci **b** and midgut **c** 2 days after precursor administration to the medium (25 µCi [³H]uridine and 25 µCi [³H]cytidine in 5 ml of culture medium for 20 h after which the animals were left in isotope-free medium for a further 48 h before killing). It can be seen that the late mRNA is only present in salivary gland cell cytoplasm. (After Edström and Tanguay, 1974)

a later RNA. The early RNA was localized on electrophoregrams between 5S and 18S RNA and migrated relatively quickly (after 30 min) into the cytoplasm, as did the heterogeneously dispersed RNA from HeLa and L cells (Greenberg, 1975). The later, highly polymeric (75S) RNA was obtained in the cytoplasm only after 3 h following the label introduction.

Both fractions differ in regard to their stability. The former fraction was no longer obtained in the cytoplasm after 1 to 2 days. The latter type of RNA could be obtained in the cells even after several days. It is, apparently, the product of the Balbiani's ring-2 (BR-2) and possibly also BR-1. The RNA may also be organospecific since it was obtained only from salivary glands and not from the intestine (Fig. 17; Edström and Tanguay, 1974). The amount of this RNA in the cytoplasm is approximately equal to 2500 pg per cell and it has a half-life of about 20 h in the cytoplasm and 65 min in the nucleus. Almost all of this RNA is transported from the nucleus into the cytoplasm (Edström et al., 1978).

In *Drosophila* tissue cultures stability differences were found among various cytoplasmic mRNA's (Lengyel and Penman, 1971).

Two fractions of poly-(A)-containing mRNA were also found in mouse blasto-cysts. One fraction has a half-life of about 7 h, the other, "longliving" fraction, of about 18 h (Schultz, 1974). During lens differentiation in chickens, the different subunits of crystallin are synthesized on mRNA's of different stability (Clayton et al., 1974). Differences in longevity were obtained for alanine aminotransferase and tyrosine aminotransferase mRNA's (T$\frac{1}{2}$ = 12–14 and 2 h respectively). It was suggested that the differential degradation rates of various mRNA molecules de-pend on the peculiarities of their molecular structure and, especially, on the part of the molecule which determines ribosomal binding and the initiation of transla-tion (Stiles et al., 1976).

Certain data have shown that the terminal cell differentiation of neuroblastoma is not accompanied by a change in the transcriptional program but occurs together with a change in the stability of some mRNA types (Croizat et al., 1977). This change in mRNA stability could be caused by its polyadenylization. The enzymatic removal of poly-(A) from the globin mRNA significantly reduced the effective translation time after the microinjection of this mRNA into *Xenopus* oocytes (Huez et al., 1974), whereupon it was shortly degraded (Marbaix et al., 1975). The addition of poly-(A) to globin mRNA, which did not contain it, restored the RNA stability (Huez et al., 1975). The same tendency was observed for histone mRNA. The addition of poly-(A) to this type of RNA increased its longevity from 10 to 40 h in the oocytes (Huez et al., 1977). Obviously there is a substance, possibly a protein, in oocytes which binds the poly-(A) sequences with the mRNA and, there-fore, stabilizes the RNA (Gurdon, 1974c). The stability of mRNA molecules in-creased when peculiar structures, caps, were formed. These structures are located on the 5'-ends, which contain the methyl bases, and the presence of caps increased the rate of translation initiation (Snatkin, 1976).

The specific gene which changed the mRNA stability in *E. coli* has been dis-covered (Kuwano et al., 1977).

Finally, the peculiarities of the mRNA stabilization in the cytoplasm could be dependent upon the number of ribosomes in existence, and the activity of ribo-somes can protect mRNA from degradation (Walker et al., 1976). During the gas-trula and neurula stages of development amphibian endodermal cells synthesized a more active RNA than did ecto-mesodermal cells of the dorsal part of the em-bryo. However, endodermal determination and differentiation (the neurula and late tailbud stages respectively) is delayed in comparison with ectodermal and me-sodermal determination and differentiation (the gastrula and neurula stages re-spectively). According to Flickinger (1971), this paradox could be explained by dif-ferences in the number of ribosomes and in the rate of ribosomal RNA synthesis. During gastrula and neurula stages the embryonic cells contain fewer ribosomes and the synthesis of ribosomal RNA is less intensive than in ecto- and mesodermal cells. In connection with this, the possibilities for stabilization of the newly synthe-sized mRNA are limited in the endoderm. This delays its differentiation in compar-ison with other embryonic layers (Flickinger, 1971). Preexisting, stable mRNA is activated by its translocation into polysomes during sprouting of the wheat embryo (Brooker et al., 1975/76). In the same manner, the poly-(A)-containing mRNA of *Artemia salina* is activated by translocation into cysts during early stages of de-velopment (Amaldi et al., 1977).

Apparently, the differential stability of mRNA molecules could be one of the mechanisms of total differentiation, since variability in mRNA stabilities affected the intensity of synthesis of their corresponding proteins.

On the other hand, regularities in translation are dependent to a great extent upon the activity of the aminoacyl-tRNA synthetases (Strehler et al., 1971) and upon the ratio of different tRNA fractions (Osterman, 1971). It was found that the ratio of these fractions was correlated with the amino acid constitution of the proteins characteristic of this type of cell. This type of correlation was obtained for tRNA in the silk secretory glands of *Bombix mori* and the reticulocytes of different animals (Lee and Ingram, 1967; Litt and Kabat, 1972; Smith et al., 1974).

It was demonstrated that the set of tRNA's in nondifferentiated amphibian tissues is similar, whereas in tissues which are differentiating in different directions, the tRNA compositions differ (Tonoue et al., 1969; DeWitt, 1971; Bielka, 1977). The heterogeneity of tRNA was also observed in the sea urchin *Paracentrotus lividus* (Molinaro and Mozzi, 1969).

In the differentiating tissues of sea urchins each of two types of soluble RNA, metyonil and lysine, may have its "own" active enzyme, aminoacyl-tRNA synthetase. During development, the activity of these enzymes is changed in proportion to the change in their corresponding RNA contents (Ceccarini and Maggio, 1969).

The tRNA complements differ in nondifferentiated tissues and in lactating milk glands. Pregnant animals synthesized mainly those tRNA molecules necessary for the intensive incorporation of their corresponding amino acids into proteins synthesized during lactation. It is possible that the correlation between the amino acid constitution of the main milk proteins, specific tRNA molecules, and their corresponding tRNA synthetases, is established under the influence of hormonal factors (Elska et al., 1971). Evidence of the direct participation of asparaginile–tRNA in the action of the asparaginile–tRNA synthetase structural gene has been obtained (Arfin et al., 1977). It is of interest that, in experiments with globin and ovalbumin mRNA, the rate of in vitro mRNA translation is higher in the presence of homologous tRNA than in the presence of heterologous tRNA. The level of protein synthesis increased in the presence of heterologous tRNA when homologous tRNA was added (Le Meur et al., 1976).

In the mealworm *Tenebrio molitor* it was demonstrated that the hormonal regulation of the phenotypic expression of genes controlling the synthesis of cuticular protein is realized on the transcriptional level, and includes the synthesis of new tRNA and the activation of their associated enzymes. The stable RNA for cuticular proteins is synthesized earlier, but is not translated until the action of the juvenile hormone induces a change in the tRNA contents (Ilan and Ilan, 1972). The existence of differential translation of some triplets suggested the possibility of regulation of cell differentiation through the selective activation or inhibition of the stable mRNA stored in the embryonic cells.

In *Tenebrio molitor* the differences in leucyl tRNA synthetase activity played an important role at the transcriptional level of regulation. It was demonstrated that these differences are determined by the presence of multiple enzyme conformation forms (Ilan and Ilan, 1975). Each conformational form demonstrated its own activity level.

Apparently, the specificity of each set of tRNA's is important at the translational level of regulation, as is the distribution of tRNA inside the cell. In unfertilized eggs of the sea urchin, *Litechinus variegatus*, one form of lysine tRNA is located in the soluble fraction, however, after fertilization it was obtained in the mitochondrial and plasma membrane fractions. This form of lysine tRNA disappeared from the soluble fraction until the pluteus stage (Yang and Comb, 1968).

The activity of numerous gene modifiers affects the system of tRNA's and aminoacyl-tRNA synthetases. An example of this is the phenogenetics of *Drosophila* mutants *Abnormal abdomen*–A^{53}g (localized on the 1st chromosome, right from white; Lindsley and Grell, 1967). The primary effect of this mutation is the distortion of histoblast differentiation and abdominal hypoderm. The penetrance and degree to which this character is expressed are dependent to a great extent on gene modifiers located on the autosomes and X chromosome, and on such environmental effects as humidity, population density, age of culture, etc.

All of these factors are realized during a two-step process: (1) intensification of soluble protein synthesis under the control of gene modifiers (cycloheximide inhibits protein synthesis and, therefore, reduced the expressivity of A^{53}g; Hillman et al., 1973); and (2) the reaction of the histoblasts to this intensification, which is mainly dependent upon the A^{53} gene (Hillman, 1973). The A^{53} gene itself demonstrates a very weak effect and the gene modifiers sharply intensified the phenogenetic expression of this trait.

It has been suggested that the A^{53} structural gene synthesizes mRNA for hypodermal proteins, and gene modifiers control the synthesis of aminoacyl-tRNA. The hypodermal protein is synthesized as a nonactive and modified compound under translation distortion conditions (Rose and Hillman, 1969, 1972).

Therefore, the selection of sets of tRNA and mRNA, which maintain the synthesis of specific proteins, could be obtained at different regulatory levels. Recently obtained data indicate that this process can take place during the formation of

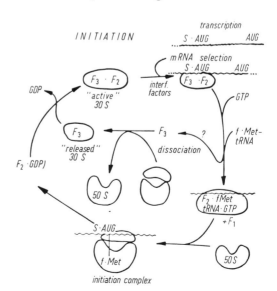

Fig. 18. Steps involved in the binding of mRNA to ribosomes and initiation of translation. (After Revel et al., 1973)

mRNA-ribosome complexes. Figure 18 represents a schematic sequence of mRNA-ribosome complex formation. It is known that the synthesis of a protein molecule on the ribosome includes three actions: initiation, elongation, and termination.

For initiation an initiatory codon (AUG or GUG) is necessary in mRNA, as well as an initiatory tRNA capable of joining with methionine under the control of methionine-tRNA synthetase. The methionine residue is formilated at the NH_2 group in the presence of transformilase. The initiation of protein synthesis requires, not only tRNA and rRNA, but also three protein compounds – F_1, F_2, and F_3, as well as GTP and the 30S subunit of ribosomes.

Both ribosomal subunits (30S and 50S in prokaryotes) take part in the elongation process. Each act in the elongation process consists of three steps: (1) binding of aminoacyl-tRNA with the decoding, acceptor region of the ribosome; (2) transpeptidation, or the formation of peptide bonds between the attached aminoacyl-tRNA and the growing peptidyl; (3) translocation, or the transference of peptidyl-tRNA and corresponding codon of mRNA inside the peptidyl bonding region of the ribosome.

The termination of translation is accomplished with the participation of terminal codons (UAG, UAG and UAA) in mRNA, the protein factor of termination, and both subunits of the ribosomes (see Kiselev et al., 1971).

In eukaryotes, the smaller (40S) ribosomal subunits are joined to mRNA during initiation. The 80S ribosomal subunit arises from the association of 40S and 60S subunits. In these organisms the total initiation complex includes GTP, aminoacyl-tRNA, protein initiatory factors, and mRNA (Ilan and Ilan, 1971).

Protein-initiating factors which were extracted from unwashed ribosomes promoted the formation of 80S complexes only when the mRNA used was extracted from individuals in the same developmental stage. The specificity of protein-initiating factors determines the control mechanism of protein synthesis by ribosomal "selection" of mRNA for translation.

Figure 19A demonstrates the formation of the entire initiation complex when all components are a part of the same developmental stage. However, if C^{14}-labeled mRNA was extracted from 7-day-old pupae, and the initiation factor was extracted from the larval stage (Fig. 19C), the initiation complex was not formed. Similar results were obtained when C^{14} mRNA was extracted from larvae and the initiation factor isolated from 7-day-old pupae. Ribosomal origin is not important, since identical results were obtained with ribosomal subunits extracted both from larvae and pupae. These results confirmed the existence of stage specificity in the initiating factor (Ilan and Ilan, 1971). Since the initiating factor is a part of ribosomal proteins, all mRNA's are not identical to ribosomes. Bacterial ribosomes can differentiate between the products of three f_2 cystrons (Lodisch, 1969). Ribosomes of *E. coli* translated T4 phage RNA more intensively than f_2 RNA after phage infection (Hsu and Weiss, 1969). Therefore, it was suggested that mRNA contains, besides the AUG codon, specific sites for the recognition of some elements of the initiation process.

The use of the specific initiation factor from rabbit reticulocytes chromatographically purified on DAE-cellulose and phosphocellulose demonstrated the stimulation of hemoglobin mRNA translation but was not effective with the

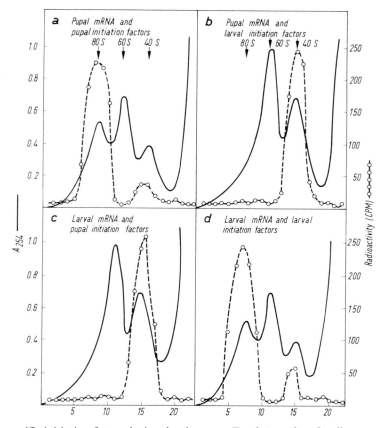

Fig. 19a–d. Stage-specific initiation factor during development *Tenebrio molitor*. In all cases seventh day pupae ribosomal subunits were used. RNP was used as mRNA, and initiation protein factors were prepared from last instar larvae of 7-day pupae and used in combinations as described above. **a** Pupae mRNA and pupae initiation factor; **b** pupae mRNA and larval initiation factor; **c** larval mRNA and pupal initiation factor and; **d** larval mRNA and larval initiation factors. (After Ilan and Ilan, 1971)

mRNA of Mengo's virus. This factor stimulates α-globin synthesis, therefore ribosomal activity could be regulated both quantitatively and qualitatively by this factor (Revel et al., 1973).

Two possible mechanisms for this phenomenon have been suggested:
1. mRNA molecules synthesized at a given developmental stage have similar specific nucleotides sequences (before the formylmethionine AUG codon) which are identified by stage-specific initiation factors.
2. mRNA's are characterized by secondary, stage-specific structures identified by the factors of initiation (Ilan and Ilan, 1971).

On the basis of observed data on identification of mRNA molecules by *E. coli* ribosomes, Revel and his co-workers found a compromise between these two alternative mechanisms. It was observed that ribosomes identified more than only the

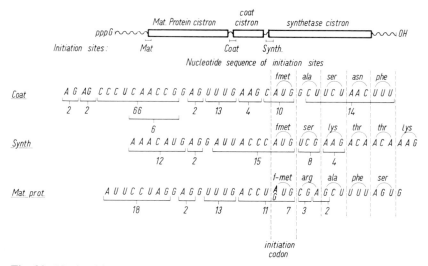

Fig. 20. Nucleotide sequence of the three initiation sites of bacteriophages MS2 or R17 RNA. The numbers refer to the T1 oligonucleotides in 2B. (After Revel et al., 1973)

AUG codon during interaction with mRNA's. Apparently, there are 6–10 nucleotides (Fig. 20; Revel et al., 1970, 1973), or the region with a specific secondary structure near the AUG codon (Steitz, 1969) which is defined as the "signal of initiation"–S. It was also demonstrated that both initiatory factors, F_2 and F_3, participated in the binding of mRNA's with 30S ribosomes.

The F_2 factor is important for the specific recognition of mRNA's. The total mechanism for mRNA "selection" is demonstrated in Fig. 21.

Tomkins and his co-workers suggested a theory on the possibility of regulation of specific gene expression through the action of repressors acting at the translational level (Tomkins et al., 1969). Recently, Heywood and others extracted two specific initiation factors, apparently of a proteinaceous nature. At the translational level one of these factors stimulated the synthesis of myoglobin, the other stimulated the production of myosin. It was possible to extract the so-called tcRNA (translation control RNA) from the fraction which contained the first factor. This RNA inhibited heterologous mRNA translation. Apparently, it is possible to induce a specific inhibition of protein synthesis using tcRNA from various types of cells. Therefore, tissue-specific sets of tcRNA exist and participate in the regulation of protein synthesis at the translational level (Heywood et al., 1974).

Later it was found that two types of tcRNA exist: (1) mRNP-tcRNA was extracted from mRNP particles. This tcRNA was found to inhibit the translation in RNP particles but did not affect the translation in polysomes. Approximately 50% of the uridyl residues which are capable of hybridizing with the poly (A) ends of mRNA were found in this type of tcRNA; (2) polysomal tcRNA, which has no effect on RNP particles but stimulates the translation of polysomal mRNA. This tcRNA contains significantly less uridile residues (Bester et al., 1975). The RNA fraction which exhibited inhibited translation was obtained from *Artemia salina* cytoplasm during the gastrula stage (Slegers et al., 1977).

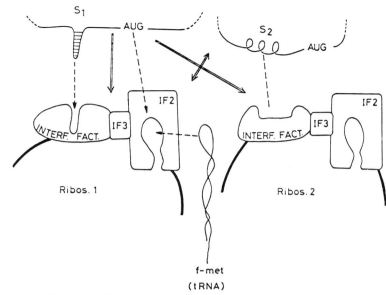

Fig. 21. Hypothesis of the mechanisms of the selective translation of mRNA. (After Revel et al., 1973)

The nuclear U_1-RNA, which contained both cap and the initiatory codon AUG, is not translated in the cell-tree system of wheat embryos, and also inhibited the translation of all mRNA's obtained from rat liver. It was suggested that U_1-RNA inhibited translation by joining with the initiatory factors or with the mRNA binding sites of ribosomes (Rao et al., 1977). The oligonucleotides which regulate translation have been recently extracted from *Artemia salina* embryos. The supernatant contained an inhibitor of translation that blocked the association, which is dependent upon the elongation factor, between aminoacyl-tRNA and the ribosomes. In this case the inhibition of initiation was less expressed. The translation activator was obtained in developing embryos, whereas this activator is absent in nondeveloping cysts. The appearance of this activator stimulated protein synthesis in developing embryos. Both of these compounds (the translation inhibitor and the translation activator) are oligonucleotides. They are, apparently, the definite products of RNA hydrolysis by nucleases of an opposite specificity (T_1 and A respectively). The appearance of the activator during embryonic development is possibly caused by ribonuclease A (Lee-Huang et al., 1977).

Therefore, cell differentiation and the peculiarities of their reaction to the action of inducers and hormones could be determined by the synthesis, distribution, and variation of specific initiatory factors (positive control) or tcRNA (negative control).

One common type of gene interaction is suppression. This interaction takes place at the translational and, possibly, post-translational levels.

In *Drosophila* the existence of a suppressor locus, su(s)–1.00 (Lindsley and Grell, 1967), is known. Mutants at this locus acted as recessive suppressors of the vermilion mutation and partially reinstated the normal phenotype.

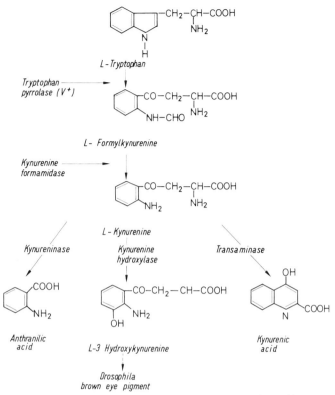

Fig. 22. Metabolism of tryptophan and the formation of brown eye pigment in *Drosophila*. (After Baillie and Chovnick, 1971)

The vermilion (v) gene is located on the same chromosome (1–33.0; Lindsley and Grell, 1967) and controls the synthesis of tryptophane pyrrolase, thereby regulating the pathway of brown pigment synthesis (Baillie and Chovnick, 1971; Fig. 22). The vermilion mutation blocks the formation of formylkinurenine from tryptophane.

There are suppressible (v^1, v^2, v^k) and nonsuppressible (v^{48a}, v^{36t}, v^{51b}, v^{51e}, v^{51d}) alleles at this locus (Green, 1955; Baglioni, 1960; Gvozdev, 1968; Baillie and Chovnick, 1971). Flies with suppressible alleles may synthesize a certain amount of brown pigment when su(s)^2 is homozygous, whereas flies with nonsuppressible alleles cannot synthesize any brown pigment and have brightly colored red eyes. The suppression of the su(s)^2 locus also had an effect on the mutations purple (II–54.5), sable (I–43.0), and speck (II–107.0). What is the mechanism for the action of the su(s)^2 gene? In this case it was demonstrated that suppression cannot be explained by the removal of an endogenous inhibitor, nor does the action of suppression affect the primary structure of enzymes (Marzluf, 1965).

It was suggested that su(s)^2 is the structural gene for tryptophane tRNA (Gvozdev, 1968) similar to prokaryotic suppressors (see Gorini, 1970). However, this hypothesis faced serious objections:

1. The gene su(s)² has a recessive character which is impossible since it directly controls tRNA synthesis.
2. Experiments with molecular hybridization demonstrated the presence of multiple copies of genes synthesizing tRNA's, at an average of 13 for each 60 tRNA's (Ritossa et al., 1966). In connection with this, the mutation of only one copy cannot significantly change the phenotype (the percentage of changed tRNA's in bacteria with nonsense-suppression is 15%, and with missense-suppression about 1%; Twardzik et al., 1971).

Tartof (1969) suggested that mutant products, which are controlled by suppressible v alleles, are monomers incapable of aggregating to form multienzyme complexes. Gene suppressors repaired this ability by changing intercellular conditions.

Perhaps more acceptable is the opinion that the su(s)² locus controls the amount of the specific tyrosine tRNA. Adult flies of this type have three chromatographically distinguishable fractions of this RNA (RNA^{tyr}), however, homozygous flies su(s)²/su(s)² lacked or contained only a little of one of the fractions tRNA$_2$^{tyr} (Fig. 23; Twardzik et al., 1971; Jacobson, 1971). The phenotypic expression of su(s)² and the chromatographic differences of tRNA are recessive and their corresponding genes are located at the same place on the left arm of the X chromosome (Twardzik et al., 1971).

The tryptophane pyrrolase of vermilion flies is associated with tRNA$_2$^{tyr}, therefore its activity is noticeably reduced. The su(s)² mutation leads to the exclusion of tRNA$_2$^{tyr} and to the restoration of enzyme activity. The treatment of homogenates from adult vermilion flies with ribonuclease induced enzyme activation, whereas the tryptophane-pyrrolase from wild-type flies is not sensitive to this treatment (Fig. 24; Jacobson, 1971).

Homozygotes su(s)²/su(s)² showed an inhibition of the production of the negative effector and therefore activated the suppressible vermilion allele. The absence in su(s)²/su(s)² individuals of an isoacceptoral form of tyrosine tRNA may possibly play the role of the negative effector (Baillie and Chovnick, 1971; Twardzik et al., 1971). This hypothesis was developed by White and his co-workers (White et al., 1973).

These authors investigated the chromatographic profiles of various tRNA's. They investigated the relative contents of different chromatographically distinguishable patterns of asparagil-, aspartil-, hystidil-, and tyrosil-tRNA's during ontogenesis and found that during the development of wild-type flies the qualitative ratio between late-eluted and fast-eluted δ-fractions of these tRNA's was changed.

In normal flies the γ-fraction predominates during all stages of development. The content of the δ-fraction in eggs and during the third larval stage is minimal. In su(s)²v, bw flies the amount of γ- and δ-fractions is equal and the content of the γ-fraction is insignificant in adult flies (Fig. 25).

Apparently the differences between the γ- and δ-fractions are determined by post-transcriptional modification of the same gene product, and not by the presence of separate γ- and δ-genes demonstrating different activity during development.

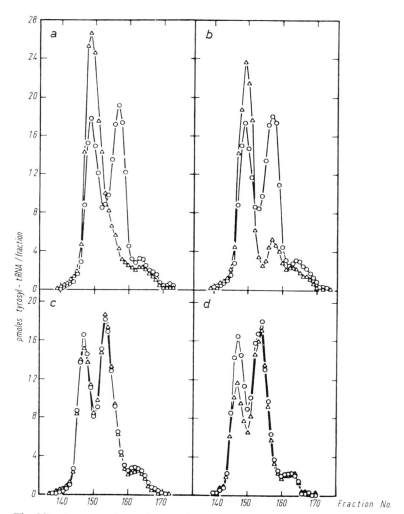

Fig. 23. a A comparison of tyrosyl-tRNA from females homozygous for the $su(s)^2$ gene with tyrosyl-tRNA from the wild-type strain. Tyrosyl-tRNA from the homozygote $su(s)^2/su(s)^2$ (—△—△—) was co-chromatographed with tyrosyl-tRNA from the wild-type strain (—○—○—). To the column was added 5 A_{260} units each of crude suppressor and wild-type tRNA aminoacylated with [^{14}C] and [^{3}H] tyrosine respectively. Chromatograph was normalized for 20 A_{260} units (1 mg) each of suppressor and wild-type tRNA added. **b** A comparison of tyrosyl-tRNA from male progeny of the first reciprocal cross ($+$/Y gene type) with tyrosyl-tRNA from the wild-type strain. Tyrosyl-tRNA from males of the first reciprocal cross (—△—△—) was co-chromatographed with tyrosyl-tRNA from the wild-type strain (—○—○—). The amounts of tRNA and elution conditions are as described in **a**. **c** A comparison of tyrosyl-tRNA from male progeny of the second reciprocal cross [$su(s)^2$/Y genotype] with tyrosyl-tRNA from the wild-type strain. Tyrosyl-tRNA from male progeny of the second reciprocal cross (—△—△—) was co-chromatographed with tyrosyl-tRNA from the wild-type strain (—○—○—). The amount of tRNA and elution conditions are as described in **a**. **d** A comparison of tyrosyl-tRNA from female progeny of the second reciprocal cross [$su(s)^2+$genotype] with tyrosyl-tRNA from the wild-type strain. Tyrosyl-tRNA from female progeny the second reciprocal cross (—△—△—) was co-chromatographed with tyrosyl-tRNA for the wild-type strain (—○—○—). Female progeny from the first reciprocal cross gave a similar column profile. The amount of tRNA and elution conditions are as described in **a**. (After Twardzik et al., 1971)

Fig. 24. Activation of tryptophan pyrrolase by RNAase T₁. *Drosophila melanogaster* were homogenized in 4 vol of 0.05 M potassium phosphate (pH 7.5) per g of frozen adult flies and centrifuged at 15,000 g for 30 min. The tryptophan pyrrolase was assayed by using a modified Bratton-Marshal assay to detect kynurenine production that results from adding 0.1 ml of homogenate to 1 ml of assay mixture consisting of 0.005 M L-tryptophan, 0.001 M 2-mercaptoethanol, 0.1 M potassium phosphate (pH 7.5) and oxygen at a concentration that results from equilibration with air. The reaction occurred at 37 °C and was stopped with 0.33 ml of 20% trichloroacetic acid at the times indicated. RNAase T₁ was either present at 2.5 U/ml (●——●) or it was absent (○----○). A homogenate of the vermilion mutant or the Samarkand strain was used. (After Jacobson, 1971)

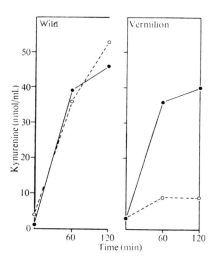

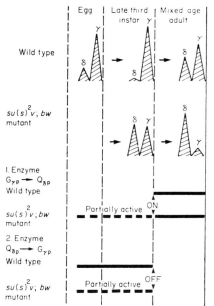

Fig. 25. Proposed models for the tRNA modifying enzyme. (After White et al., 1973)

Two models were suggested in explanation of the observed data:

1. An enzyme modified tRNA by converting the nucleotide designated as Gγp into Qδp (which is similar to the minor nucleozide found in *E. coli*). This enzyme is not active usually from the egg stage until the late third larval stage. Therefore, tRNA synthesized during this time period belongs to the γ-fraction. During the prepupal stage this enzyme became active and the synthesized tRNA began to be modified into the Qδp form. The form of the tRNA, synthesized before the prepupal stage, was not modified due to one of three causes: (a) the enzyme could identify

only newly synthesized tRNA; (b) the enzyme might be localized inside the nucleus and, therefore, would not be able to modify tRNA located in the cytoplasm; (c) the enzyme could be active in only a certain cell type during the prepupal stage. The su(s)², bw mutant differs in that the enzyme is partially active during embryogenesis and larval development.

2. The enzyme which modified tRNA controls the $Q\delta p \rightarrow G\gamma p$ reaction. It is active during embryogenesis, from the larval stages until the prepupal stage when it is inactivated. The su(s)²v, bw mutant can synthesize this enzyme only in the partially active form, therefore the majority of tRNA's remained in the unmodified δ-form (White et al., 1973).

As has been mentioned, the su(s) locus also suppressed alleles of the purple, speck, and sable loci. In this case it is possible that the reduction in content of tRNA^{asp}, tRNA^{aspart}, tRNA^{hys}, and tRNA^{tyr} during *Drosophila* development is responsible for suppression. Jacobson et al., (1975) suggested that the su(s)$^+$ locus produced an enzyme which modified the second isoacceptor of tRNA^{tyr}, and when such modification is absent [e.g., in the su(s)² mutant], tRNA is not able to accept tyrosine from one of the fractions of tyrosyl-tRNA ligase.

This very elegant hypothesis requires further proof. Mischke et al. (1975) could not induce tryptophane pyrrolase activity in fresh homogenates of the vermilion mutant of *D. melanogaster* and *D. hydei*, although the treatment was done in exactly the same manner as by Jacobson (1971). At the same time, enzyme activity could be induced almost to the wild-type level by homogenation of vermilion flies in the presence of 0.02M EDTA, the standard component in mixtures of ribonuclease. The authors suggested that the tryptophane pyrrolase activity, stimulated by EDTA, is possibly derived from another enzyme (peroxidase, for instance), but not from the tryptophane pyrrolase molecule.

In addition, Wosnick and White (1977) found that the conditions of growth of *D. melanogaster* significantly influenced the amount ot the nucleozide Q found in certain fractions of tRNA's. This effect was demonstrated in both wild-type and su(s)² flies. Flies with the su(s)²v bw genotype have 78% of their tyr tRNA4 in the γ-form and brown eyes, and are not distinguishable from su(s)²v bw flies in which only 5% of their total tRNA lacks Q nucleozide. The suggestion that these tRNA's are the specific inhibitors of tryptophane pyrrolase in flies with the γ-form and that their absence determined the suppression, is in disagreement with observed data. There is no evidence that su(s)² directly controls enzyme synthesis, as was originally supposed.

An interesting suppression mechanism was described for Pgd lethal mutants *D. melanogaster* with a significantly reduced 6-phosphogluconate dehydrogenase activity (Gvozdev et al., 1977). Suppressor mutations su₁Pgd, su₂Pgd, etc., have been obtained. These mutations are located at the Zw locus which controls the glucoso-6-phosphate dehydrogenase enzyme and do not restore 6-phosphogluconate dehydrogenase activity. Therefore, the Pgd⁻ mutation could be suppressed by the Zw⁻ mutation. The following hypothesis was used to explain this suppression.

Normally, glucoso-6-phosphate is used in glycolysis and in the pentose cycle. The mutation in the Pgd gene blocks the oxidation pathway in pentose biosynthesis and, hence, 6-phosphogluconate is accumulated. This situation is lethal for an organism. If there is another termination of the pentose cycle by a mutation in gene

Zw, glucoso-6-phosphate will be mainly metabolized in glycolysis. This, in turn, will lead to the anaerobic biosynthesis of pentoses from glycolytic metabolites with the participation of transaldolase and transketolase enzymes. It is also possible that the lethal effect on the Pgd⁻ mutation is aggravated because the accumulation of 6-phosphogluconate is inhibitory to the first steps of glycolysis. Similar data were obtained by Hughes and Lucchesi (1977).

Finally, the various interactions of genes were described at the post-transcriptional level for different stages of development. In some cases these relations between genes allowed a rough approximation of the action mechanisms of different gene-modifier groups.

The complexity in this situation lies in that we cannot talk about the functional activity of separate genes (with rare exceptions) on the basis of individual mRNA production. The necessary methods have not been developed, therefore in experimental genetics indirect methods based on final product determinations are used.

What is known of the transcription process during ontogenesis allows the suggestion of the presence of temporal and spatial differences in the phenotypic expression of biochemical traits. A large amount of data, obtained mainly by electrophoretic and immunochemical methods, has confirmed this suggestion.

1.2.2 Spatial Heterogeneity of Certain Isoenzymes and Proteins

The investigation of isozyme changes during ontogenesis at the organ, tissue, and cellular levels began to be widespread after their discovery by Markert (Hunter and Markert, 1957).

Initially, the concept of isoenzymes had only an operational character (Markert, 1968). Isoenzymes (isozymes) were considered to be any multiple fraction of the same enzyme (Shaw, 1969; Serov, 1968). But later, with the collection of various experimental data, this concept acquired a genetic meaning, and after this the concept of isozymes was given to the genetically determined variants of the same enzyme in the same organisms, as characterized by their similar substrate specificity. (*The Nomenclature of Multiple Forms of Enzymes*, 1971) Enzyme fractions which were not genetically determined were termed multiple forms of enzymes.

It is possible to distinguish two groups of isozymes in regard to the character of their genetic control: (1) isozymes controlled by a number of multiple alleles; (2) isozymes controlled by nonallelic genes. This last group may be further subdivided into: (a) isozymes that produce hybrid forms (a typical case is that of lactate dehydrogenase, LDH); and (b) isozymes that do not produce hybrid forms, as, for instance, soluble and mitochondrial malate dehydrogenase (MDH). For isozymes controlled by alleles at one locus the term allozymes was proposed (see review by Shaw 1965, 1969; Serow, 1968; Markert and Whitt, 1968). Isozymes typically differ in their mobility in an electrical field and may be easily separated with the aid of electrophoresis in starch, polyacrylamide or agarose gels, or on acetate cellulose strips followed by specific histochemical staining of suitable blocks (Wilkinson, 1965; Maurer, 1971).

Initial data in the study of isozymes were received in the investigation of the isozymes of lactate dehydrogenase (LDH) whose genetic control had been sufficiently studied. This enzyme plays and important role in glycolysis, and catalyzes the reversible reaction.

$$CH_3—CH—COOH + NAD \rightleftarrows CH_3—C—COOH + NADH + H^+$$
$$\underset{\text{Lactic acid}}{\overset{|}{OH}} \qquad\qquad \underset{\text{Pyruvic acid}}{\overset{||}{O}}$$

Details of the physicochemical peculiarities of isozymes are described in Jakovleva's review (Jakovleva, 1968). It is known that the LDH molecule is a tetramer, composed of two types of polypeptides, subunits A and B. The synthesis of each of them is controlled by a separate gene. Because any combination of subunits in a tetramer is possible, there are five isozymes of LDH separated by the aid of electrophoresis: A_4 (LDH-5, the slowest isozyme), A_3B, A_2B_2, AB_3, and B_4 (LDH-1, the fastest isozyme). Often the A subunit is designated as M (muscle-type, prevalent in the muscle) and the B subunit as H (heart-type, prevalent in the heart) (see review by Wilkinson, 1965). In mice the genes controlling synthesis of the A and B polypeptides are located in separate linkage groups (Shaw and Barto, 1963; Markert and Whitt, 1968; Shaw, 1969; Ruddle et al., 1970). There are differences in the amino acid composition between the two subunits of LDH (Wachsmuth et al., 1964) that cause antigenic differences.

Antibodies against the isozyme LDH-1 do not give a cross-reaction with LDH-5 and vice versa (Table 12). Isozymes 2, 3, and 4 react with both antiserums (Cahn et al., 1962; Kaplan and White, 1963).

Figure 26 demonstrates the organospecific features of the distribution of the isozymes of LDH (Markert, 1964). These differences in distribution are also apparent at the cellular level. For instance, LDH-1 was identified in frog egg yolk plates and in the connective tissue of the oviduct. LDH-3 was found in the cytoplasm of these eggs, certain cells of the oviduct epithelium, and, also, in the oviduct secretory glands. LDH-2 was not found in the egg, but appeared in some of the epithelial cells of the oviduct (Nace et al., 1960; Nace, 1963).

Table 12. Antigenic characteristics of the chicken LDH.
(After Kaplan and White, 1963)

LDH type		Anti-A	Anti-B
LDH-1	B_4	0^a	$+$ [b]
LDH-2	B_3A_1	$+$	$+$
LDH-3	B_2A_2	$+$	$+$
LDH-4	B_1A_3	$+$	$+$
LDH-5	A_4	$+$	0

[a] Reaction is absent
[b] Positive reaction

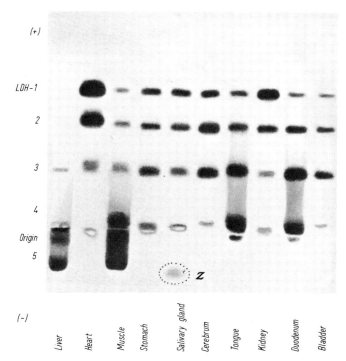

Fig. 26. Zymogram of LDH isozymes from tissues of the rhesus monkey. Note that each tissue contains distinctive proportions of the isozymes. All five are present in every tissue although quantities of LDH-5 in some of the tissues are too small to be evident on this photograph. (After Markert, 1964)

Neurons and glial cells in hippocamp and the medula of the rat differ in their LDH spectrum; in the former fast isozymes are prevalent, and in the latter slow isozymes are the most common form (Alexidze and Chaglid, 1970; Korochkin, 1972 b; Pevsner, 1972). The fast isozymes of LDH are in the majority in the ascending part of the loop of Henley in the kidney and in the distal convoluted tubules. Slow isozymes are prevalent in the collecting ducts and vasa recta (McMillan, 1967).

Parenchyma cells comprising 60% of the total weight of the liver and Kupfer's cells (a variety of reticular endothelium) were purified using differential centrifugation from mouse and rat livers. Electrophoresis of these extracts in polyacrylamide gels demonstrated only the LDH-5 form in the parenchymatous cells, but in the Kupfer's cells LDH-1, LDH-2, LDH-3, LDH-4, and LDH-5 appeared (Fig. 27; Palmer and Kjellberg, 1967; Berg and Blix, 1973).

In our laboratory a variation of microelectrophoretic technique was developed. This method, in contrast to other microelectrophoretic methods (Alexidze and Chaglid, 1970; Rosenberg, 1970), makes possible the determination of LDH isozymes in any size cell, not only in large cells but in small cells as well (Korochkin, 1972 b, c; Korochkin et al., 1972 b). The principles of this technique are demon-

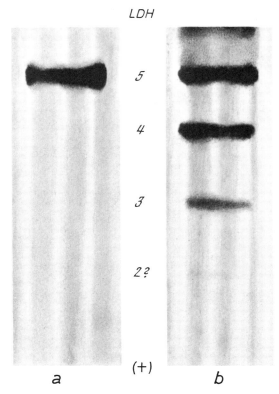

LDH

5

4

3

2?

(+)

a   *b*

Fig. 27. Lactate dehydrogenase patterns obtained by disc electrophoretic treatment of the enzyme from highly purified fractions of rat liver parenchymal **a** and Kupffer cells **b.** (After Berg and Blix, 1973)

strated in Fig. 28. With the aid of this method it was discovered that parenchymatous cells in the rat liver differ in their spectrum of LDH isozymes from that found in the stromal cells (Fig. 29). In the parenchymatous cells the slow fraction is prevalent, to all appearances this is LDH-5, although in some cases the weak activity of LDH-4 appeared. In the stromal cells there are two or three fast migrating fractions which appear along with LDH-5 and LDH-4. It is possible that these are LDH-3, LDH-2, and LDH-1. The heterogenetity of liver cells (parenchyma cells and stromal cells) was determined very clearly on these kinds of electrophoregrams (the patterns appearing on staining gels after electrophoresis) through quantitative ratios of different isozymes.

Some authors (Wachsmuth et al., 1969), using specific inhibitors for the A and B subunits of LDH, demonstrated histochemically that isozymes of LDH are distributed unevenly within liver lobes. In the middle portion of the lobe LDH-5 is prevalent, and on the periphery LDH-1, LDH-2, and LDH-3 also appeared in substantial concentrations. These data were confirmed in our laboratory with the immunohistochemical method of Coons. Characteristically, various mouse strains demonstrated considerable variability in this parameter, especially in the activity of the LDH fast fractions (Leonova, 1974).

Specific peculiarities in the subcellular distribution of LDH isozymes were found. The fast fractions were obtained mainly in the mitochondria, however the

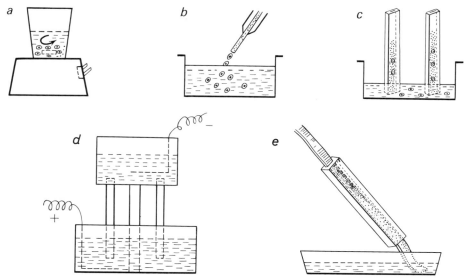

Fig. 28 a–e. Micromethod for the isozyme determination in single isolated cells. **a** Preparation of cell suspension. **b** Mixture of isolated cells and the gel components. **c** Gel polymerization in microcapillaries. **d** Electrophoresis. **e** Extraction of a gel and its staining

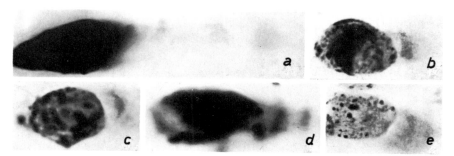

Fig. 29 a–e. Electrophoregrams of LDH from single cells of rat liver. **a** Connective tissue cell with fast fraction of LDH. **b,c** Parenchimal liver cell with only one fraction of LDH. **d,e** Cells with two LDH fractions (Korochkin, 1973)

slow ones were found in the supernatant, microsomal, and nuclei (Agostini et al., 1966; Nemchinskaya et al., 1968; Pokrovsky and Korovikov, 1969; Baba, 1970; Baba and Sharma, 1971).

The discovery of additional organospecific fractions of LDH, evidently under separate genetic control, generates special interest. These fractions are LDH-X and LDH-E.

Isozyme LDH-X is a homotetramer with a molecular weight of 150,000 (Stambaugh and Bucley, 1967) and was discovered in the testis of mammals and man. In regard to its electrophoretic mobility this isozyme takes an intermediate position between LDH-3 and LDH-4 in man, between LDH-4 and LDH-5 in the rabbit,

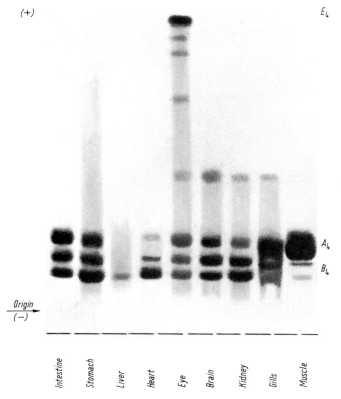

Fig. 30. LDH isozymes from fish *Lepomis cyanellus*. (After Whitt et al., 1973 b)

and in the mouse its mobility is slower than that of the LDH-5 isozyme. Two fractions of LDH-X were found in homogenates of guinea-pig and rat testis with mobilities between LDH-3 and LDH-4, and between LDH-4 and LDH-5 respectively. In homogenates of bull testis three fractions were found. These fractions of LDH-X were not found on the electrophoregrams of drake, chicken, pig and cat testis (Goldberg, 1963, 1965; Blanko and Zinkham, 1963; Zinkham et al., 1964).

There is a third locus, C, which controls the synthesis of the isozyme LDH-X. This locus is closely linked with the B locus in pigeons (Blanco et al., 1964, 1975; Zinkham et al., 1964). This isozyme was mainly obtained from testis tissues, particularly from the heavy mitochondria fraction (Blanco et al., 1975). Erickson and co-workers (Erickson et al., 1975) used biochemical and immunological methods to investigate the LDH localization in mouse spermatozoa. It was found that 62% of the total LDH activity is localized in the supernatant and the main component of this LDH is LDH-X. An analysis of the fine structure of LDH localization showed that a significant amount of this enzyme was found on the plasma membrane segment located behind the acrosome. The immunocomplex was not found on the membrane of the middle and main parts of the spermatozoa. Apparently, LDH-X played some role during spermatogenesis (Blanco and Zinkham, 1963) since this enzyme was absent when aspermia and testicle atrophy occurred (Szeinberg et al., 1966). LDH-E_4 is the isozyme which demonstrates the largest anodic mobility in comparison with the other fractions of LDH (Fig. 30). This isozyme was discovered in the retina of teleosteins (Markert and Faulhaber, 1965) and was designated as E_4 (Massaro and Markert, 1968).

There is serious reason to believe that this isozyme is related to LDH-B_4 and that there is a separate locus controlling the synthesis of subunit E (Whitt, 1969). In the brain and eyes of bony fish a heteropolymer was discovered which contained B and E subunits (B_3E_1). The absence of heteropolymers between A and E polypeptides could be explained through either temporal (asynchronous involvement in function) or spatial (synthesis in different cells or in separate regions of the same cell) differences in gene function, synthesizing mRNA for the formation of these polypeptides (Whitt et al., 1973 a). The gene which controls the synthesis of the E polypeptide acts mainly in the photoreceptor cells and the synthesis itself occurs mainly inside segments of these cells, primarily in the ellipsoidal regions where very high concentrations of mitochondria were found.

What is the specific functional significance of the LDH-E_4 isozyme that is inherent to only one type of cell? This might be understood from the following scheme for the transformation of visual pigments in the vision cycle (after Whitt et al., 1973 a, b).

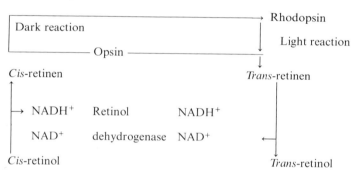

It is supposed that LDH-E_4 plays a role in the regulation of this process in the following ways: (1) if NAD^+ becomes a limiting factor in rhodopsin regeneration, LDH may play a role in converting NADH to NAD^+; (2) the high substrate affinity of LDH-E_4 makes possible the utilization of low concentrations of lactate as an energy resource by photoreceptor cells (Whitt et al., 1973 a, b).

Isozymes LDH-X and LDH-E_4, therefore, demonstrate the most extreme cases of the differential expression of nonallelic genes, whereby the function of these isozymes is limited to only one or two types of cell.

Similar qualitative and, more often, quantitative features of the spatial distribution of isozymes in tissues and cells is peculiar not only to isozymes of LDH. Many, if not all, isozymes and enzymes, controlled by both allelic and nonallelic genes, display differences of this type.

In a number of papers the organ and tissue specificity distribution was described, for esterases (Markert and Hunter, 1959; Paul and Fottrell, 1961), acid phosphatase (Lundin and Allison, 1966), mitochondrial (M) and soluble (S) forms of malate dehydrogenases (Conclin and Nebel, 1965), fructosodiphosphate aldolase (Penhoet et al., 1966) intra- and extramitochondrial isocitrate dehydrogenase (Lowenstein and Smith, 1962), and other enzymes (see review by Serov, 1968; Brumberg and Pevzner, 1975).

As was the case in LDH, the heterogeneity of other isozymes was explored not only among various organs but also within organs. Specifically, in the rat liver it is possible to detect six fractions of esterases using α-naphthylacetate as a substrate. The first, third, and sixth fractions isolated, which belong to a group of carboxylesterases, were found to be located in the parenchyma cells and hepatocytes, mainly in the endoplasmic reticulum. The fraction 1 isozyme predominated in embryo during the prenatal period and disappeared after birth. The activity of fraction 3 gradually increased, and band 6 appeared only after delivery. Fractions 2, 4, and 5 (arylesterases) are specific for reticuloendothelial cells and their activities do not change during ontogenesis (Kaneko et al., 1972). Just as for esterases, one fraction (B) of the three fractions (A, B, and C) of acid phosphatases is detectable in hepatocytes, another fraction (C) is found in the reticuloendothelium, and the third (A) is specific for nondifferentiated hepatic cells (Kaneko et al., 1970). Beckman and his collaborators found a distinct specificity in the isozyme spectrum in different types of human blood cells for a number of isozymes (Beckman et al., 1973).

As a whole, the impression is gained that various isozymes form nonrandom combinations specific for different cells and tissues, and possibly for different individuals or populations, as was proposed in an interesting hypothesis by Grossman et al. (1970).

The various tissue and cell types demonstrate this feature also in their total protein spectrum. The distribution of certain proteins between tissue and cell types was investigated in fairly great detail using the immunohistochemical technique of Coons.

Among the parenchyma cells of the liver it is possible to find cells which contain serum albumin (10% of the cells) or fibrinogen (1% of the cells), however, among Kupfer's cells these percentages are 18% and 33% respectively (Hamashima et al., 1964).

The organospecific protein S-100 exists in the neurons of the brain as well as in glial cells, but its localization in these cells is different. Neurons contain this protein in the nucleus and nuclear membrane, however in glial cells it is concentrated in the cytoplasm (Hyden and McEwen, 1966; Korochkin et al., 1972b; Sviridov et al., 1972).

Organospecific antigens were obtained in D. melanogaster larvae using two-dimensional immunoelectrophoresis. There are three specific antigens found in the salivary glands: two found in the fat bodies surrounding the salivary glands; three antigens located in the ganglias; two in the hemolymph; and one in the Malpighian tubules (Karakin et al., 1977).

This considerable quantitative (differences in the amount of the various fraction of isozymes) and qualitative (organospecific proteins) variability develops as a result of complicated ontogenetic processes.

1.2.3 Temporary Specificity in the Establishment of the Isozyme Spectrum

The successive phenotypic expression of the various fractions of an enzyme reflects the activation sequence of their respective genes during development. However, this correlation is approximate since there is a temporary gap between the transcription and translation processes. The inhibition of RNA synthesis by actinomycin demonstrated that the structural gene of one of the hemolymph proteins in *D. melanogaster* larva was transcribed between 54 h and 57 h of development, whereas the protein itself was immunochemically obtained between 57 h and 60 h (Roberts, 1971).

Occasionally, this temporary gap is very large. For instance, the mRNA used in hemoglobin synthesis is synthesized in chicken nuclei during the primitive streak stage (Wainwright and Wainwright, 1966, cited after Ingram et al., 1974). It is then transported into the cytoplasm where it is stored in the stable form. γ-aminolevuline acid ($10^{-3}M$), which is the precursor of heme, stimulated hemoglobin synthesis in the developed blastoderm even in the presence of actinomycin D (10 µg/ml). The addition of γ-aminolevuline acid apparently stimulated globin synthesis on the translational level (Levere and Granick, 1965). Just the same, the RNA used in procollagen synthesis is synthesized in the embryonic nuclei during the neurula stage and then is stabilized in the cytoplasm. RNA extracted from nuclei during the neurula stage and injected into oocytes is capable of synthesizing collagen. Nuclear RNA extracted at later developmental stages lacks this ability (Rollins and Flickinger, 1972). These facts hinder the explanation of the control of cell specialization as a simple process of differential gene transcriptions organized in time and space. Biochemical investigations (Havkin, 1969, 1974) have demonstrated that, during different stages of plant development, and in various plant tissues, the same proteins and enzymes are present, but in different quantitative proportions. Therefore, is this case, tissue specialization should be explained, not through differential gene transcriptions, but through the presence of a specific regulating system, organizing the various gene expressions in different tissues at numerous levels. However, animals demonstrated qualitative differences as well, i.e., the presence of organospecific proteins.

Independently of these circumstances, functional activization could deal with nonallelic genes controlling isozyme synthesis as well as allelic genes of homologous loci. Let us consider both of these cases.

1.2.3.1 Nonallelic Genes Controlling the Synthesis of Isozymes Which Form Hybrid Patterns

Among these isozymes lactate dehydrogenase (LDH) has been the most carefully studied (Masters and Holmes, 1972).

The activity of this enzyme gradually decreased during the preimplantation period in mammals, e.g., mouse, which has been genetically well investigated (Brinster, 1965; Epstein et al., 1969). After implantation LDH activity increased again

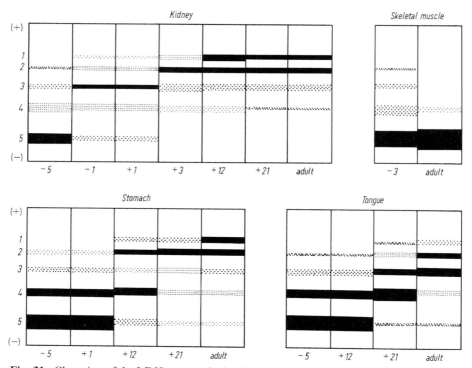

Fig. 31. Changing of the LDH pattern during development of mouse tissues. The LDH patterns changed asynchronously. *Numbers*, days before (−) and after (+) birth. (After Markert, 1963)

and, during embryogenesis, a change in the isozyme spectrum was registered (Wolf and Engel, 1972). This transformation is especially evident during the postnatal period.

The LDH-5 fraction predominates in embryonic tissues. Apparently this could be explained physiologically by the fact that LDH-5 can function intensively under anaerobic conditions. This condition is typical for embryonic tissues and for the skeletal muscles. The activity of this enzyme is maximal with pyruvate concentrations that inhibited LDH-1. Apparently, the predominance of LDH-5 affects the rapid conversion of pyruvate to lactate and, therefore, leads to the depletion of oxygen under anaerobic conditions.

However, in some mammalian skeletal muscles, e.g., m. cambaloideus, an LDH isozyme distribution identical with that found in cardiac muscle was obtained. This could be explained by the fact that these muscles contain mainly red filaments. The primary physiological role of these muscles is to fix the body in a static position and, therefore, they are able to maintain contraction for long periods of time (Wilkinson, 1965).

A gradual change to LDH-1, affecting various tissues to a differing extent, is obtained during tissue and cellular differentiation (Fig. 31; Markert, 1963). For instance, this transformation is not significant, and LDH-5 predominates in tissues of the skeletal muscles and liver, where specialization is complete. On the other hand, a notable transformation from a spectrum with LDH-5 predominance to one in which LDH-1 predominated was obtained in cardiac muscles (Markert and Ursprung, 1962; Masters and Hinks, 1966). Apparently this has important metabolic significance, since the pyruvate reduction to lactate is catalyzed by LDH-1, and is significantly inhibited by a small pyruvate concentration. It was suggested that the rapid accumulation of lactate cannot occur in tissue which is rich in LDH-5 (e.g., heart), and it is then possible to expect the total oxidation of glucose through the citric acid cycle (Cahn et al., 1962). Naturally, the following questions arose: (1) whether the change in the LDH isozyme spectrum is combined with the withdrawal of differentiating cells from the mitotic cycle; (2) whether these changes differ in cells originating from different embryonic layers; (3) how the special peculiarities of the individual affect this differentiation.

Data exist which allow definite answers to each of these three questions.

It was demonstrated that, during intensive cell proliferation, LDH activity rapidly decreased before every cell division. The fluctuation in enzyme activity depended mainly upon the A subunits. The addition of phytohemagglutinin to lymphocyte and granulocyte cultures induced a proportional increase in the polypeptide A contents after 4 h. The LDH activity, as well as the activity of a number of other enzymes (fumarase in KB cells, deoxycytidine monophosphate deaminase, thymidilate kinase in HeLa cells, ornithine transaminase in liver cells, and glucoso-6-phosphate dehydrogenase in hepatome cells) fluctuated during the cell division cycle, reaching a peak at the end of the S period or during the early G_2 phase (Brent et al., 1965; Bello, 1969; Martin et al., 1969; Sellers and Granner, 1974).

It is possible that these fluctuations are dependent upon a change in the ratio of synthesis and degradation rates of the LDH subunits since actinomycin apparently does not affect the course of the process (Ruddle and Rapola, 1970). Precisely the same fluctuations in the concentration of subunit A were obtained in in vitro cultures of skeletal and cardiac muscle. Under these conditions, a gradual increase in the contents of polypeptide A was registered (Fujisawa, 1969). The same tendency for a reduction in LDH activity during proliferation, followed by a gradual reorganization of the isozyme spectrum, was described during the triton retina regeneration from the pigment epithelium in the whole organism (Mitashov and Korochkin, 1974).

Therefore, cellular withdrawal from the mitotic cycle and entry into the differentiation phase is apparently accompanied by certain changes in LDH activity and in the ratio of the formed LDH subunits. These changes could develop divergently in predominantly A or B polypeptide directions, depending on the peculiarities of the differentiating tissue. The peculiarities of the embryonic layers, from which this tissue system is developed, could affect this development. For instance, in our laboratory demonstrated, using immunohistochemical methods, that the mesoderm of the rat embryo contained more A subunits than did the endoderm and ectoderm (Posnachirkina, unpublished work).

The kidney glomeruli, as well as the mesenchyme from which they are formed, have a relatively low total LDH activity. In both of these cases the A type (M-type) of polypeptides predominate. The total enzyme activity and that of its slow isozymes increased during differentiation. At the same time, the tubules which are formed from mesodermal segments are characterized by an increase in the B subunit concentration during differentiation (Smith and Kissano, 1963). It is possible to suppose that the oxygen supply in the developing cells played an important role. However, the glomeruli, which are rich in oxygen, contained almost only A polypeptides, similar to papillae which are comparatively poor in oxygen.

Generally, the cellular origin from either embryonic layer left an imprint on the LDH spectrum.

Finally, interspecific differences in LDH patterns were obtained during ontogenesis when the differentiation of parenchymatic liver cells and the connective tissue cells of the liver were investigated. For instance, the LDH-4 and LDH-5 fractions predominate in both of these cell types at various developmental mouse stages, whereas LDH-4 predominates in the rabbit, LDH-4 and LDH-5 both predominate in the cat (the amount of LDH-5 noticeably increased during aging), and LDH-3 is the major fraction in the guinea pig (Karlsson and Palmer, 1971).

The LDH spectrum remained highly stable after its establishment during cell differentiation, and during long up to 180 days cell cultivation in vitro. This effect was demonstrated for aneuploid cell cultures as well as for diploid cell cultures, although an initial, nonspecific change in the pattern of the LDH spectrum was observed. This nonspecificity in the transformation of the LDH spectrum consisted of a relative decrease in the fast migrating isozyme (Vesell et al., 1962).

How does the activation of nonallelic genes, which control the synthesis of their respective products, proceed during development–synchronously or nonsynchronously? Is the predominance of a certain isozyme established at the transcriptional level or during the subsequent steps of realization of the hereditary information?

Data obtained by Whitt indicated that various fish tissues have LDH isozyme specificity and that both synchronous and nonsynchronous activation can occur in genes controlling LDH subunit synthesis (Whitt, 1970).

However, this problem will be discussed in the next chapter, after a consideration of the peculiarities in activity of nonallelic genes which control isozyme synthesis and nonrelated products. The peculiarities in the action of allelic genes will be examined as well.

1.2.3.2 Control of the Synthesis of Isozymes and Nonrelated Enzymes by Nonallelic Genes

The action of these genes should be examined using the example of two organisms: mammals, with a regulation type of development; and *Drosophila*, with an egg development of the mosaic type.

Mammals. The basic processes of morphological and biochemical differentiation occurred after implantation in the mouse embryo. However, during the preimplantation period, changes in the activity of some enzymes were also obtained. For instance, the paternal glucosephosphate isomerase already demonstrated activity during the eight blastomere stage (Brinster, 1973). Similar data were obtained for β–glucuronidase (Chapman and Wudl, 1975). In our laboratory, the activity of the paternal isozyme of 6-phosphogluconate dehydrogenase was demonstrated in the rat embryo at the eight blastomere stage (Chlebodarova et al., 1975). Apparently, at least some of the paternal genes began to be active during development at very early stage–before implantation. It appears that protein synthesis is sufficiently high during this period (Epstein and Smith, 1974). Similar data were obtained in the rabbit embryo (Van Blerkom and Manes, 1974). In any event, differences between the protein patterns of the internal cell mass and the protein patterns of the trophoblast were obtained during the blastocyst stage in mouse embryos (Van Blerkom et al., 1976). In connection with this, it is possible to talk about the significant activation of biochemical processes and enzyme-protein synthesis in mammalian embryos during the preimplantation period (Epstein, 1975; Wales, 1975).

In general, three groups of mouse enzymes, differing in the character of enzyme activity which changed during ontogenesis, could be roughly distinguished.

1. Enzymes with a comparatively high and constant activity in the zygote, followed by a gradual decrease in activity during cleavage.

To this group belong the enzymes LDH, glucose-6-phosphate dehydrogenase (G-6-PDH), glucose–phosphate isomerase (GPI), and phosphofructokinase (PPK). LDH, G-6-PDH, and GPI demonstrated highly specific activity on the 8–16 cell embryo stage, whereupon their activity gradually decreased (Brinster, 1965, 1966; Epstein et al., 1969; Chapman et al., 1971). Only LDH retained noticeable activity until the implantation stage, but immediately before implantation its activity disappeared almost completely. PPK demonstrated a high activity in mature eggs before fertilization. During development from the morula to the blastocyst stage the activity of this enzyme decreased and then increased once again (Brinster, 1971 a).

2. Enzymes with an initially low activity which gradually increased during embryonic development. To this group of enzymes belong fructose-1,6-diphosphate aldolase (FDP-aldolase) and NAD-dependent malate dehydrogenase (NAD-MDH). The activities of these enzymes increased gradually until the morula-blastocyst transition, whereupon the enzyme activity increased more rapidly (Epstein et al., 1969).

3. Enzymes which display an interrupted increase in activity. A rapid increase in enzyme activity occurs during blastocyst formation after the morula stage from a minimal level of enzyme activity. To this group belong 6-phosphogluconate dehydrogenase (6-PGDH; Brinster, 1971 b), hexokinase (HK; Brinster, 1968), and hypoxanthineguaninephosphoribosile transferase (GHPRT; Epstein, 1970). Similar behavior was demonstrated by adenine phosphorybosile transferase (APRT), although its activity increase occurs at earlier stages (8–16 cells; Epstein, 1970).

Specific tendencies were also obtained for various isozyme activities in mouse embryos (Wolf and Engel, 1972; Table 13).

A definite sequence of activation for various nonallelic loci has been described. It was possible to distinguish electrophoretic variants of enzymes existing in the fertilized egg before implantation (LDH-1,G-6-PDH and PGI), after implantation [LDHα(LDH$_s$), G-PGDH, and PGM$_1$], and during postimplantational development.

It is possible that the differences in LDH-1 and LDH-5 behaviors are imaginary and that activation of the genes controlling their synthesis occurs synchronously.

Table 13. The temporary sequence of appearance of various electrophoretic fractions in C3H mouse embryos. (After Wolf and Engel, 1972)

Enzyme	Mature oocyte	The first obtaining of enzyme after fertilization, h											
		18-45	97-116	192	218	228	232	248	281,5	301,5	335,5	398,5	420
LDH-β	+	+	+	−	+		+						
G-6PDH	+	+	−				+	+					
Glucosophosphate isomerase	+							+					
LDH-α	−	−	−	+	+		+						
6-PGH	−	−	−	+	+		+						
Phosphoglucomutase	−	−		+	+		+						
Soluble isocitrate dehydrogenase NADP-IDH-S	−	−	−				−	+	+				
Mitochondrial NADP-IDH-M	−	−	−					−	+	+			
NADP-MDH-S	−	−				+							
NADP-MDH-M	−	−					−	−	−	+	+	+	
α-glycero phosphate dehydrogenase	−									+	+	+	
Succinate dehydrogenase	−							−	−	−	−	−	
ADH	−	−						−	−	−	−	−	+

However, Engel and Kreutz (1973) suggested that, during mouse embryo development, the activation of the locus which codes for M polypeptide subunits (LDH-A) occurs first. Kolombet (1977) came to the same conclusion based on the investigation of the LDH pattern during mouse embryogenesis. She suggested that the locus which codes for M subunits is activated before 5 ½ days after fertilization, whereas the locus which codes for H subunits is only activated 7 days after fertilization. As an alternative, Kolombet postulated synchronous LDH gene activations that produced different LDH subunits. If this is the case, the presence of LDH-1 in 5–7-day-old embryos could be explained by the presence of maternal isozymes synthesized on preexisting maternal mRNA.

Genes controlling the synthesis of esterases began simultaneous activation in some tissues of developed mice and rat embryos. Fractions of all esterases were obtained at the same time in developed postnatal brains of these animals. The patterns observed for esterases are characteristic of adult animals, although some qualitative differences were noted (Sviridov et al., 1970, 1971; Serov and Korochkin, 1971).

In general, it is possible to suggest that, for the synchronous activation of genes which control the synthesis of their relative products, it is not important whether these genes are located on different chromosomes (LDH) or are closely linked (esterases) (Popp, 1965; Ruddle and Roderick, 1965; Petras and Biddle, 1967).

On the basis of the data given in Table 13, it is possible to distinguish various groups of genes that began to be active at various time periods during mouse embryonic development.

Drosophila. The *Drosophila* enzymes could be divided into two groups in regard to their changes in activity during ontogenesis (Rechsteiner, 1970a).

The first group of enzymes includes ADH, MDH, and NADP-dependent isocitrate dehydrogenase. The activities of these enzymes were practically unchanged during embryogenesis and began to increase after hatching.

The second group of enzymes includes α-glycerophosphate dehydrogenase and β-hydroxybutyrate dehydrogenase. The activities of these enzymes increased from 5- to 20-fold during embryonic development. However, Wright and Shaw (1969) did not find fluctuations in α-glycerophosphate dehydrogenase activity during embryogenesis. Therefore, data concerning this enzyme cannot be considered incontrovertible. The second group also includes LDH, which demonstrated an unusually great increase in activity during embryogenesis. It was suggested that in this case enzyme activation is not dependent upon the intensive synthesis of new molecules, since within the first 20 hours of development no noticeable modification in the protein content was found (Rechsteiner, 1970b).

In contrast to the second group of enzymes, the increase activities of the first group of enzymes is correlated with larval growth and with protein content (Church and Robertson, 1966; Dewhurst et al., 1970). The investigation of enzyme activities for amylase (Doane, 1969; Ursprung et al., 1968), ADH, IDH, 6-phosphogluconate dehydrogenase, G-6-PDH (Wright and Shaw, 1970), and aldehydoxidase (Dickinson, 1971) demonstrated that activities began to increase shortly before hatching. The change in aldehydoxidase activity is characterized by certain peculiarities. As in many other enzymes, ADH activity rapidly increased after

Table 14. Activity of isoamylases during development: Amy[3,6]. (After Doane, 1969)

Stage	Time	Total activity [a]	Relative activity in bands [b]	
			Iso-No. 3	Iso-No. 6
	h ± 1 from hatching			
Larva:				
1st Instar	1	1.8	0.88	0.12
	22	5.5	0.76	0.24
2nd Instar	26	3.3	0.88	0.12
	46	20.2	0.72	0.28
3rd Instar	50	14.0	0.60	0.40
	72	69.6	0.46	0.54
	84	110.4	0.42	0.58
	92	250.2	0.46	0.54
	96	247.7	0.41	0.59
Pupa (mid-):	144	37.5	0.45	0.55
	Days from emergence			
Adult				
Female	0	26.2	0.60	0.40
	4– 5	237.1	0.36	0.64
	8–10	235.6	0.29	0.71
	12–14	229.4	0.28	0.72
Male	0	23.4	0.51	0.49
	4– 5	222.3	0.31	0.69
	8–10	221.8	0.26	0.74
	12–14	220.2	0.23	0.77

[a] Total activities are expressed as 10^{-4} mg starch/min/individual at 25 °C, pH 7.4
[b] Relative activities of isozymes No. 3 and No. 6 were determined by scanning procedures. Extracts used included 20 or more individuals raised on standard medium

hatching and reached a maximum in the third larval stage, sharply decreasing during the pupal stage, and then increasing and reaching a plateau in the imago. In contrast to the fluctuation in activity of ADH, aldehydoxidase activity is not high during the larval period, it increased gradually during the pupal stage, and did not demonstrate sharp fluctuations during ontogenesis (Ursprung et al., 1968; Wright and Shaw, 1969; Dickinson, 1971).

The activity of acetylcholine esterase (ACHE) increased intensively in *D. melanogaster* during the first larval stage, whereas choline acetyltransferase activity increased extensively until the second larval stage (Dewhurst et al., 1970). Phenoloxidase activity increased during the third larval stage and reached its maximum before pupation (Peeples et al., 1969). Other species of *Drosophila* apparently showed similar tendencies in the changes of their enzyme activities. For instance,

Fig. 32. Comparison of esterase patterns from adult flies and third stage larvae from various inbred strains of *Drosophila virilis. Even numbers*, adult flies; *odd numbers*, larvae (micromodification of starch gel electrophoresis, Korochkin et al., 1973b)

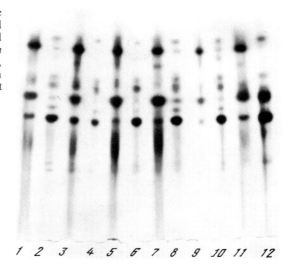

increase in esterase activities was obtained in *D. virilis* shortly after hatching (Matveeva and Korochkin, 1974).

A number of enzymes demonstrated a substitution of the larval by the pupal type of enzyme and then with the imaginal type of enzyme. These stage type enzymes could differ in their electrophoretic mobility and/or the relative activity of the isozymes (Wagner and Selander, 1974). This phenomenon was found in *D. melanogaster* during analysis of the changing isozyme spectrum of alkaline phosphatase (Schneiderman et al., 1966; Sena, 1966; Nilsson, 1967; Wallis and Fox, 1968), amylase (Table 14; Doane, 1969), leucine aminopeptidase (Sakai et al., 1969), and for enzymes of the α-glycerophosphate cycle (O'Brien and MacIntyre, 1972a). Some of the esterases in *D. virilis* demonstrated the same tendency (Fig. 32; Korochkin et al., 1973b; Korochkin and Matveeva, 1974).

As follows from Table 14, almost all the amylase activity in newly hatched larvae is restricted to isomylase 3, but shortly thereafter isoamylase 6 began to be very active and 24% of the total amylase activity is concentrated in this isozyme in larva of the late first stage. A similar change in the isozyme spectrum was found in *Drosophila* larva of the second moulting. The relative activity of isozyme 6 began to increase and, therefore, the activity of isozyme 3 was approximately equal to 40% of the total esterase activity at the end of the larval period. Significant changes in the isozyme spectrum do not occur during the pupal stage. Isozyme 3 predominates in the newly hatched flies, especially in females, but the activity of this isozyme constantly decreased and the relative activity of isozyme 6 correspondingly increased. In connection with this observation the important question of whether or not this tendency is characteristic of all individuals came into being. According to Doane (1969) the intestine and hemolymph are rich in amylase. Amylase activity is also detectable in a number of other organs: salivary glands, fat bodies, Malpighian tubules, and ovaries. Moreover, the amylase distribution within any organ could differ in different species of *Drosophila*. It was observed that amylase activity

Table 15. Relative activities of isoamylases in tissues[a] *D. melanogaster*, Amy[3.6]. (After Doane, 1969)

Stage	Age	Hemolymph		Midgut	
		Iso-No. 3	Iso-No. 6	Iso-No. 3	Iso-No. 6
Larva	72 h ± 1	0.56	0.44	0.42	0.58
	92 h ± 1	0.52	0.48	0.44	0.56
Adult (♀)	8– 9 days	0.44	0.56	0.25	0.75
	12–13 days	0.50	0.50	0.17	0.83

[a] Relative activities were determined by scanning the banding patterns produced by the blood and midgut extracts of individual larvas and adults reared on standard food. Six to ten individuals were assayed in each group and the results averaged for the table. Variability in any category was very low, less than 2%

in various tissues differs during development. Amylase 3 activity is higher than amylase 6 activity in the hemolymph of late larva of Amy $^{3.6}$ stock. In the intestine of the same larva the activity of isozyme 6 is higher than that of isozyme 3 (Table 15). Similar organ specificity was found during *D. virilis* development for the distribution of 6^F and 6^S, 4^F, and 4^S esterases. In this case, those fractions of esterases which increased in activity during the pupal stages and predominated in the imago are synthesized mainly in the imaginal discs during differentiation (Korochkin et al., 1973b).

What is the mechanism for the described redistribution of total enzyme activity among various isozymes?

It was demonstrated that the 3 and 6 fractions of isoamylases are controlled by closely linked nonallelic genes (Bahn, 1967). Therefore, it was suggested that these two closely linked genes are either regulated independently of each other or this regulation occurs at the post-transcriptional level and depends on the activity of gene modifiers.

Two alternative hypotheses were suggested for the fluctuation in activity of alkaline phosphatase and esterases (Wallis and Fox, 1968; Korochkin et al., 1973b).

The first hypothesis suggested that there are two separate genes for esterase-4 and esterase-6 (and only one for alkaline phosphatase). Those fractions of esterases controlled by these genes could exist in cells in two different states–free, and combined with cell membranes or other structures and compounds. The combined fraction of esterase is obtained together with membrane fragments after the extraction of proteins for electrophoretic separation. As a result, these "combined" esterases appeared as a separate fraction on electrophoregrams with a slower mobility than the "free" esterases (Fig. 33). The presence of such "free" esterases and those combined with various membrane–containing fractions (mitochondrial, lysosomal, microsomal, nuclear, and cellular) were obtained in *D. virilis* and *D. texana* (Korochkin et al., 1973b). Electrophoretic mobility differences in lysosomal and microsomal glucuronidases in mice were found not to be caused by differences in the primary structure of the enzyme molecules, but to be dependent upon the qual-

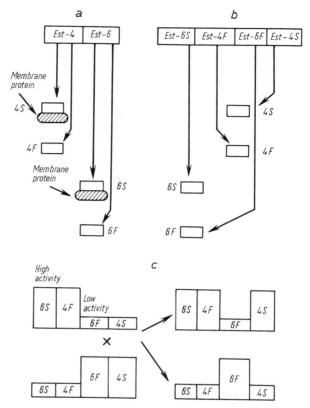

Fig. 33a–c. Two hypotheses of the F- and S-esterases subfractions formation. The *A* hypothesis supposed that S- and F-subfractions differ in their ability to associate with proteins of the cell membranes. The *B* hypothesis supposed that S- and F-subfractions are controlled by separate genes. *C*– the possible explanation of the origin of *D. virilis* strains with high and low esterase-4 activity as a result of crossing over (see text)

ity and quantity of the membrane component which remained bound to the enzyme after extraction (Ganschow and Bunker, 1970). These data supported the hypothesis of the presence of free and combined forms of esterases in *Drosophila*.

The second hypothesis proposed another possible variant of genetic control (Fig. 33b) based on the suggestion that the production of each of the subfractions of esterase-4 and esterase-6 (4F and 4S, 6F and 6S) is controlled by a separate structural gene. These genes are the product of tandem duplication (Roberts and Baker, 1973). The same explanation holds for the larval and pupal fractions of alkaline phosphatase.

A large amount of observed data supports the second hypothesis (however, it does not contradict the first hypothesis either). Among these findings are: (1) the presence of independent null mutants for each of the subfractions in various individuals from different strains; (2) the specificity of intracell localization, e.g., the S-subfraction of esterase-4 in *D. virilis* of the 140(va) stock is located mainly in the mitochondrial and lysosomal fractions; (3) the organospecificity of the subfraction distribution, e.g., the 4S and 6F fractions predominate in those organs which are formed during the larval period (intestine, Malpighian tubules), whereas the 4F and 6S fractions predominate in organs developed from the imaginal discs (genitals, eyes, thorax, etc.); (4) the presence of independent subfraction variants which differed in their electrophoretic mobility; (5) the presence of a specific coordination in the redistribution of the total activity among different subfractions. This activity redistribution during ontogenesis is peculiar to different strains of flies, e.g., the 4S and 6S esterase subfractions predominate during the larval stage in the 103-B$_2$ inbred strain of *D. virilis*, whereas the 4F and 6F subfractions predominate in the imago.

According to the data mentioned above, the organization of the block which contains esterase genes can be schematized as in Fig. 33 B. This scheme explains the rare appearance of individuals in some strains, who displayed an equal expression of the 4S and 4F subfractions. The existence of 4S and 4F fractions with similar (both intensive and weak) activity expressions could be a consequence of the crossing-over between corresponding, closely linked genes (Fig. 33. B). A similar mechanism was described for the amylase isozymes in *D. melanogaster* (Bahn, 1967).

Therefore, the distribution of total activity is characterized not only by the cellular and tissue arrangement of isozymes, but also by fluctuation in the isozyme spectrum as well. For instance, larval isozymes are replaced by pupal isozymes during the pupal stage. This tendency was established for a number of enzyme systems and could be regulated on different levels:

1. It is possible that, at the transcriptional level, a change in the multiple fractions of RNA-polymerases acts as a regulating factor. Acceptance of this mechanism involved differences in gene promoters controlling the synthesis of larval and "adult" enzymes (Fig. 34). The existence of multiple forms of RNA-polymerases, able to discriminate between different types of promoters, was postulated by Travers (1976). He suggested a model for their synthesis and discussed their role in the regulation of growth. The functional activity of RNA-polymerases could be modified by some proteins, among them those controlled by the nutritional regime (Barbiroli et al., 1977). The synthesis and functional condition of some RNA-polymerases can be controlled by various factors: hormones, cyclic AMP, albumin, etc. (Rutter et al., 1973). The possible importance of different forms of RNA-polymerase in the regulation of development was demonstrated in mice embryos by Warner and Versteegh (1974) and in *Misgurnus* embryos by Ekisashvili (1974).

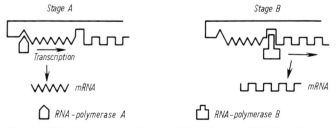

Fig. 34. The hypothetic role of multiple forms of RNA-polymerases in the regulation of enzyme synthesis during cell differentiation

2. Gene modifiers play a role in the regulatory mechanism at the post-transcriptional level. These modifiers realized their control either at the level of selective translation (as described above), or through the conformational change of the "larval" enzymes through binding them with different cell structures or compounds (Fig. 33. A).

These are the basic tendencies in the actions of nonallelic genes which code for different enzymes.

1.2.3.3 Allelic Genes

A large number of papers have dealt with the sequences in the phenotypic expression of differential activity of homologous loci during ontogenesis. Analysis of the development of multiple forms of various enzymes is the most effective way to expose the tendencies of subsequent allele activations during ontogenesis.

Animal populations often demonstrate polymorphism in the electrophoretic mobility of some enzymes. In this case it is possible to investigate the sequence in the appearance of maternal and paternal enzymatic forms in the heterozygous, hybrid offspring during its developmental differentiation. Cases of synchronous activation of homologous loci are well known, at least during the investigation of the phenotypic expression of a trait.

There are two types of NADP-dependent isocitrate dehydrogenase in mice, one with a fast electrophoretic mobility (F) and the other with a slow (S) mobility. These electrophoretic variants are controlled by a pair of allelic loci (Henderson, 1965). Only the maternal fraction of this enzyme is obtained in 6–9 day old embryos. Beginning on the 12th day of development, the heterodimer form of the enzyme began to demonstrate enzymatic activity. The full spectrum of IDH isozymes (both parental and hybrid forms) appeared after the 16th day. The fact that the hybrid isozyme (Fig. 35) was obtained before homodimer isozymes indicated the synchronous activation of parental alleles (Donahue and Stern, 1970). A similar picture was obtained for glucose phosphate isomerase, where heterodimer activity appeared as early as the late blastocyst stage, e.g., during the 5th day of embryogen-

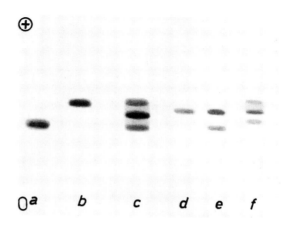

Fig. 35 a–f. Electrophoretic samples of S-fraction of NADP-IDH from C3H **a** and CBA **b** mice liver, adult hybrids between them **c**, and hybrid embryos after 248 h **d**, after 278 h **e** and after 365 h, **f**. (After Wolf and Engel, 1972)

Table 16. 6-PGD phenotypes of the 36-h–48-h embryos *Coturnix coturnix japonica*. (After Ohno et al., 1968)

Mating type	Embryo phenotypes				
///////////////////////////	A/B	A/D	B/−	D/−	Unfertilized
♂A/A × ♀B/D	21	11	0	0	4
♂A/A × ♀B/D	33	17	0	0	10
♂A/A × ♀B/D	17	7	0	0	4
///////////////////////////	B/D	A/D	B/−	A/−	Unfertilized
♂D/D × ♀A/B	8	6	0	0	2
♂D/D × ♀A/B	11	15	1	0	6

esis (Chapman, et al., 1971). NADP-dependent malate dehydrogenase is already present in heterodimer forms when it is first obtained electrophoretically (the 11th day after fertilization) (Engel and Wolf, 1971). The synchronous appearance (on the 12th day after fertilization) of the parental forms of mitochondrial NADP-dependent malate dehydrogenase was demonstrated during mouse development (Engel, 1973). The same type of phenotypic expression of parental alleles coding for glucose phosphate isomerase was found in interspecies reciprocal hybrids of *Lepomis gulosus* and *L. cyanellus* (Champion and Whitt, 1976).

There are three alleles (A, B, and D) of the autosomal locus coding for electrophoretic variants of 6-phosphogluconate dehydrogenase (6-PGD) in the Japanese quail, *Coturnix coturnix japonica*. The males, homozygous for one of these alleles, were crossed with females heterozygous for two other alleles. The activity of the maternal enzymatic forms, which are characteristic of an unfertilized egg, gradually decreased during the first 24 h of incubation (until the primary streak stage) whereupon the genes controlling this enzyme began to be active. Between 36 h and 48 h of incubation (the first heart contraction stage in quail embryos) the synchronous activation of paternal and maternal loci was observed and, as a result, the hybrid 6-PGD phenotype developed (Table 16; Ohno et al., 1968). A similar situation was observed for alcohol dehydrogenase (ADH) during ontogenesis. This locus has three alleles (A, B, and C) which affect the electrophoretic mobility of the enzyme in starch gels. The development of the B/C heterozygous embryo is characterized by: (1) the synchronous activation of both parental genes and (2) the specific sequence of this activation in different tissues–found in the liver during the 9th day of incubation, and, a few days later, in the kidney (Ohno et al., 1969).

The activation of paternal genes in ♀*Coturnix coturnix jap.* × ♂*Gallus gallus domesticus* was dependent upon the combination of paternal alleles. If a female quail has A or C alleles of ADH, the hybrid (19th day, hatching stage) demonstrated only the maternal forms of ADH, and total ADH activity is twice as low in the original maternal form. Apparently, the genes existing in the hybrid genome did not start to function before hatching. Thereafter the activity of both forms of ADH (quail and chicken) began to increase more or less simultaneously. At this point the activity of the allodimeric hybrid is detectable on gels, together with an increase in maternal isozyme activity and the appearance of the paternal autodimer fraction of ADH. These observations are usually interpreted as an activation of both alleles in the hybrid genome.

Fig. 36 a–c. LDH- and ADH-isozyme patterns of trout liver, 150 *(left)* and 192 *(right)* days after fertilization; **a** *Salmo trutta;* **b** hybrid; **c** *S. irideus.* (After Hitzeroth et al., 1968)

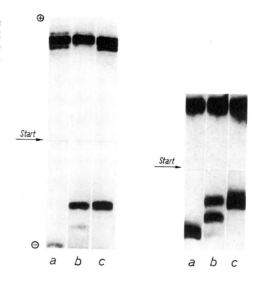

It is possible to obtain only B homodimer activity and, sometimes, weak allodimer activity when chicken ADH is combined with the B allele of quail ADH. These data suggested the weak activation of the paternal ADH gene or some disturbance in the translation of the synthesized mRNA. It was suggested that the B gene (or another, closely linked locus) is able to suppress the activity of the corresponding homologous locus (Castro-Sierra and Ohno, 1968).

On the other hand, numerous data have been obtained demonstrating the asynchronous activation of paternal and maternal enzymatic forms in the ontogenesis of fish and, recently, in *Drosophila* development. The electrophoretic spectrum of LDH and phosphoglucomutase in the interspecific trout hybrid *Salmo irideus* × *S. trutta* remained constant until the eye pigmentation stage ("Augenpunkt") and is similar to the isozyme pattern of the unfertilized egg. Later in development, an additional fraction characteristic of the maternal organism appeared. This reflected an earlier activation of maternal rather than paternal genes (Goldberg et al., 1969; Klose et al., 1969). The phenomenon is illustrated particularly well during the development of C forms of LDH characteristic of eyes and in the liver form of ADH. The C subunit of LDH could be obtained from embryo head extracts at approximately the 95th day of development. The maternal fractions of the enzyme (C^R) were obtained first, and the paternal fractions (C^B) did not appear on the electrophoretic slabs. The parental fractions of the enzyme were not obtained until the 134th day of development. The ADH^R fraction was obtained from hybrids on liver electrophoregrams after the 150th day after fertilization, ADH^B and hybrid (allodimer) bands were obtained after the 192nd day of development (Fig. 36; Hitzeroth et al., 1968). The asynchronous activation of maternal and paternal genes controlling different electrophoretic variants of 6-phosphogluconate dehydrogenase was also found in the embryogenesis of another fish, *Rutilus rutilus*, heterozygous for this enzyme. The same tendency was obtained for polar fox × red fox hybrids (Serov and Zakijan, 1974).

In the ♀*Drosophila texana* × ♂*D. virilis* (varnished mutant) hybrids the paternal fraction of esterases is obtained before the maternal fraction during ontogenesis (Korochkin et al., 1973b). Schmidtke et al. (1976) obtained temporary hemizygosity of α-glycerophosphate dehydrogenase parental alleles in interspecific trout hybrids. These observations supported the suggestion of asynchronous parental gene activation.

However, direct proofs of asynchronous parental gene activations are absent from those publications cited. It is impossible to exclude the possibility that paternal and maternal genes may begin to function simultaneously, but that the asynchrony of their phenotypic expression is determined by the system of post-transcriptional regulation (see above). Wright and Shaw suggested that the differential activity of maternal autosomal alleles is characteristic only of interspecies hybrids, whereas for intraspecies offspring this phenomenon is less likely (Wright and Shaw, 1970).

The late appearance of the paternal product (as a result of the maternal effect) should be distinguished from the phenomenon of asynchronous phenotypic expression of parental isozymes in the hybrids. The delay in the parental gene expression allowed for an understanding of the biochemical basis of the maternal effect, which was extensively studied by Sokolov (1959) on the morphological level.

1.2.4 Isozymes and the Maternal Effect

Wright and Moyer (1966) investigated LDH isozymes obtained from tissues of different frog species–*Rana palustris*, *R. pipiens*, and *R. p. sphenocephala*. These three species are different in regard to the gene controlling the B subunits of LDH. It was demonstrated that the paternal and hybrid LDH fractions could be obtained in the hybrid embryo until the heart contraction stage. Triploid hybrids with vvm genotypes were obtained from *R. pipiens* from Vermont, (USA) with the v genotype crossed with *R. pipiens* from Mexico with the m genotype. To this end the method of nuclei transplantation was used. The tendencies in the development of the isozyme spectrum for mitochondrial malate dehydrogenase (MDH) and 6-PDH are similar to those in diploid hybrids. The parental form of isozymes are first detectable in the triploid hybrid at the same stages as in the hybrid fractions– the 18th stage for 6-PDH and the 19th stage for MDH and IDH (Subtelny and Wright, 1969). A disturbance in the formation of the normal isozyme spectrum was obtained when interspecific and interracial crosses produced lethal haploid hybrids (Subtelny and Wright, 1969).

A delay in the appearance of the paternal forms of some dehydrogenases (glutamate oxaloacetate transaminase and glucoso-6-phosphate isomerase) was found in *R. pipiens* × *R. palustris* hybrids (Johnson and Chapman, 1971a, b). Similar data were obtained in *Drosophila* hybrids from stocks differing in their electrophoretic variants of various enzymes: octanol dehydrogenase (Pipkin and Bremner, 1969); 6-phosphogluconate, α-glycerophosphate, alcohol, isocitrate, and glucose-6-phosphate dehydrogenases (Wright and Shaw, 1970); aldehydoxidase (Dickinson,

Table 17. Number of embryos characterized by different LDH patterns in hybrids ♀ sph-4 × ♂ sph-1. (After Wright and Moyer, 1968)

Stage of hybrid development	Type LDH-B^b/LDH-B^b	Type LDH-B^a/LDH-B^b
20	1	4
24	4	4
25	6	3
Tadpole, 3 days	3	1
Tadpole, 11 days	3	1
Adult, 3 months	3	1
Total	20	14

1968); esterases (Korochkin et al., 1973 b). The "maternal" spectrum of the previously listed enzymes was already formed in the oocyte. These genes are nonactive until a certain stage of development subsequent to their activity in the oocyte stage.

There are several possible causes which could affect the differential phenotypic expression of enzymes in early development– the maternal effect. Among these are: (1) the stability of enzyme accumulated during oocyte maturation; (2) the stability of mRNA's which code for the maternal subunits of enzyme; (3) the comparatively early activation of maternal genes as opposed to paternal genes.

Wright and Moyer (1968) tried to find the answer to this question based on the detailed analysis of LDH isozymes during development of different species of *Rana*. Especially clear results were obtained when the intraspecies hybrid *R. p. sphenocephala* was investigated. Parents in these crosses differed in genes controlling the B subunits of LDH. One of these was heterozygous (LDH-B^a/LDH-B^b)– type 4, and the other was homozygous for this locus (LDH-B^b/LDH-B^b)–type 1. When a heterozygous mother was used in these crosses the embryo demonstrated the maternal phenotype B^a/ B^b, until the 18th stage (muscle contraction) and began to change during the 19th stage (heart contraction). Among the developing embryos it was possible to find two types of LDH spectrum (Table 17).

The potentiality for direct synthesis of B^a subunits will remain in differentiated embryo tissues if mRNA's for B^a subunits are synthesized, then masked during oogenesis. However, the analysis of crosses involving 3 alleles B–♀ sph 4(LDH-B^a/ LDH-B^b) × ♂ pip(LDH-B^c/LDH-B^c) put this suggestion in doubt. It is known that B subunits combine randomly with other subunits, forming a number of B homopolymers. The mRNA synthesized during the maturation of the oocyte could be activated in the developed hybrid and, therefore, serve as a template for B^a and B^b subunits. In this case, the simultaneous observation of B^a and B^b subunits would be possible during the initial establishment of the isozyme spectrum of LDH in offspring with the LDH-B^b/LDH-B^c genotype. However, this was not observed. Therefore it was suggested that the maternal effect is caused by the stability of proteins synthesized during oogenesis (Wright and Moyer, 1968). This suggestion is very possible, since stability was demonstrated for a number of enzymes in tissue cultures (Yagie and Feldman, 1969). It is known that polysome formation in the frog, as well as noticeable intensification of protein synthesis, were obtained during

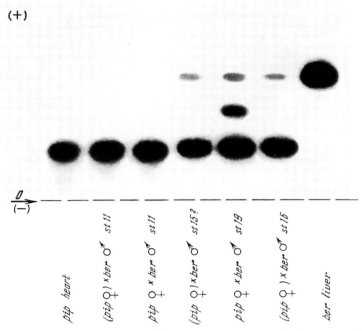

Fig. 37. 6-Phosphogluconate dehydrogenase patterns of embryos resulting from the diploid cross pip ♀ × ber ♂ and the androgenetic haploid cross (pip ♀) × ber ♂. The haploids (pip ♀) × ber ♂ at Shumway stage 15 and 16 are the same age as pip ♀ × ber ♂ diploid controls at stage 18½ and 19, respectively. (After Wright and Subtelny, 1971)

precisely the 18th stage of development (Wallace, 1963). Finally, using nuclei transplantation among races of *Rana*, Wright and Subtelny (1971, 1972) found that, during early stages of embryogenesis, only the "maternal" form of enzymes is present. These races differed in their electrophoretic mobility of various dehydrogenases. After activation of the transplanted nucleus genome (which takes place at the same time after fertilization as in the control diploid hybrids) only two fractions of the enzyme are detectable upon electrophoresis and no hybrid band activity was observed. These results yielded (1) absence of synthesis of "maternal" enzymes, and (2) the conclusion that the maternal effect is caused, in this case, by the stability of their corresponding proteins (Fig. 37).

These data also indicated that activation of "paternal" alleles occurs, automatically, during a given period of time, independently of the activity of "maternal" isozymes (although this activity is high), and independently of the morphological stage of embryonic development. This activation was dependent only on the amount of time following fertilization. The paternal form of 6-phosphogluconate dehydrogenase was found in embryonic haploid hybrids of nucleo-cytoplasmic origin ♀ *R. pipiens* × ♂ *R. berlandieri*. This fraction is obtained at the moment when the control diploid embryos are at the 19th stage (first heart contraction), although the haploids did not reach this same developmental stage during this time. However, there are known correlations between morphogenesis and the activation of paternal genes (Neyfach, personal communication).

Another question remains: what is the origin of the maternal type molecules which exist in the oocytes? Are they accumulated during the activity of the oocyte nucleus, or do they migrate from the maternal organism through the trophic cell system or another type of circulating system? If a female *Drosophila*, heterozygous for the ma-1 gene [inhibited xanthine dehydrogenase (XDH) activity] was crossed with a male hemizygous for this gene, all offspring, despite expectations, were similar to the mother, i.e., the progeny demonstrated the wild-type phenotype (high activity of XDH). The suggestion that XDH molecules migrated from the maternal organism into the oocyte was disproven by the following experiment. Females with the $\dfrac{ma-1}{+}\dfrac{ry}{ry}$ genotype were crossed with males of the ma-1 $\dfrac{+}{+}$ genotype (where ry is a mutant allele of the XDH structural gene causing the absence of the active enzyme). Although the maternal organism cannot pass active XDH molecules, since it is homozygous for the ry gene, all progeny displayed the active enzyme (Ursprung et al., 1968; Markert and Ursprung, 1973). The mechanism of maternal influence is not clear in this case. The most likely explanation is that both systems, the activity of the oocyte genome itself and the interaction between the oocyte and the maternal organism, played an important role (see Chapt. 1.4).

Apparently, either mechanism could predominate in different species and in different stages of evolutionary development.

Recently Gerasimova obtained new data connected with the previously discussed problem (personal communication). She investigated the activity of the maternal isozyme of 6-phosphogluconate dehydrogenase in *Drosophila* progeny from crosses

$$\female \overset{y\,Pgd^A}{\underset{y\,Pgd^A}{\Longrightarrow}} (XX/Y) \times \male \underline{\underline{Pgd^- Zw^-}}\;.$$

The maternal isozymes of 6PGD remained in F_2 males of the $Pgd^- Zw^-$ genotype until the early pupal stage. If the male parent had the slow isozyme (Pgd^B/Y), the hybrid isozyme was obtained in the progeny. During the 22nd h of development the paternal isozyme B occurred, and then hybrid band activity appeared, although comparatively weaker. Hybrid isozyme activity disappeared before isozyme A (maternal) activity vanished. On the basis of this experiment it could be supposed that the maternal effect is caused by the presence of stable mRNA, from which the maternal isozyme 6PGD is translated during all stages investigated. This could occur, since the subunits that appeared as a result of enzymatic degradation underwent further degradation and did not participate in metabolism. The presence of stable mRNA is corroborated not only by hybrid enzyme formation, but also by an increase in the total 6PGD activity per male individual of the $Pgd^- Zw^-/y$ genotype, at the expense of the maternal fraction, until the prepupal stage. The activity of this enzyme is equal to zero during the pupal and imago stages. If these data can be repeated and corroborated, it would be possible to suggest an active role for the stable fraction of maternal mRNA as an additional cause of the maternal effect.

1.2.5 The Influence of Gene Modifiers on the Phenotypic Expression of Enzymes

It was demonstrated that the levels of the phenotypic expression of various enzymes and proteins is determined during ontogenesis not only by the activity of the structural genes, but also by the influence of various gene modifiers. For instance, the influence of autosomes on the expression of the glucoso-6-phosphate dehydrogenase gene located on X chromosome was demonstrated in mice (Hutton, 1971). A similar situation was observed during analysis of the expressions of immunoglobuline M autosomal genes which are dependent upon the action of genes located on X chromosomes in man (Grundbacker, 1972). In the *virilis* group of *Drosophila* the degree of expression of the esterase isozymes depended not only upon the second chromosome where the structural genes are located, but also upon other chromosomes (Korochkin et al., 1973 b). Later, the role of the X chromosome was also demonstrated, for the regulation of organospecific esterase activity. This esterase is coded by the Est-6 structural gene which is located on the third chromosome (Korochkin et al., 1974). The regulation of alcohol dehydrogenase activity (the structural gene is located on the second chromosome of *D. melanogaster*) by gene modifiers linked with the X chromosome was demonstrated by Pipkin and Hewitt (1972) and later corroborated by Barnes and Birley (1975). However, there is no significant influence of gene modifiers located on the X chromosome on the total regulation of alcohol dehydrogenase activity. Almost 80% of the total ADH activity differences between strains with high and low ADH activity are dependent upon gene modifier(s) on the second chromosome, and 20% of these differences are dependent upon gene modifiers on the third chromosome (Ward and Hebert, 1972; Ward, 1975; Barnes and Birley, 1978).

The activity of phenoloxidase in *D. melanogaster* is regulated by several genes located on the X chromosome (Peeples et al., 1969) and also by genes linked with II and III chromosomes (Lewis and Lewis, 1963). Particularly, the various lozenge (lz) mutants are located on the X chromosome and affected the reduction in size of the eyes. In these mutant flies the partial or total absence of one or two fractions (A_1 and A_2) of phenol oxydase was observed. At the same time, a noticeable correlation was observed between the degree of activity reduction and various morphogenetic disturbances (Peeples et al., 1969).

It was also demonstrated that in some of the lz mutants, the activities of the fast migrating esterase isozyme differed. The differences in esterases patterns between wild-type *D. melanogaster* (Oregon R) and mutant flies were detectable at 120–144 h of development and continued through 24 h for adults. This period covered the time of incomplete eyes, claw, and genital development in the mutant flies (Peeples, 1966).

MacIntyre (1974) compared the activity of lysosomic enzymes in euploid *D. melanogaster* strains and in strains containing a duplication of various regions of the second and third chromosomes. It was found that numerous regions of these autosomes affected each investigated enzyme to different degrees. For example, the activity of arylsulphatase increased by more than 50% when the additional region 74A–74D (III chromosome) was present. It is possible that the localization of the structural gene occurs in this region. At the same time, other parts of the second (38C–41F, 54F–57B) and third chromosomes (79D–83CD, 88C–91B) significantly (but to a lesser degree) affected the enzyme activity.

The activity analysis of the X chromosome-linked G-6-PDH, α-GPDH and IDH genes which are located on the autosomes of *D. melanogaster* demonstrated that the absence of various parts of the 2nd and 3rd chromosomes affected the phenotypic expression of the above-mentioned enzymes (Rawls and Lucchesi, 1974). On the basis of these experiments it was supposed that the enzyme activity differences are caused by the dose effect of regulatory loci that are located inside the aneuploid segments. The fluctuations in enzymatic activities reflected the differences in the number of enzyme molecules as a function of the effect of regulatory loci on enzyme synthesis or degradation. The influence of controlling elements on the isozyme pattern was also found in some fish species (Anders et al., 1973 a, b).

On the basis of the investigations that were performed in our laboratory, we suggested that all members of the *Drosophila* genus have two structural genes controlling the synthesis of α-esterases. All variation in the esterase spectrum of the various subgenera and species of *Drosophila* is caused by the influence of different gene modifiers, but not due to the subsequent tandem duplication of structural genes, as was earlier suggested (Roberts and Baker, 1973). These modifiers determine the binding of α-esterases with the membrane proteins, with inhibitors, and with activators (Korochkin et al., 1974).

The variations in gene activity obtained in such systems could be detected by disturbance in the interaction between corresponding structural and regulatory genes. The concrete mechanisms of these interactions can, apparently, affect transcriptional and post-translational processes.

Courtright (1976) suggested two mechanisms for the controlling of phenotypic expression of isozymes in *D. melanogaster*:

1. Regulation with the aid of gene regulators, acting, probably, on the transcriptional or translational levels of the specific structural gene. Examples of such genes could be the lxd (low activity of xanthine dehydrogenase) and lozenge (lz) genes. The product of the first gene reduces the number of xanthindehydrogenase molecules and acts in the cis and trans position. The gene lz controls phenoloxidase activity. Because the electrophoretic mobility of the enzyme in mutant flies is not changed, it is possible to suggest that this control is a result of the interaction between structural and regulatory genes on various levels. Concrete mechanisms for these interactions could occur on the transcriptional as well as the post-transcriptional level.

There are few actual experimental data concerning the transcriptional level of gene action regulation and these data are restricted, in general, by observations of the dose compensation mechanism (see above).

Another example of regulation of isozyme expression on the transcriptional level is the position effect. In *D. melanogaster* an inactivation of the Pgd gene (the structural gene at 6-phosphogluconate dehydrogenase which is located on the left arm of X chromosome) was obtained when this gene was translocated to the heterochromatic region of the same chromosome as a result of the Dp (l;f) R rearrangement. Then dramatic locus inactivation that was obtained as a result of Y chromosome removal or a reduction in the temperature leads to a reduction of the total Pdg activity. There was an observed reduction in the activity of the fast iso-zyme, controlled by the Pgd^A alleles, in the locus which was translocated close to the heterochromatin region by the duplication Dp (l; f) R. At the same time, the enzyme activity produced by the homologous locus of the X chromosome in the same individual did not change (Gvozdev et al., 1973).

It was demonstrated that in the Dp (l; f) R chromosomal mutation induced in the giant chromosomes of the salivary glands of *D. melanogaster* a transcriptional rate decline and a disturbance in the regular replication of the euchromatic regions occurred (Ananiev and Gvozdev, 1974 b). These results suggested that the Pdg gene inactivation occurs, in this case, on the chromosomal level.

The influence of positional effect on the phenotypic expression of isozymes was also demonstrated in the case of amylase from *D. melanogaster* (Bahn, 1971).

2. Regulation through the aid of gene regulators acting on the post-transcriptional level. In this case only some models, which have been studied comparatively well, can be discussed.

In two cases a significant complexity in the genetic control of isozyme expression during ontogenesis was demonstrated. The first case deals with the interaction of genetic factors determining the definite realization of the β-glucuronisade activity in the proximal tubulus of mice kidneys. This avtivity is dependent upon the following genes (Paigen et al., 1975): (a) the structural gene (GuS) which is codominant, located on the end of the 5 chromosome, and determines the structure and catalytic properties of the enzyme. Labarca and Paigen (1977) extracted poly-A-containing RNA's from mice kidneys. Among these RNA's was the GuS gene product which was used as a template in the test system of the amphibian oocyte. It was demonstrated that in the amphibian oocyte 25 pg of β-glucuronisade per day could be synthesized if 60 ng poly (A) RNA was injected in the oocyte; (b) the processing genes, which determine the functions of the metabolic apparatus controlling the post-transcriptional modification of the enzyme (structural modifications, the binding with corresponding cellular structures, and the degradation of the enzyme). An albuminous factor, egosine, was detected that is coded for the structural gene Eg (localized on the eighth chromosome). This factor is necessary for the binding of the enzyme with the membranes of the endoplasmic reticulum and lysosomes; (c) the gene regulators (Gur) that controlled the rate of glucuronisade synthesis as a response to the physiological action of hormones; (d) temporal genes (Gut) which determined the program of structural gene expression during the growth and development of the organism; (e) the bg gene which is located on the 13th chromosome and affected the enzyme secretion throught the proximal tubules; (f) the Tfm gene (located on the X chromosome) that is responsible for the synthesis of the receptor which changes the structural gene activity.

The structural gene is not linked with the processing gene Eg, however the regulatory genes Gur and Gut are linked with the structural gene. The gene regulators act, perhaps, in the cis position and control the amount of the specific RNA synthesized by the structural gene. A similar system of the genetic regulation was demonstrated for the β-galactosidase expression in mouse liver (Paiger et al., 1976; Breen et al., 1977).

Another example of the post-transcriptional regulation of isozyme activities could be a system of genes controlling the phenotypic expression of the organospecific isozyme S of esterases in the ejaculatory bulbs of the *virilis* group of *Drosophila* (Korochkin et al., 1978 a, b). This enzyme came into the female spermatheca during copulation and, through the destruction of fat, promoted fertilization.

Determination of the corresponding cells of the genital discs for the synthesis of this esterase occurs very early, 10–12 h after the second moulting (Korochkin et al., 1976 a). However, the active transcription of the corresponding gene and the synthesis of the enzyme itself are obtained only at various periods after the hatching of the imago. With the aid of the mutagen and the selection of mutant individuals from natural populations, a number of various *D. virilis* stocks were obtained. These stocks differ in initiation of enzyme synthesis. The structural gene Est-S is located on the second chromosome (192.1 map units). The phenotypic expression of esterase-S, controlled by this gene, is the result of a number of gene interactions which are realized on different regulatory levels.

The following genes are involved in the control of the phenotypical expression of this enzyme (Korochkin et al., 1978a, b): (1) The Est-S structural gene which codes for the S isozymes. This gene is closely linked with three genes which code for three other esterase isozymes. A more detailed analysis demonstrated that these genes are located within the 2F8–2G2 region. For this purpose the *D. virilis* stocks were used in which various segments of the second chromosome of *D. lummei* were incorporated. Finally, the direct localization of the esterase genes was done on a cytological map of the salivary gland (Korochkin, 1977). For this purpose, the poly-(A)-containing RNA's were extracted from the ejaculatory bulbs during the period of maximal synthesis of this S-esterase. Among the various extracted mRNA's was the mRNA which codes for the organospecific esterase. This observation was corroborated by the injection of the pooled mRNA's which were obtained into *Xenopus* oocytes whereupon the synthesis of S-esterase was obtained. The poly(A) RNA's were labeled by radioactive iodine and subsequently hybridized with DNA on salivary gland chromosome preparates. It was demonstrated that the label is obtained only on the bands (but not between them), and that four adjacent bands on the second chromosome in the $2G_1$ region were labeled. These bands are located in the zone where the esterase genes were located cytogenetically. This is the sequence of the esterase genes that was determined with the aid of genetic analysis: Est-α_2, Est-α_1, Est-S, and then Est-β. Therefore, the Est-S structural gene is located in one of the bands at the $2G_1$ zone. (2) The Est-S[RA] gene regulator is located on the X chromosome. This gene is closely linked with the Crinkled eyes (Ce) gene that is adjacent to the bobbed locus. This gene controlled the level of S-esterase activity and affected the amount of S-esterase molecules synthesized. (3) The Est-S[RT] gene regulator is located on the fifth chromosome in the 5A5–5B2 region. This gene determined the time of translation initiation of the structural gene. It was suggested that this gene is responsible for the synthesis of the specific fraction of tRNA which allowed the translation of the organ specific esterase isozyme (Korochkin et al., 1978 a, b). (4) The Est–S[MB] gene is located on the fourth chromosome and controls the binding of S-esterase with the cell membranes, and, therefore, regulates the ratio between the free and the membrane bound fractions.

Complex regulations by numerous gene regulators were described for another isozyme system. The phenotypic expression of A and C isozymes of the fructosodiphosphate aldolase is regulated by a number of genes in the tissues of various vertebrates. The distribution of total enzymatic activity between these isozymes is regulated on the level of (1) transcription, (2) the rate of translation, (3) the processing of the newly synthesized, but functionally not active, predecessors of the enzyme, (4) the cellular condition, through the regulation of metabolite concentrations, and (5) the conditions which affect the association–dissociation rate of enzymatic subunits (Lebherz, 1975).

In some cases the gene modifiers could, apparently, change the dominant–recessive interaction between isozymes (which is regularly of a codominant character). For instance, a gene modifier was found in *D. melanogaster* which controlled the dominant expression of the slow isozyme of alkaline phosphatase in heterozygous flies. In the F_2 a segregation was obtained wherein the ratio of individuals with slow and fast isozymes was equal to 3:1 (Schneiderman et al., 1966).

Paigen and Ganschow (1965) described structural, regulatory, temporary, and architectural groups of genes. To the last group of genes belong those which control the localization of the enzyme in some of the cell organelles. The spatial distribution of enzyme molecules and, therefore, the distribution of total enzyme activity inside the cell is controlled by genetic factors which are closely linked with gene or are identical to the structural gene. This spatial organization of total enzymatic activity significantly controlled cell metabolism, the functional condition of cells and, in turn, is a significant factor in post-transcriptional regulation (Ward et al., 1973). Tompkins demonstrated that in salamanders the recessive lethal gene g affected abnormalities in gill development, hyperpigmentation, the reduction of yolk metabolism, and mortality after hatching. On the biochemical level the action of this gene is expressed in the disturbance of the binding of at least one of the esterase fractions with cell membranes. It was suggested that this gene controls the synthesis of the membrane component which is necessary for the binding of certain proteins (Tompkins, 1970).

Genetically determined differences in the ability to bind various enzymes often served as the basis for intraspecies differences and, also, as a basis of the organospecific peculiarities in the isozymic spectrum (Korochkin et al., 1973 a). Electrophoretic variants of glucuronidase in mice apparently reflected differences in membrane elements which are bound to the extracted enzyme molecules (Ganschow and Bunker, 1970; Lalley and Shows, 1974). The gene Eg is located on the eighth chromosome in mice. This gene was found to control the incorporation of β-glucuronidase (its structural gene is located on the fifth chromosome) into microsomes. Therefore, the gene which controls the processing of the polypeptide is not linked with the structural gene (Karl and Chapman, 1974).

The dependence of glucoso-6-phosphatase activity upon its interaction with membranes has been clearly demonstrated. This enzyme is located on the membranes of the smooth and rough endoplasmic reticulum and remains associated with isolated microsomes. This multifunctional enzyme demonstrated glucoso-6-phosphohydrolase activity (hydrolase) and pyrophosphate glucosophosphotransferase activity (transferase):

Hydrolase: glucoso-6-phosphate $+ H_2O \rightarrow$ glucose $+$ Pi
Transferase: glucose $+$ PPi $\rightarrow$ glucoso-6-phosphate $+$ Pi

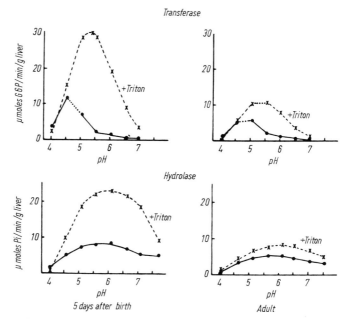

Fig. 38. The effect of pH on the transferase and hydrolase activities of isolated microsomes of rat liver, before and after treatment with Triton X-100. (After Duck-Chong and Pollak, 1973)

Protein synthesis and membrane formation in embryonic and postnatal rat liver occurs faster than in the liver of adult individuals. It was demonstrated that an increase in transferase activity coincided with an increase in hydrolase activity. However, the ratio between these two activities was not constant. This ratio increased from 0.8 (day of delivery) to 2.6 (4–7 days after delivery) and subsequently decreased to 1–2 (adult rat). These fluctuations were caused by the binding of enzyme molecules with microsomal membranes, and differences in the expression of catalytic properties were caused by the relative proportions of membrane-bound and free fractions. It was demonstrated with the aid of enzyme extractions with the nonionic detergent Triton X-100 and sodium deoxycholate (Duck-Chong and Pollak, 1973; Fig. 38). There are bacterial mutants with reduced esterase activity. In some of them esterase activity is reduced because of low levels of enzyme synthesis. Other mutants have low esterase activity due to the weakening of the association of the membrane–enzyme compound (Frehel et al., 1974).

The mechanism of regulation of tyrosinase activity during pigment cell differentiation is multiphasic. The synthesis of this enzyme is controlled by the structural gene, however three fractions, T_1, T_2, and T_3, of this enzyme were discovered in mice. Apparently these fractions are different conformers of tyrosinase. The enzyme could be synthesized as a protyrosinase which is activated by protease. The activation of protyrosinase leads to an increase in the T_1 fraction but not in T_2 or T_3. In the free condition tyrosinase regularly bound with cytoplasmic inhibitors. After incorporation into the melanosoma this enzyme began to be inaccesible to inhibitors. When the activity of the T_2 and T_3 fractions is reduced or absent the melanosomic matrix demonstrated noticeable defects in ultrastructure. Therefore, mouse tyrosinase is directly or indirectly incorporated into the formation of the melanosomic matrix (Rittenhouse, 1968; Hamada and Mishima, 1972; Quevedo, 1973; Ikejima and Takeuchi, 1974). Consequently, the total enzymatic activity and the distribution of the activity among its numerous fractions are controlled by a series of gene modifiers.

With the aid of radio-immunoprecipitation it was established that tyrosine-DOPA-oxydase (TDO, tyrosinase) is not synthesized in frog embryos until the early neurula stage. However, the enzyme is inactive during this period. It is possible that it is associated with inhibitors of a protein nature, but could be activated by soft proteolyzation (trypsinization). Until the hatching stage, natural stimulation of tyrosinase activity occurs. This process is accompanied by melanin formation in the pigment cells. At the same time, the DOPA-decarboxylase activity of this enzyme is obtained initially during the neurula stage and is accompanied by catecholamine synthesis. The change in the substrate preference of TDO during development is not dependent upon transcription and is perhaps determined by the synthesis of various enzyme conformers or by the interaction of this enzyme with inhibitors under the influence of environmental stimuli (Benson and Triplett, 1974 a, b).

The discovery of the previously cited mechanisms for enzyme activity regulation made possible the revision of the traditional conceptions of the biochemical-genetic nature of the formation of some traits during ontogenesis. It was suggested that the "c" locus, which affects mouse albinism, is the structural gene for tyrosinase and that homozygous c/c individuals have lost enzyme activity. However, recently, with the aid of biochemical analysis, it was demonstrated that homogenates of albino mouse tissues showed tyrosinase activity when treated with Triton X-100 detergent, which affects the dissociation of enzymes and membranes. Moreover, the homogenate which was prepared from albino mouse tissues induced the inhibition of enzyme activity in homogenates of black mice when these two types of homogenate were mixed. This observation resulted in the conclusion that the c locus is not the structural locus for tyrosinase but is the regulatory locus which controls the synthesis of a product which binds tyrosinase and inhibits enzyme activity (Hearing, 1973). Mutants of the albino locus (c^{65K}, c^{112K}, c^3, c^{14cos}, etc.) are known. These mutations inhibited glucoso-6-phosphate dehydrogenase activity in mice. The additional pleutropic effect (the morphogenesis disturbance and the embryonic mortality) and the absence of the dose effect in heterozygous individuals are difficult to explain by a mutation in the G-6-PDH structural gene. In connection with this it was suggested that the mutant gene controls the synthesis of membrane proteins. Such mutant membranes are not able to bind both glucose-6-phosphate dehydrogenase and tyrosinase. This, in turn, affected various morphogenetic abnormalities, such as the absence of pigmentation, the disturbance of thymus and kidney development, and the disturbance in size and shape of amelanotic pigment granules (Gluecksohn-Waelsch and Cori, 1970).

Sorenson and Scandalios found a complex of catalase and inhibitors in the corn scutellum extracts. This inhibitor was possibly of a protein nature and is very specific to catalase since it inhibited the activity of this enzyme in liver but did not affect peroxidase activity in corn, although peroxidase is related to catalase (Soren-son and Scandalios, 1974, 1976).

If the above cited data are corroborated in the future, it could serve as one the demonstrative examples of the originality of gene regulation mechanisms and of the action of gene modifiers in eukaryotes on the post-translational level.

Such mechanisms could, apparently, play an important role in the formation of stage-specific spectrums of isozymes, as demonstrated by the electrophoretic in-vestigation of LDH during mouse and rat ontogenesis. It was shown that there is only one LDH-1 (B_4) isozyme in the embryo before implantation (Auerbach and Brinster, 1967; Brinster, 1967, 1971 b; Epstein et al., 1971; Engel, 1972; Glass and Doyle, 1972; Wolf and Engel, 1972; Engel and Kreutz, 1973). On the other hand, Gibson and Masters (1970) believed that LDH is represented by isozymes 4 and 5 in the mature egg and during the first cleavage stages. The presence of the LDH-1 isozyme, described by other authors, is determined by the adsorption of this en-zyme on the outer ova membrane during migration along the oviduct.

Poznakhirkina and co-workers found isozymes LDH-4 and LDH-5 in rat ova using an extraction with Triton X-100. They suggested that these are highly stable fractions of LDH which are associated with membranes (Poznakhirkina et al., 1975). Rauch demonstrated that the fifth isozyme is present and continues to dem-onstrate activity for at least 6 days after the cow LDH is injected into frog eggs (Rauch, 1969). These experiments support an idea of the high stability of LDH isozymes.

However, it was recently observed that the appearance of additional LDH isozymes in the ova of rats and mice is the result of oviduct liquid adsorption on the surface of ova (Lubimova and Korochkin, 1978). This liquid contained all of the LDH fractions. It is characteristic that the main portion of LDH-4 and LDH-5 bound with protein inhibitors (Fig. 39). Triton X-100 treatment of ova associated

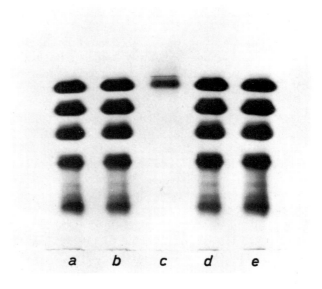

Fig. 39a–e. LDH patterns from kidney and ova of mouse. Only LDH-1 was obtained in ova (**c**). Micro-electrophoresis by T. Lubi-mova

a   b   c   d   e

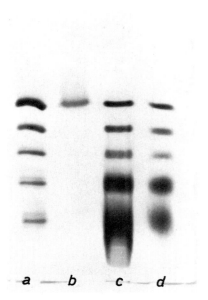

Fig. 40a–d. Microelectrophoresis of LDH iso-zymes of mouse. **a** Kidney; **b** ova, Triton X-100 extraction; **c** oviduct fluid, Triton X-100 extraction; **d** oviduct fluid, water extraction. Preparations by T. Lubimova

with zona pellucida released LDH-4 and LDH-5 and these fractions were subsequently obtained on electrophoregram. A thoroughly designed set of experiments using all kinds of electrophoresis, especially micromethods, allowed the final conclusion to be made that the ovum itself possesses only the isozyme LDH-1 (Lubimova and Korochkin, 1978). Occasionally, subfractions of this isozyme were obtained. These subfractions had an intermediate electrophoretic mobility as compared to LDH-1 and LDH-2 (or slightly faster than LDH-1) (Fig. 40) (Monk and Ansell, 1976; Monk and Petzoldt, 1977). The LDH-5 fraction appeared only after implantation and, apparently, occurs initially in the inner cell mass (Monk and Ansell, 1976).

It is of interest that the various intercellular liquids contained inhibitors in high concentrations. These inhibitors associated selectively with different isozymic fractions or prevented the free association of monomers. The liquid in the oviduct of some mammals could serve as a good example.

There is a gene regulator, Ldr-1, which controls the presence of the A4 isozyme at LDH in erythrocytes of some mouse strains (Shows and Ruddle, 1968). Apparently this gene regulator affects the differential expression of one of the LDH isozymes through selective binding with an inhibitor, cell membrane, or by the differential degradation of one of the synthesized polypeptide subunits (B), e.g., on the post-translational level (Kaplan and Everse, 1972). A similar regulatory system was found in our laboratory for fox LDH (Serov and Chlebodarova, 1973). It was observed that LDH-2 predominate in the erythrocytes of some animals. Genetic analysis of the progeny obtained in crosses between individuals with different erythrocyte LDH spectrums made possible the suggestion that the autosomal gene Ldr-1[a] brought about the low level of LDH-2 isozyme activity. Animals which were homozygous for the Ldr-1[b] allele or heterozygous Ldr-1[a]/Ldr-1[b] demonstrated a normal level of LDH-2 isozyme activity. Nagamine (1974) found a specific inhibitor of the A subunit of LDH in patient plasma. This inhibitor is possibly of a proteinaceous nature and migrated during electrophoresis to the region between β- and γ-globulins of plasma. The ability of this inhibitor to inactivate the A polypeptide is reduced after heat treatment of 65 °C. Another inhibitor, specific for polypeptide B, was described in the plasma, saliva, and hemolyzates of humans (Nagamine, 1974). There is the possibility of a nonrandom association of subunits and, therefore, the development of a specific spectrum of LDH in the various organs of animals (Lebherz, 1974).

Genes which affected the peculiarities of the LDH spectrum and controlled inhibitor synthesis could demonstrate differential activity in different organs. Interstrain differences in the LDH pattern were demonstrated, not only for mouse erythrocytes (Shows and Ruddle, 1968), but also for mouse liver (Chlebodarova et al., 1978). Two phenotypes were obtained for both tissues. In mice of the C3H/He, DBA/1J, and C57BL/6J strains the LDH pattern in erythrocytes is represented only by the LDH-5 isozyme, i.e., they do not have isozymes with B subunits. It is customary to designate this phenotype as B_1^-. In mice of the CBA/Lac and AKR/J strains both LDH-5 and LDH-4 isozymes are present (phenotype B_1^+). Individuals with B_1^- and B_1^+ phenotypes are homozygous for a and b alleles, respectively, of the Ldr-1 gene. Individuals homozygous for the a allele (B_1^- type of erythrocytic LDH) have five isozymes of LDH in the liver. The activity of these isozymes gradually decreased from the LDH-5 fraction to the LDH-1 fraction. This pattern of the LDH spectrum, designated as the B_2^+ type, was demonstrated in CBA/Lac and AKR/J strains. In mice of the C57BC/6J, C3H/He, DBA/IJ, and DBA/2J strains only three isozymes, LDH-5, LDH-4, and LDH-3, were found (type B_2^-). The isozyme LDH-3 demonstrated very low activity; this indicates a reduction in activity of isozymes containing B subunits.

The question then arose, are these peculiarities of LDH patterns in erythrocytes and liver affected by the same gene or by different genes with organ specificity? If these differences in the LDH spectrum are inherited by one gene, test cross progeny should demonstrate an even proportion of $B_1^+ B_2^+$ and $B_1^- B_2^-$ phenotypes. If these two types of LDH pattern are the result of the action of two genes located in different linkage groups, the test cross should result in four phenotypes, $B_1^+ B_2^+$, $B_1^+ B_2^-$, $B_1^- B_2^+$, and $B_1^- B_2^-$, in even proportions. In the test cross progeny these four phenotypes were observed. However, the proportion of the $B_1^+ B_2^-$ and $B_1^- B_2^+$ phenotypes was 19.6%. These data allow the suggestion that the peculiarities of LDH isozyme patterns in mouse liver and erythrocytes are inherited as separate genes with a distance between them of 19.6 ± 4.1 map units. The F_2 data is in good agreement with this hypothesis. The gene which controlled the specificity of the LDH pattern in mouse liver was designated as Ldr-2 (Chlebodarova et al., 1978).

During ontogenesis the activity of this Ldr-2 gene is noticeable, beginning 6–8 days after birth, and for Ldr-1 starting from 12–14 days of postnatal life. During this period it was possible to obtain interstrain differences in the LDH spectrum. Consequently, gene regulators began to function after structural genes and, therefore, their control was realized on the post-translational level rather than on the transcriptional level. In this case the mechanism of action could be the association of isozymes with some type of inhibitor or with proteins of cell membranes.

An interesting system of alcohol dehydrogenase activity was obtained from plants during ontogenesis. Two alleles controlled Adh_1^f and Adh_1^s variants which had fast and slow electrophoretic viability. The first variant is apparently activated early, the second in the scutellum and endosperm. There is another gene, Adh_r, which regulates the level of enzymatic activity only in the scutellum. Its allele, Adh_r^N, specifically increases the relative activity of Adh_1^s. This effect is averted by the dominant allele Adh_r. It is possible that the gene Adh_r is also potentially active in the endosperm, but prior to the moment of its being "switched on" in the endosperm, the synthesis of ADH had already been terminated. Therefore, the gap in time of activation of nonallelic genes prevented the interaction between these genes in this plant organ. The authors have listed some genetic data which support the suggestion that the Adh_r gene synthesizes a limiting factor which is necessary for the transcription of the structural ADH genes. There is competition between the ADH genes for this factor. However, direct proof of the transcriptional level (but not other levels) of regulation have not yet been established (Schwartz, 1971, 1973 b; Efron, 1971).

The multiple fractions of ADH found in *Drosophila* are formed at the post-translational level by the binding of enzyme molecules with unknown and, possibly, proteinous, thermostable components (Gerace et al., 1973). A model, explaining the tissue specificity distribution of dimeric esterase-2 isozymes in heterozygous butterflies, *Ephestia kühniella*, was proposed (Leibenguth, 1973). He suggested that there are regulatory alleles $RG^{a,b,c}$ which differ in their sensitivity to some inducer compounds. These regulatory genes control the proportions of isozymes in tissues. The dominant gene which controls esterase activities in microsomes of rat liver has been described (Zemaitis et al., 1974).

Finally, it was suggested that the actual activity of some of the mitochondrial enzymes is affected, not by the number of molecules synthesized in the cell, but by the number of acceptors on the intramitochondrial membrane where the enzymes are located during activity. Therefore, the enzyme activity is controlled, not only by its structural gene, but by gene modifiers as well. These gene modifiers regulate the number of acceptor sites on the intermitochondrial membrane (O'Brien and Gethman, 1973).

The influence of particular chromosomes on the phenotypic expression of biochemical traits has been observed experimentally. For instance, somatic hybridization of cells of mouse liver adenocarcinoma with a human (or mouse) fibroblast led to a reduction in the esterase-2 activity of heterokaryons. This enzyme is characteristic of adenocarcinoma cells. The elimination of human C_{10} chromosomes from the cloned heterokaryons induced a reinstatement of enzyme activity. It was suggested that this chromosome regulates the phenotype of the corresponding esterase isozyme (Klebe et al., 1970; Ruddle, 1971). However, the mechanism of this regulation and the level of its action (transcriptional or post-transcriptional) remains unknown. Also still to be explained remain a number of events which were obtained in experiments with somatic hybridization. Among these are: (1) There is a reactivation of one of the esterase fractions which is characteristic of the hamster in heterkaryons made of hamster cells and human fibroblasts. This secondary activation of the hamster gene is induced by a gene activator which is linked with the AdeB gene on a chromosome of the B group (Fa Ten Kao and Puck, 1972);

Fig. 41. LDH isozymes of medaka. *XX* is the standard pattern from the eyeball of XX females. The other patterns are derived from skeletal muscle. In *YY* the LDH zymograms of skeletal muscle are distinctive and are designated types *I*, *II*, or *III*. Note the absence of these three distinctive types in zymograms of the eyeball of XX females (*XX*), and of skeletal muscle of XX females (*XX*) or XY males (*XY*). (After Matsuzawa and Hamilton, 1973)

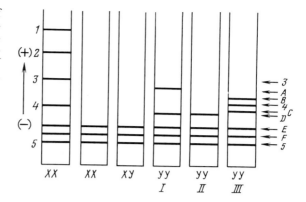

(2) The reactivation of rat hypoxanthine phosphoribosyltransferase was obtained in heterokaryons of rat hepatoma and human fibroblast cultures under the control of human chromosomes (Croce et al., 1973).

It was demonstrated that, in some cases, certain chromosomes of *Drosophila* controlled the synthesis of membrane components which are able to associate with esterases, and, as a result, to control the distribution of various isozymes inside the cell. The mutation of corresponding loci determined the rearrangement of enzyme fraction activities among various cell structures (Korochkin et al., 1973 b). In the fish *Oryzias latipes*, a system of gene modifiers was discovered. These genes are located in the sex chromosome and affect the spectrum of LDH isozymes in muscles (Fig. 41; Matsuzawa and Hamilton, 1973).

Recently it has been demonstrated that the incorporation of individual chromosomes of rye into the wheat genome influences the phenotypic expression of "wheat" malate dehydrogenase, possibly through a modificatory effect of gene regulators (Flavell and McPherson, 1972).

A classical example of gene interaction on the post-transcriptional level is the genetic control of xanthine dehydrogenase (XDH) activity in *Drosophila*. The XDH activity is regulated by at least four loci: the structural gene of XDH, rosy (ry), which is located on the third chromosome (52 map units); the gene lxd (low enzyme activity) which is located on the third chromosome; and two genes, maroon-like (ma-l) and cinnamon (cin) which are located on the X chromosome (Courtright, 1976).

There are two hypotheses which explain the interaction of these genes. According to the first hypothesis (Fig. 42A), XDH is a polymer made up of polypeptide subunits which are coded for by the ry gene and other loci. It was suggested that the antigenic properties of XDH molecules are stipulated by subunit structures which are controlled by the structural gene. In very elegant work from the Chovnik laboratory it was proven that the rosy locus is the continuous sequence of nucleotides which codes for the polypeptide molecule which, in turn, forms the dimer molecule of XDH. Using various strains of *D. melanogaster* differing in the electrophoretic mobility of XDH (also including the null mutant) the authors, on the basis of recombination analysis, calculated the size of the structural site of LDH

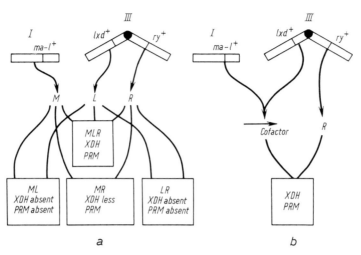

Fig. 42 a, b. Models of the genetic control of xanthine dehydrogenase in *Drosophila melano-gaster* **a** "subunits" hypothesis; **b** "cofactor" hypothesis. (After Markert and Ursprung, 1973)

which codes for the polypeptide. The length of this site was estimated as 3×10^3 nucleotide pairs. This is in good agreement with the molecular weight of the XDH molecule (150,000 daltons, equal to 1,000 amino acids per subunit) (Gelbart et al., 1974).

According to the second hypothesis (Fig. 42B), the genes ma-1$^+$ and lxd$^+$ control the synthesis of a cofactor which binds with a protein that has antigenic properties to XDH. After the protein–cofactor association has occurred the enzyme begins to be active. Experiments done when extracts of different fly strains were mixed supported this suggestion. Active XDH is formed when ry and ma-l extracts were combined. Therefore ma-l flies produced the factor which is complementary to the ry factor and, vice versa, the ry flies contained the complementary factor to the ma-l flies. ldx flies also produced a factor which is complementary to ry, but not to the ma-l factors. There is some evidence that these factors are of a proteinous nature (Glassman and Mitchell, 1959; Chovnik et al., 1964; Glassman et al., 1966, 1968; Courtright, 1967).

Courtright suggested a scheme of interaction of genes involved in the expression of XDH and related enzymes.

1. Interaction of cin and ma-1
   cin + ma-l→precursor

2. Activation of precursor
   precursor $\xrightarrow{\text{1xd gene} + \text{Mo}^{+6}}$ complementing factor

3. Formation of active enzyme
   A1DOX-CRM —————→ aldehyde oxidase
   XDH-CRM $\xrightarrow[\text{factor}]{\text{complementing}}$ xanthine dehydrogenase

The interaction of structural and regulatory genes occurred, in this case, following a scheme on the post-translational, but not the transcriptional level (Courtright, 1975, 1976).

It is obvious that such processes are included in the mechanisms which determine the dependence of character realization. The expression of a character, therefore, is affected by the structural gene, gene modifiers, and the "genetic" environment where these genes act.

Differences in the synthesis and degradation rates played an important role in the differential expression of various isozymes.

In the natural population of *D. melanogaster* there are two variants of ADH which differ in their electrophoretic mobility. Characteristically, flies homozygous for the fast variant (Adh^F/Adh^F) of the isozyme demonstrated higher ADH activity than flies homozygous for the slow (Adh^S/Adh^S) isozyme. It was demonstrated, with the aid of immunological methods, that, for complete inhibition of the enzyme activity in flies homozygous for the fast isozyme, it is necessary to use twice as much of the specific antiserum than in the case of Adh^S/Adh^S flies. The heterozygous Adh^F/Adh^S needs an intermediate amount of antiserum. Therefore, the genetically controlled differences in ADH activity are dependent upon differences in molecule numbers (Gibson, 1972). It is possible that gene modifiers regulate the amounts of ADH molecules which are controlled by alternative alleles (Day et al., 1974).

A similar phenomenon was discovered in our laboratory. It was demonstrated that the parental alleles of the Est-6 locus have unequal expression in the ejaculatory duct of *D. melanogaster* × *D. simulans* hybrids. This differential gene expression is associated with different numbers of enzyme molecules (similar to the case of ADH) since the rate of "simulans" esterase synthesis is higher than that of "melanogaster" synthesis (Table 18; Aronstam, 1974; Aronstam et al., 1975).

The rate of synthesis of the histidase enzyme in mice of the C57BL/6 strain is higher than that in the C3H/He strain and, therefore, the first strain has higher histidase activity in the liver. This difference between two mice strains is dependent upon differences in the activity of a two-allelic autosomal locus (Hanford and Arfin, 1977).

Table 18. C^{14} incorporation in esterase-6 isozymes in hybrid males *D. melanogaster* × *D. stimulans*. (After Aronstam, 1974)

Incubation time, h	D. sim Est · 6^s/Est · 6^s X D. mel Est · 6^s/Est · 6^F Isozyme		D. mel Est · 6^s/Est · 6^{s}x D. sim Est · 6^F/Est · 6^F Isozyme	
	Slow	Fast	Slow	Fast
6	1643	931		
	1314	743		
24	2816	1770	1840	2640
	2557	1412	1665	2494
48	3503	2002		
	3450	1857		

In some cases the specific genes were found which controlled the activity differences of allelic genes. For instance, δ-aminolevulinate dehydratase in the liver of AKR mice (allele Lv-a) demonstrated higher activity than C57BL/6 mice (allele Lv-b). Heterozygous mice are characterized by an intermediate level of enzyme activity. Purified enzymes, extracted from both types of homozygotes (Lv-a/Lv-a and Lv-b/Lv/b), are identical in a number of physicochemical properties (Coleman, 1966; Doyle and Schimke, 1969; Doyle, 1971). It was demonstrated, with the aid of isotopic techniques, that the locus Lv regulates the rate of enzyme synthesis, however it is not the structural gene for δ-aminolevulino-dehydratase, nor does it affect the enzyme degradation rate. Therefore, the enzyme activity in tissues is regulated by a number of enzyme molecules (Schimke, 1973). Oganji (1971) suggested that there are special genes which control the rate and initiation of octanol dehydrogenase structural gene activity in *Drosophila*.

The enzyme synthesis of β-glucuronidase and β-galactosidase in mouse are controlled by two unlinked genes. Their activities change coordinately during mouse development. However, the ratio of activities of these two enzymes differs in various tissues (0.65 in brain, 6.1 in liver). These differences in enzyme activity ratios are caused by organospecific genetic systems which regulate the rates of synthesis and degradation of enzyme molecules (Meisler and Paigen, 1972).

It is also known that mutations which change the degradation rate of catalase in mouse liver do not affect its degradation in the kidneys (Ganschow and Schimke, 1970). Catalase activity is regulated by two genetic factors. The first factor controls catalytic enzyme activity such that the specific activity of purified catalase from the liver of C57BL/6 and C57BL/Ha mice is 60% of that found in the DBA/2 strain of mice. The second genetic factor controls the content of catalase activity in the liver through the regulation of the degradation rate. In the liver of C57BL/Ha mice the degradation rate of catalase was found to be half that of DBA/2 and C57BL/6 mice. As a result, in the liver of C57BL/Ha mice there are three times as many enzyme molecules (Rechcigl and Heston, 1967; Ganschow and Schimke, 1970). Similar systems, apparently, occur in *Drosophila* (Rawls and Lucchesi, 1974). The predominance of LDH-5 in the R3230AC tumor of rat mammary glands is caused by its low degradation rate in comparison with other fractions of LDH (Williams and Fritz, 1973). At the same time, the tissue-specific spectrum of LDH depends upon the rates of degradation and synthesis of A and B polypeptides. It was observed that LDH-5 in rat liver is synthesized 32 times as fast as in cardiac muscle, and 13 times as fast as in skeletal muscle. The degradation rate of this isozyme is 10 times higher in cardiac muscle than in liver and 22 times higher than in skeletal muscle (Table 19; Fritz et al., 1969). Later, corresponding constants were also determined for other LDH isozymes in various organs (Table 20; Fritz et al., 1973).

As can be seen from Table 20, the homopolymer A_4 in the heart demonstrated a higher level of degradation than any other tetramer containing the B polypeptide. However, in liver the tetramers A_4 and A_3B have the same rate of degradation.

Don Michele and Masters (1977) pointed out that the lowest degradation constant was obtained in the rat for muscle LDH (0.2 day^{-1}), and the highest for cardiac muscle (0.72 day^{-1}). A high rate of synthesis (65 mg protein/g tissue/day) was observed for liver LDH. The slowest turnover of LDH was obtained in muscle tis-

Table 19. Protein turnover in normal rats. (After Fritz et al., 1969)

Protein	Rate constants		Half-life (days)
	Synthesis (k_{AP}) (pmol/day/gm)	Degradation (k_{PA}) (day^{-1})	
Heart muscle LDH-5	2.2	0.399	1.6
Liver LDH-5	65.0	0.041	16.0
Skeletal muscle LDH-5	5.2	0.018	31.0
Heart muscle total soluble	8.7×10^5	0.669	1.0
Liver total soluble	16.7×10^5	0.310	2.2
Skeletal muscle total soluble	0.34×10^5	0.031	22.0

Table 20. Constants of the LDH isozymes synthesis and degradation in rat tissues. (After Fritz et al., 1973)

Tissue and isozyme	Degradation Kd, day^{-1}	Concentration [P] picomol/g	Synthesis K, picomol/g/day
Heart			
5	0.077	$42 \pm 3 \ (15)^a$	3.2
4	0.015	$116 \pm 11 \ (14)$	1.7
3	0.021	$560 \pm 24 \ (14)$	11.8
2	0.022	$1801 \pm 88 \ (13)$	39.6
1	0.028	$1902 \pm 78 \ (12)$	53.2
Skeletal muscle			
5	0.005	$2692 \pm 76 \ (15)$	13.5
4	0.009	$68 \pm 7 \ (5)$	0.6
3	0.018	$42 \pm 4 \ (5)$	0.8
2	0.031	$41 \pm 4 \ (4)$	11.3
1	0.052	$47 \pm 7 \ (5)$	2.4
Liver			
5	0.036	$2199 \pm 146 \ (17)$	79.2
4	0.028	$27 \pm 6 \ (5)$	0.8

[a] Number of investigated animals

sue. The fastest synthesis of total LDH and LDH-5 occurs in the liver and muscle of the heart and uterus.

Uracil degradation in mice is controlled by one allelic pair of the Pd locus. It was suggested that this locus regulates the activity of all three enzymes of the pyrimidine-degradation system: dihydrouracil dehydrogenase, dihydropyrimidinase, and 3-ureidopropionase (Dagg et al., 1964). This was corroborated for 3-ureidopropionase by Sanno et al., (1970).

The differential expression of Ct_1 and Ct_2, which controls the synthesis of catalase in corn, is mainly dependent upon the ratio between synthesis and degradation rates (Quail and Scandalios, 1971). It is characteristic that the differential rate of translation is an important regulatory mechanism in hemoglobin synthesis in mammals, wherein the α-chain is translated faster than the β-chain. Therefore, two different types of mRNA are translated at a different rate inside the same cell (Hunt et al., 1969).

On the basis of their own observations, Fritz and co-workers proposed a hypothesis to explain the control of LDH isozyme concentrations in animal cells. This hypothesis explained how these concentrations could be changed if the rates of transcription and translation remained constant (Fritz et al., 1971). They proposed the following scheme:

Amino acid pool

$$\downarrow K_{AP}^{A_4} \text{(synthesis)} \qquad \downarrow K_{AP}^{B_4}$$

$$A_4 \quad + \quad B_4 \underset{K_{-1}}{\overset{K_1}{\rightleftharpoons}} 2A_2B_2$$

$$A_2B_2 + B_4 \underset{K_{-2}}{\overset{K_2}{\rightleftharpoons}} 2AB_3 \xrightarrow{K_{PA}^{AB_3} \text{(degradation)}}$$

$$A_2B_2 + A_4 \underset{K_{-3}}{\overset{K_2}{\rightleftharpoons}} 2A_3B \quad K_{PA}^{A_3B}$$

It was suggested that homopolymers A_4 and B_4 are synthesized as a result of first-order reactions, which are characterized by rate constants – $K_{AP}^{A_4}$ and $K_{AP}^{B_4}$.

The heteropolymers A_2B_2, AB_3, and A_3B are formed during second-order reactions. The fast intermediate reaction could also occur when monomers and/or dimers are formed. According to this model, only the asymmetric heteropolymers A_3B and AB_3 are degradated with the succeeding elimination of subunits from the reaction system. On the other hand, it was demonstrated that free exchange between LDH fractions does not occur, and that the separated subunits (produced as a result of enzyme degradation) are disintegrated (Rauch, 1969; Nadal-Ginard, 1976).

Finally, the enzyme organization in multienzyme complexes or polyenzyme systems also significantly affects the phenotypic expression of various enzymes. The assembling of such systems in the cell is apparently one of the basic mechanisms of intergenic interactions since the enzymes comprising the multienzyme complex are products of different loci (De Moss et al., 1967; Chalmers and De Moss, 1970; Maletsky, 1972).

The multienzyme complex in *Neurospora* is composed of anthranilate synthetase, N-(5-phosphoribosyl) anthranilate isomerase, and indol-3-glycerol phosphate synthetase, and is controlled by tryp-1 and tryp-2 genes. The association of two subunits, which are controlled by these loci, is necessary for the expression of activity of anthranilate synthetase. Mutations of the tryp-1 and tryp-2 genes led to various disturbances in complex formation and enzyme activity expression (Chalmers and De Moss, 1970). Lipid-containing multienzyme complexes of various constitutions were obtained in the plasma membranes of rat liver (Blomberg and Raftell, 1974). It was suggested that *Drosophila* enzymes, controlling different stages of xanthommatine formation (brown eye pigment), are integrated into multienzyme complexes. Some of the classical eye mutants (v, cn, st, ltd, cd, w) reduced phenoxasinone synthetase. This enzyme catalyzes the condensation of 3-hydroxy-kinurenine into xanthommatine and distorts its association with pigment granules, therefore apparently distorting the function of the multienzyme complex wherein it is associated. Two of the previously cited mutants (v, cn) also affected the activity of other enzymes, possibly by binding with the same complex (Phillips et al., 1973). These kinds of mutations could serve as a basis for pleiotropic gene action.

There are numerous variants in the conditions of assembly of multienzyme complexes since loci which are responsible for enzyme syntheses could be heterozygous. If all of these loci are heterozygous, the number of variants in complexes is determined by the following product (Maletsky, 1972):

$$\prod_{i=1}^{m} C_i ,$$

Where C is the number of molecular variants for i enzyme from the total number m-enzymes associated in the multienzyme complex. For each enzyme this number could be calculated from the following formula (Shaw, 1965):

$$C_k^n = \frac{(n+k-1)!}{n!(k-1)!} .$$

Where k is the number of various polypeptides which are synthesized by the cell and from which the enzyme molecule is formed; n is the number of polypeptides in the enzyme molecule (i.e., for LDH $k=2$, $n=4$).

Each variant of the enzyme assembly of the multienzyme complex is realized with a certain degree of probability which is dependent upon numerous conditions and differs in various tissues and cell types.

The activity of either isoenzyme, naturally, depends, to a great degree, upon which variant of a given multienzyme system will be realized. Since in some cases some of the enzyme molecules cannot associate, this could occur, for instance, as a result of a lack of conformational accordance. Thus, the complex effect of differences in translation rate and also post-translational events (assembly with membranes or inhibitors, degradation, formation of multienzyme complexes, etc.) is an important factor in the realization of differential expression of biochemical traits in the phenotype.

There are specific regulating gene systems which control the maintenance of the processes which occur on all of these levels.

1.3 The Phenomenon of Allelic Exclusion

1.3.1 The Concept of Allelic Exclusion, Its Phenomenology

In the previous chapters the various degrees of phenotypic expression of maternal and paternal biochemical traits were discussed. However, extreme cases are known when one of the parental characteristics is not expressed at all. Such an extreme case of differential gene expression is allelic exclusion. In this case, one of the homologous alleles is inactivated and does not participate in the function of the gene.

The possibility of differential conditions of homologous alleles was initially indicated by Prokofieva-Belgovskaja (1946). She obtained differences between maternal and paternal chromosomes in the salivary glands of heterozygous y ac v/sc⁸, and homozygous sc⁸/sc⁸ flies of *D. melanogaster*. Later, Lyon (1961, 1968) demonstrated that in female mice, heterozygous for the sex-linked affecting hair color, some of the regions of hair possessed color typical of the father, whereas other regions demonstrated the maternal color. In XO females this mosaicism was absent. Data obtained suggested the inactivation of one of the X chromosomes of mammalian females.

However, only a few years ago no serious evidence of the inactivation of either whole of partial autosomes had been demonstrated. It was the convention that both of alleles of the autosomal locus were active in the heterozygous cell. The classic example of such a situation is hemoglobin. Both alleles of each locus, controlling the synthesis of whole molecules, are active in the same erythrocyte (Bennett and Owen, 1966). If the cell was heterozygous for one of the genes, the products of both alleles were obtained in this molecule accordingly to the generally adopted theory (Finger et al., 1966; Smithies et al., 1968; Watkins, 1966).

Evidence of the functional inactivation of autosomal alleles was initially observed when the synthesis of immunoglobulins (Ig) was studied. Genes controlling Ig synthesis are located on autosomes and are codominant, since in heterozygotes the products of both parental alleles (allotypes) are present. However, each mature plasmic cell in the heterozygous organism is specialized for the synthesis of only one of the two allotypes of light (L) chain Ig or heavy (H) chain. The second allotype can be synthesized only in another cell (Gell and Sell, 1965; Oudin, 1966).

Allotypic exclusion of Ig was indirectly demonstrated when patients with polymyeloma were inspected (Harboe et al., 1962; Martensson, 1963; Martensson and Kunkel, 1965; Cohn, 1967). It was demonstrated that in heterozygotes the heavy chains of myelomic proteins have only one allotype, whereas in the blood of the same individuals both allotypes are present (Table 21).

Analogous data were demonstrated for light chains (Harboe et al., 1962). Direct evidence of allelic exclusion was demonstrated in the example of immunoglob-

Table 21. Gm-types of whole serum and isolated 7S-myelomic proteins. (After Harboe et al., 1962)

Gm-types of isolated 7S myelomic proteins	Gm-types of whole serum			
	a^+b^-	a^+b^+	a^-b^+	a^-b^-
a^+b^-	2	11	0	0
a^+b^+	0	0	0	0
a^-b^+	0	1	1	0
a^-b^-	3	3	14	5

ulin synthesis when methods of immunohemolysis (Weiler, 1965) and immunofluorescence (Pernis, 1966, 1967) were used.

Allotypic localization in the cell was analyzed in a number of experiments in the lymphoid tissue of rabbits (Pernis et al., 1965). With the aid of antiserums to allotypes which are controlled by allelic genes (the combinations A_a^1 and A_a^2 for locus Aa; Ab⁴ and Ab⁵, Ab⁴ and Ab⁶, Ab⁴ and Ab⁹ for locus Ab were investigated) the presence of two populations of plasma cells was demonstrated. Each of these populations contained only one of the two allotypes. In general both types of cell are distributed randomly. However, there were fields which showed the predominance of one of them.

Successive precipitation of radioactively labeled globulins of heterozygous rabbit progeny showed that all Ig of the blood serum is a mixture of parental homozygous-type molecules. Hybrid molecules were not found, although these kinds of molecule were synthesized in in vitro conditions and no structural limitation was demonstrated for their formation in vivo (Dray and Nisonoff, 1963). There is good agreement between percentages of cells containing a certain allotype and the relative amounts of this allotype. Thus, the ratio of cells containing A4 and A5 in the spleen is equal to 61/39 for CL 262 rabbits (Pernis et al., 1965). According to papers dealing with serum proteins in heterozygotes Ab⁴/Ab⁵, the ratio of Ig which reacted with anti-A4 and anti-A5 is equal to 64/27 (Dray and Nisonoff, 1963); 63/36 (Leskowitz, 1963) 62/38 (Bornstein and Oudin, 1964). In another rabbit (V22) with the Ab⁴/Ab⁵ genotype, the predominance of cells with the A4 allotype was smaller (the A4/A5 ratio was equal to 53/47). However, these numbers were in good comparison with the allotypic ratio for the serum.

This coincidence satisfied the hypothesis that the number of molecules of different allotypes is dependent upon the number of cells which synthesize these allotypes rather than the rate of molecule synthesis within the cell (Pernis et al., 1965; see review Baranov and Korochkin, 1973).

The presence in heterozygous cells of only one of two types of parentally inherited Ig allotypes was also found in humans (Pernis et al., 1970) and mice (Weiler, 1965). At the same time, it is possible that each of the allelic genes of the a and b loci in rabbits (as well as in other cases) is represented by a number of closely linked genes which control the synthesis of the variable part of the H-chain. These sections of H-chains demonstrated both common and different antigenic determinants (Landucci-Tosi et al., 1975; Mage et al., 1977). Also in populations of lym-

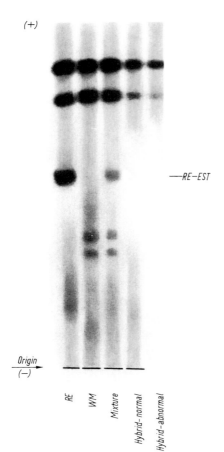

(+)

—RE-EST

Origin
(−)

RE
WM
Mixture
Hybrid-normal
Hybrid-abnormal

Fig. 43. Esterase isozymes in the brain of the interspecific sunfish hybrids: ♀ *Lepomis gulosus* × L. microlophis ♂: *RE*, male redear sunfish; *WM*, female warmouth; mixture, mixture of the redear extracts of male and female; hybrid normal, morphologically normal hybrid; hybrid abnormal, hybrid with a malformation of the jaw. Note that the esterase isozyme unique to the paternal redear sunfish (*RE-EST*) is absent in both hybrids. (After Whitt et al., 1972)

phoid B-cells some cells have maternal and paternal active alleles which code for Ig allotypes (Loor and Kelus, 1977).

Recently published papers have demonstrated the allelic exclusion which occurs in the Lpp locus of pigs (Rapacz and Haster, 1968), the Ac locus of Ig (Vice et al., 1970), and the Lpg locus of lipoprotein allotypes (Albers and Dray, 1969). Analysis of the Ess locus of α-arylesterase (Albers et al., 1969) and A locus γA Ig (Conway et al., 1969) in rabbits allowed the suggestion that functional haploidy of autosomal genes could occur comparatively often.

In connection with this, the analysis of isozymes of interspecific crosses is demonstrative. In interspecies fish hybrids of ♀*Lepomis gulosus* × ♂*L. microlophus* the inhibition of one of the paternal esterases was observed concomitant with asynchronous gene activation (Fig. 43; Whitt et al., 1972). In ♀*Micropterus salmonides* × ♂*Lepomis cyanellus* fish hybrids the expression of maternal glucoso-6-phosphate dehydrogenase and one of the liver esterases is inhibited (Whitt et al., 1973 a, b). Some of the fractions of paternal esterases are not active in ♀*D. virilis* × ♂*D. littoralis* hybrids (Korochkin et al., 1973 a).

1.3.2 Suppressive Action of Allotypic Antibodies on the Synthesis of Immunoglobulins of Corresponding Specificity

Lymphoid tissue of animals heterozygous for either Ig locus is represented apparently by two competitively developed cell populations as a result of allelic exclusion. These populations are possibly under selection caused by certain factors. Thus, variations in the proportion of cells with different allotypes could occur. This selection is the best natural explanation of variation in the quantitative ratios of serum Ig's with the different allotypic constitutions in heterozygous individuals (Oudin, 1966).

It was experimentally shown that the selective factor in lymphoid tissue is the antibodies, which are specific to allotypes and consequently suppressed the synthesis of corresponding Ig's. Dray (1962) induced allotypic suppression in rabbits by treatment of animals during fetal and neonatal periods with antibodies to the paternal allotype. The idea of this experiment, which deals with the change in the quantitative expression of light chains of Ig allotypes, was next (Fig. 44). F_1-heterozygous offspring $b^4 b^5$ were obtained as a result of reciprocal crosses between homozygous rabbit individuals $b^4 b^4$ and $b^5 b^5$ genotype. Before mating the $b^4 b^4$ females were immunized with b^5, and mothers with $b^5 b^5$ genotypes were immunized with b^4 allotypes. The antibodies synthesized in the maternal organism affected the embryo, since the placental barrier is permeable to γ G-globulins. γ G-globulins induced a paternal allotype suppression which remained during the entire postnatal life of the offspring. At the same time the total concentration of γ G-globulins did not change, because the synthesis of the maternal allotype was increased compensatorily. One of the rabbits chosen for the experiment had a ratio of b^4:b^5 that was equal to 300:1 at 5 months of age, 70:1 at 1 year, 63:1 at 2 years, and 19:1 at 3 years. These ratios occurred instead of the regular ones, when normal ratio of the b^4 allotype concentration in blood serum is approximately twice as high as the b^5 allotype concentration.

The suppression of the paternal allotype was induced in $b^4 b^5$-progeny, not only through in utero (through the mother) immunization, but also when the progeny

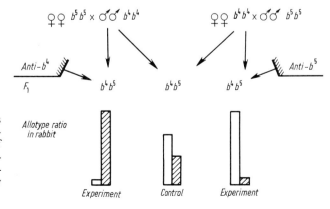

Fig. 44. Scheme of Dray's experiments of the suppression of allotypes of light chains of immunoglobulins in $b^4 b^5$-heterozygous rabbits. (After Baranov and Korochkin, 1973)

of the same genotype as the nonimmunized mother, b^4b^4 or b^5b^5, were injected with antibodies immediately after birth (b^5 and b^4 respectively) (Mage and Dray, 1965; Dubiski, 1967 a, b).

Later, similar experiments were performed in the suppression of the α-locus of Ig heavy chains in the rabbit (Mage, 1967).

The significant interest is the obtaining of allotypic suppression in homozygous individuals. Theoretically it could be possible to expect that hypogamma globulinemia, experimentally induced with antibodies in homozygous individuals, should be accompanied by a compensatory increasing in the Ig synthesis which are controlled by other loci. It is impossible to induce this allotype suppression with in utero methods in the homozygous progeny of homozygous mothers, because these offspring inherited only the maternal allotype. It is also practically impossible to induce suppression by the neonatal injection of antibodies, because the newborn homozygous progeny which went through the placental period of prenatal life has no maternal γ G-immunoglobins. The maternal Ig's specifically reacting with the injected allotype antibodies, will therefore prevent the effect of antibodies on the lymphoid tissue.

Dubiski avoided these difficulties by injection of anti-b^5-antibodies to b^5b^5-progeny immediately after birth. This progeny suppressed the b^5 of heterozygous b^4b^5 mothers. The synthesis suppression of the b^5 allotype in b^5b^5 progeny was accompanied by a large increase in the concentration of Ig which does not have this allospecificity. At the same time, normally in rabbits only 5% to 30% of γ G Ig's do not have specificity of the b locus (b-negative molecules). The same proportion was found for molecules which do not have allotypes of the a locus (a-negative molecules). Therefore, as a result of b^5 suppression in rabbits homozygous for this allele, the synthesis of light chains, which are controlled by another locus, is increased (Dubiski, 1973 a, b).

The second known locus for the light chains of Ig γ-type is the c locus. Vice and co-workers studied the role of this locus in b^5 suppression and the possible participation of this locus in the synthesis of light chains controlled by other loci. They improved the suppression method and transplanted b^5b^5 zygotes into the uterus of b^4b^4 recipient mothers. As a result of this, they totally excluded the appearance of maternal antibodies with b^5 specificity in the embryos. The suppression of b^5 Ig in newborn b^5b^5 rabbits was induced by the injection of anti-b^5 antibodies. The total amount of γ G Ig in rabbits with the suppressed allotype was not changed due to a compensatory increase in the concentration of Ig's with (b)-light chains. Immunological specificity of the c locus was demonstrated by 44% of b-negative γ G molecules of the original 73%, whereas the rest of the molecules (29%) did not have allospecificity for the c and b loci in the 24-week-old rabbits. Therefore it was demonstrated that the synthesis of light chains is controlled by a third or a number of other loci. The result of an experiment using individuals heterozygous for the c locus ($b^5b^5c^7c^{21}$) suggested that light chain synthesis could take place, if not through a third locus, then minimally through another, unknown third allele of the c locus (Vice et al., 1969 a, b).

The inhibition of allotype synthesis was obtained when zygotes homozygous for the α heavy chain locus were transplanted into the uterus of immunized (or nonimmunized) female rabbits, followed by injection of antiallotypic antibodies into the newborn rabbits. The observed data demonstrated that, in suppressed a^2a^2 homozygotes, the other locus, responsible for heavy chain synthesis, begins to function. Thus, molecules (a^-) and (a^+) are the subclasses of γ G-globulins (David and Todd, 1969).

The first attempt to suppress the synthesis of Ig allotypes in mice was unsuccessful. The majority of progeny from immunized mice mothers died, and surviving offspring demonstrated a normal concentration of the paternal allotype. However, Herzenberg successfully suppressed Ig synthesis with Ig-1b specificity, initially for short periods of time and then for the entire life of the mouse (Herzenberg, 1970). Recently, short-term suppression of Ig synthesis in mice was induced with the aid of antibodies to some of the determinants of the variable part of light polypeptide chains (Ruffini et al., 1970).

The mechanism of Ig allotype suppression is not clear. The lack of experimental data concerning this problem forces us to limit ourselves to the more or less likely suggestions. In the first paper dealing with suppression of Ig synthesis by allotype antibodies it was reported that those rabbits suppressed to b^5 having b^4b^5 genotypes did not produce the b^5 allotype in the lymphatic node and spleen, but possessed only the b^4 allotype. These preliminary histoimmunochemical investigations allowed, even then, the conclusion that suppression occurs on the cellular level and is based on a more complicated process than the simple neutralization of allotypes in the blood by injected antibodies (Dray, 1962).

The presence of Ig allelic exclusion served as a basis for the suggestion of selective action of antibodies on the heterogeneous lymphoid cell population. Apparently, allotypic suppression reflected a change in the number of cells specialized for the production of suppressed and alternative allotypes in heterozygotes. A correlation was obtained between the ratios of b^4:b^5 allotypes in the serum of b^5 suppressed b^4b^5 allotype rabbits and the proportion of spleen cells, stained according to Coons, with antibodies to b^4 and b^5. Antibodies, therefore, directly affected lymphoid cells and stimulated, as was demonstrated in in vitro experiments, their blast-transformation. The presence on the lymphocyte surface of the antigenic determinants of Ig, including allotypic determinants, suggests the conclusion that these determinants are the receptors for antibodies which induce suppression (Sell and Gell, 1965 a, b; Jones et al., 1970).

Two hypotheses were suggested in explanation of the suppression mechanism (Dubiski, 1967 a, b). According to the first hypothesis, the fixation of antibodies on the surface of immunocytes, which are competent to synthesize their corresponding allotype, determines the cytotoxical effect and the death of cells. As a result, the cell clone with the alternative allotype has the advantage in competitory development (Fig. 45a). It is realistic to suggest that the inactivation of lymphoid cells by antibodies occurs in a manner similar to the mechanism for the establishment of immunological tolerance (Bretscher and Cohn, 1968).

The second hypothesis suggested that all immunocytes are in a resting condition, until the moment of specific antigenic stimulation. The antibodies which lock one of two allotypic receptors on the cell surface commit the immunocytes to synthesize the alternative allotype (Fig. 45b). According to this hypothesis, which suggests the influence of autoallotypic antibodies on the development of immunocytes, the antibodies do not kill these cells.

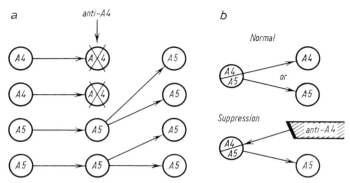

Fig. 45. Two hypotheses (**a** and **b**) represented the mechanism of suppression action of antiallotypic antibodies. Explanation in text. (After Baranov and Korochkin, 1973)

Therefore, for the realization of the suppressive action of antibodies it is apparently necessary to have: (1) the presence of the cloning cells according to differential activities of their allelic or nonallelic genes; (2) the presence of cell surface antigenic determinants – the receptors of antibodies. It is possible that the further investigation of protein models which satisfy these two conditions will result in an understanding of the intimate mechanisms of genome activity during the cellular differentiation of ontogenesis.

The mechanism which determines the phenotypic restriction of allotypes appears to act only in vivo since there is an absence of allelic exclusion in in vitro clones of lymphoid cells (Bloom et al., 1971).

1.3.3 Examples of True Allelic Exclusion

In most cases, allelic exclusion was registered on the phenotypic level when either trait was absent in development. As follows from the observations discussed in Chapts. 1.1 and 1.2, this absence could be due to an event at any level of regulation. Moreover, differential gene expression could occur because of the somatic segregation of the whole genome or by somatic crossing over (Smithies, 1964; Bečak et al., 1966; Ohno, 1967; Stern, 1968). At the same time, true allelic exclusion indicates only those cases where the inactivation of one of the pairs of alleles was proven. In this case, the absence of a trait is caused by the cessation of transcription of the corresponding piece of DNA. In connection with this, we will first discuss the theory of differential inactivation of homologous loci. Subsequently, some actual data on allelic exclusion will be considered in the light of certain theoretical prerequisites.

1.3.3.1 The Differential Activity of Parental Genomes

The most illustrative example of the differential activity of whole parental sets of chromosomes is the phenomenon of inactivation of one of the chromosomal com-

Table 22. Data on several characters of newly hatched larvae of *Pseudococcus gahani, Ps. obscurus,* and their male hybrids. (After Nur, 1967)

	Distance, pore to eye	Diameter of large spine (L)	Diameter of small spine (S)	Ratio S/L	Hind legs: tibia + tarsus	Anal lobes with 3 spines
Pseudococcus gahani	7.9 ± 0.3	4.2 ± 0.1	3.2 ± 0.1	0.76 ± 0.02	127.8 ± 0.5	0.7% (150)[a]
Ps. obscurus	1.0 ± 0.1	4.7 ± 0.1	2.8 ± 0.1	0.59 ± 0.01	138.1 ± 0.9	1.8% (110)
Ps. obscurus × gahani	1.4 ± 0.1	3.7 ± 0.1	2.0 ± 0.1	0.54 ± 0.01	119.0 ± 0.6	11.5% (122)

[a] Number of lobes examined
Linear dimensions are given in microns and are based on 50 measurements. Spines from anal lobes with three spines were not measured

plexes in males of some species of *Coccoidea*. In this case, cytological observations are supported by genetic data (Nur, 1967; Brown, 1969; Chandra, 1971). During early embryogenic stages in these males, there is total heterochromatization (and, as a result, inactivation) of the chromosomes of the father (Schrader and Hughes-Schrader, 1931). Evidence of genetic inactivation was obtained when the expression of some mutations in males was investigated. It was demonstrated that only the maternal genes are active (Table 22). Paternal dominant genes were not expressed (i.e., eye color genes and dominant lethal genes which were induced in males before mating). The genetic inactivation and heterochromatization of the paternal genome is correlated with its late replication in S-phase, with its exclusion from reproduction in endomitotic nuclei, and with a dramatic reduction (as compared to euchromatic chromosomes) in RNA synthesis during interphase (Berlowitz, 1965a, b; Chung-chin-Huang, 1971). Therefore, the cytological criteria of chromosome inactivation are present. It was demonstrated that histones do not affect chromosome heterochromatization (Palotta et al., 1970).

In some tissues of the male, deheterochromatization occurs during a certain period of development. It is precisely these tissues which do not develop normally if the individual has either an irradiated paternal genome or the genome of another species. Decondensation does not take place in all cells and, as a result, these tissues are functionally mosaics.

To explain the mechanism of paternal genome inactivation in *Pseudococcus*, Nur (1973) postulated that females produce two types of eggs: those with a compound that induces heterochromatization of the paternal chromosomes and those without this compound. After fertilization the first type of egg develops into males and the second into females.

1.3.3.2 The Differential Activity of X Chromosomes in Mammals

As mentioned previously, the hypothesis of differential activity of X chromosomes was developed by Lyon (1961) and was used to explain gene dosage compensation in mammals. The inactivation of one X chromosome in XX females is accompanied by the heterochromatization of this chromosome and the subsequent formation of the heteropycnotic body. According to this hypothesis: (1) the heteropycnotic X chromosome could be of paternal or maternal origin in various cells of the same organism; (2) this chromosome is genetically inactive (Lyon, 1961). There is also biochemical evidence which indicates that at least the main part of the X chromosome is inactive. It was noted that when an enzyme controlled by an X chromosome-linked locus is not expressed, then another enzyme, which is controlled by a gene located at a significant distance from the first locus, is also absent (Gartler et al., 1972 a, b). This inactivation was shown to be stable. For instance, a cell clone with the fast fraction of glucose-6-phosphate dehydrogenase was established from patients heterozygous for electrophoretic variants of this enzyme. After the hybridization of cells from this clone with cells of a mouse fibroblast culture, the resulting heterokaryon had only the fast G-6-PDH fraction. Hence, the other X chromosome remained inactive (Migeon, 1972). However, in somatic hybrids between mouse strain D7 cells and human fibroblasts, derepression of the inactivated X chromosome can occur (Kahan and DeMars, 1975). Cooper et al. (1975) found hybrid fractions of G-6-PDH in kangaroo fibroblast cultures, indicating that in these cells both X chromosomes are active. Generally, X chromosome inactivation occurs randomly. A case of nonrandom X chromosome inactivation, however, has been described in mice. In this experiment, the wild-type allele of the sex-linked Ta gene was translocated to an autosome. As a result, the expression of motled color was inhibited. This obscured phenotypic effect allowed the supposition that the X chromosome gene segment which was translocated to an autosome was active in all cells, while these genes on the intact X chromosomes were inactive (Lyon et al., 1964; Searle, 1968). In the kangaroo, only the paternal X chromosome is inactivated (Cooper et al., 1971; Van de Berg et al., 1977). The preferential inactivation of the paternal X chromosome was observed in the yolk sac of mice (Takagi and Sasaki, 1975; Takagi, 1976; West et al., 1977), and in rats (Wake et al., 1976). The mechanism of such preferential inactivation remains unknown.

There are numerous well-known cytological observations of differences in paternal X chromosome behavior (Ohno et al., 1959; Ohno and Hauschka, 1960). Such data demonstrate particularly well the asynchrony of chromosome replication, since one chromosome is labeled by H^3-thymidine significantly earlier than the other (German, 1964 a, b; Zibina and Tichomirova, 1965; Zacharov, 1968). An interesting question is during what developmental period does inactivation of one of the X chromosomes begin (or, on the contrary, when does activation begin)? Data suggest that this process occurs very early in embryonic development. Both X chromosomes are active in oocytes and during early cleavage (Gartler et al., 1972, b, 1973). Adler et al. (1977) have suggested that dose compensation in mammals is caused by the inactivation of one of the two active chromosomes and not by the activation of one of two inactive X chromosomes. They studied preimplantation changes in α-galactosidase which is controlled by a gene located on the X chromosome.

One index of inactivation is the formation of sex chromatin bodies. In man and macaque, these bodies were identified in trophoblast cells, in 10–12-day-old blastocysts and in 16–19-day-old embryos (Glenister, 1956; Park, 1957; Liberman, 1966; Zacharov, 1968) – i.e., at the 2600–5000 cell stage (Hamerton, 1964). At the same time, sex chromatin was observed in 14% of the trophoblast cells on the 13th day of embryonic development, but is absent in the mesoderm of the chorion.

Similar results were obtained with cats. In this case, sex chromatin was first found in trophoblastic cells of the late blastocyst stage. In embryonic cells, it was first observed during the postimplantation stage of development (Austin and Amoroso, 1957; Austin, 1962). One of the rabbit X chromosomes was heteropycnosed 2 days before implantation, at about the 400 cell stage. In the rat, this does not occur until after embryo implantation. It occurs on the 6th day of development of trophoblast cells and on the 7th day in cells of the egg cylinder (Zibina, 1964).

Experiments have been done where one cell of a mouse blastocyst heterozygous for an X-linked gene that determines hair color was injected into a blastocyst of a mouse homozygous for a third allele for hair color. It was demonstrated that X chromosome inactivation occurred on the 3th–3.5th day of the blastocyst stage (Gardner and Lyon, 1971). There is evidence that inactivation occurs after the isolation of the embryoblast cells – significantly earlier than mesoderm and ectoderm delimitation and at the same time as implantation (Nesbitt, 1971). Apparently, the time of X chromosome inactivation differs in various animals. This time can cover stages from the morula with only 50 cells (the pig) to the late blastocyst or early primitive streak stages with several thousand cells in humans or monkeys (Lyon, 1974).

It is possible that X chromosome inactivation does not occur along the entire chromosome at the same time. Rather, it occurs gradually and spreads along the chromosome. Observations of the inactivation of one X chromosome during the cultivation of mouse teratocarcinoma cells suggested this. In this case, the reduction of G-6-PDH activity as a result of X chromosome inactivation always preceded the reduction of another X-linked enzyme-hypoxanthine phosphoribosyltransferase. These temporary differences in activity could be caused, however, on the post-transcriptional level as well, as a result of differences in mRNA stability (Martin et al., 1978).

Sex chromatin formation does not occur in all cells of the embryo. In 3-months-old human embryos, the sex chromatin body is present in the nuclei of somatic cells, while it is absent in oogonia and oocyte nuclei (Ohno et al., 1961).

Is the inactivation of the X chromosomes during development synchronous? Results of an analysis of the distribution of pigmented melanocytes in ciliary bodies and iris and choroid of allophenic and chimeric mice (for details, see Part II) suggest asynchrony of inactivation. In cell precursors of retinal melanocytes, this inactivation is apparently going on by the 7th day of pregnancy (Deol and Whitten, 1972). Cattanach's results, however, differ from those of Deol and Whitten. Cattanach described a translocation with respect to the time of X chromosome inactivation. In this translocation, and autosomal segment of chromosome 7, which caused the wild-type alleles of albino, pink eye and ruby-2, has been inserted into the X chromosome. Variegation occurs in animals which are heterozygous for the translocation and homozygous for the recessive alleles on the intact chromosome 7. In cells with an active translocated X chromosome (Cattanach-X), spots of wild-type were observed. When this translocated chromosome was inactive, then albino, pink eye or ruby-2 spots were formed. Some of these spots had small wild-type spots inside. This was thought due to a tendency of the autosomal pigment that was inserted into the X chromosome to avoid inactivation during X chromosome differentiation (Cattanach, 1974).

In general, the following two characteristics of differential activity of the X chromosomes have been observed:

1. Asynchrony of X chromosome inactivation in various tissues and, possibly, cells of the same tissues.
2. Inactivation occurs during early stages of embryogenesis. In connection with this, large parts of the developed organism demonstrate mosaicism when this organism is heterozygous for a sex-linked trait. In addition, death can occur during early stages of development if the X chromosome contains lethal genes (Zibina and Tichomirova, 1965).

1.3.3.3 Cases of Functional Inactivation of Homologous Autosomal Loci

There are many papers reporting the asynchrony of DNA replication in homologous autosomes. (Lima-de-Faria et al., 1961; German, 1964a, b; Prokofieva-Belgovskaja and Gindilis, 1965; Bianchi and Molina, 1967; Kofman-Alfaro and Chandley, 1970). This asynchrony is visible when the action of distinct segments of chromosomes is compared (Perondini and Dessen, 1969; Iordansky and Pavulsone, 1970; Kiknadze, 1972).

There are a number of reports of differences of length of homologous autosomes in metaphase (Prokofieva-Belgovskaja and Gindilis, 1965). The heteromorphism in animal and plant homologous chromosomes of nonnucleolar secondary constrictions is well documented (Prokofieva-Belgovskaja, 1963; Palmer and Funderburk, 1965). However, at the present time all these facts are difficult to interpret from the point of view of functional heteromorphism because the nature of changes in chromosome morphology and its relation to chromosomal action are still unclear.

The best arguments for possible differential activity of autosomal genes are cases of heteromorphism of chromosomal regions for which the relation between cytological and functional expressions have been clearly demonstrated. These include puffs of polytene chromosomes, lateral loops of lampbrush chromosomes and nucleoli.

1.3.3.4 Heteromorphic Puffs

Heteromorphism of puffs has been found in many *Diptera* and the gallfly (Scherbakov, 1968; Ashburner, 1969a, b; Kiknadze, 1972). There is a sharp asymmetry of RNA synthesis at loci with heteromorphic puffs (Serfling et al., 1969) in addition to different transcriptional activities of homologous loci during asymmetric puffing. In this chapter we will discuss only cases of heteromorphism caused by normally occuring differences in gene expression. The stability of a band structure in the cell is not strong enough evidence for heteromorphism because cytologically undetectable changes can totally inhibit the expression of either locus. More convincing criterions are as follows: (1) puffing differences in homologous loci that are affected by genotype and environment (i.e., in various stocks or hybrids); (2) the different degree of puff heteromorphism in various cells of the same organism.

The behavior of several chromosomal regions from various strains of *D. melanogaster*, *D. simulans*, and hybrids between them were investigated (Aschburner, 1967, 1969a, b). Puff 64C was found only in the *vg6* strain and never for other *D. melanogaster* strains. A puff developed by both homologous chromosomes during synapsis was found in F_1 hybrids between *vg6* and any other strain of flies. The heteromorphic puff found only on the *vg6* homologous chromosome was visible during asynapsis. Similarly a 3B3 puff was found in *D. virilis* × *D. texana* hybrids (Poluektova, 1970a, b). In both cases a region usually nonfunctional, showed puffing. As a result, what is known as the correction of puffing has occurred. This correction, however, is not always total, because, for instance, the common puff in *vg6* × *Oregon R* hybrids is regularly smaller than in homozygous *vg6* flies.

A hypothesis was proposed to explain this phenomenon. Aschburner (1970a, b) suggested that there is a regulatory gene which activates adjacent regions for puffing and, hence, the structural gene that is located in this region. The product of the gene regulator is transported to the second homolog when there is synapsis. This homologous region cannot form a puff because of a mutation in the gene regulator. In this way, puff correction takes place. During asynapsis, when the homologous chromosomes are separated, the product of the gene regulator cannot reach the homolog that has the inactive puff region (Fig. 46).

The differences in the characteristics of DNA replication could play an important role in differential transcriptional activity of homologous chromosomes, especially in interspecific hybrids. It was demonstrated that in polytene chromosomes of *D. melanogaster* there is positive correlation between transcription intensity and the rate of replication of the corresponding X chromosomes (Berendes, 1966a, b; Ananiev and Gvozdev, 1974a, b).

Differences in DNA syntheses in homologous regions of polytene chromosomes were also demonstrated for *Anopheles atroparvus* × *A. labranchiae* hybrids (Tiepolo et al., 1974).

The correction of the structurally damaged locus was not obtained when *vg6* flies were crossed with flies which carried the pericentric inversion, in (3LR) C165, with breakage points in the 64C region. Almost total inhibition of puffing in the 64C region of *vg6* chromosomes was demonstrated in *D. melanogaster* × *D. simu-*

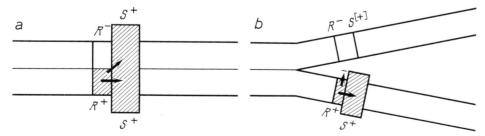

Fig. 46a, b. A formal explanation of the synapsis dependent nonautonomy of 22B:8–9 and of 64C in heterozygotes. In **a** the homologs are synapsed and the signal from the intact regulating gene of the puff active parent is able to activate both structural gene alleles. In **b** the homologs are asynapsed and this chromosome-limited signal transfer is unable to occur – only the cis structural gene is active. (After Ashburner, 1970a, b)

lans hybrids when conjugation of homologous chromosomes had occurred. However, in those nuclei where conjugation was not present, the *vg6* chromosomes always showed a puff at 64C.

Puff 22B is characteristic for Pacific but not for Oregon strains of *D. melanogaster*. Total reparation of activity of the 22B locus of the Oregon strain was obtained, in Pacific × Oregon interstrain hybrids. This event was demonstrated regardless of conjugation of the homologous chromosomes.

The expression of homologous regions varies in various cells of an organ during the different steps of development. Apparently, all cases of differential development of puffs in conjugated and nonconjugated chromosomes can be considered an appropriate example because the degree of synapsis can vary significantly from cell to cell. In the genetic literature, suggestions that the degree of conjugation affect the expression of allelic and nonallelic genes were proposed years ago (Ephrussi and Sutton, 1944; Lewis, 1950). At the present time, however, it is not known how asynapsis affects puffing. Moreover, the cause and effect relationships between these events are very complicated (Evgeniev, 1970).

Finally, there is much interest in the asynchrony of homologous puff formation. An example is the behavior of the 63E region *Drosophila*. Both the *melanogaster* and *simulans* species have this puff, although in *simulans* it is smaller. In hybrids the conjugated homologous chromosomes have a symmetrical puff while in nonconjugated chromosomes the puff is formed firstly on the *melanogaster* chromosome and then on the *simulans* chromosome. Here, the *simulans* puff retains its autonomy in puff size (Fig. 47). Similar behavior was demonstrated for the 2B puff in larvae heterozygous for the "half way" mutation (Rayle, 1967) (see Fig. 47). Asymmetry of the puff as a result of asynchrony of homologous loci function was described for several regions of *Sciara ocellaris* (Pavan and Perondini, 1967; Perondini and Dessen, 1969; Fig. 47). In separate loci (for instance, C12), the differences in puffing time on the homologous chromosomes were accompanied by asynchrony of replication of DNA in this region: the homolog with early puffing had a maximal rate of replication in the G larval stage while the other homolog had a maximum rate in the H stage. Asynchrony of reduction of the homologous puffs was described for the gallfly *Harmandis laevi* (Zhimulev and Lichev, in press).

1.3.3.5 Heteromorphism of the Lateral Loops in "Lampbrush" Chromosomes

The lateral loops of lampbrush chromosomes are active chromosomal regions, similar to puffs. These chromosomes have been studied less than polytene chromosomes from the point of view of the differential activity of homologous chromosomes. However, there are examples of the functional heteromorphism in the behavior of homologous loops (Callan, 1963, 1965).

There is a giant granular loop on chromosome XII of *Triturus cristatus cristatus* which never fuses with other loops. In hybrid females obtained from crosses with *T. cristatus carnifex*, the tendency to conjugate was demonstrated. In some oocytes both of the sister loops of one homolog are very small; in some oocytes one loop is small and the other is reduced; on another homolog both of the loops are reduced

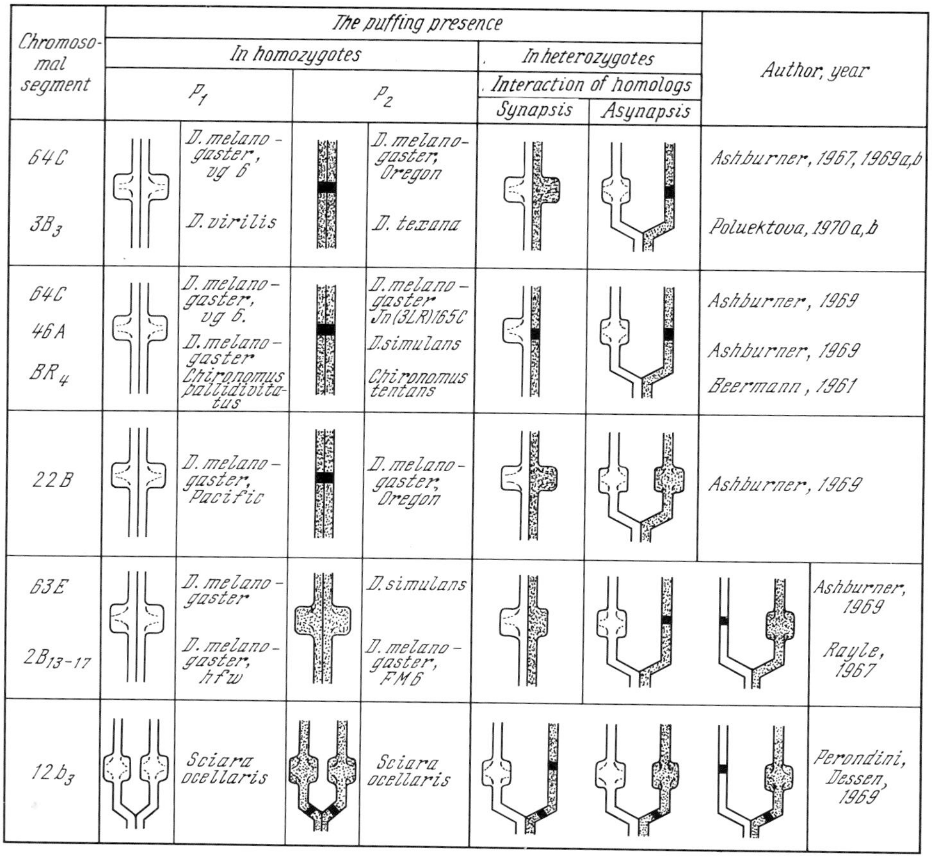

Chromosomal segment	The puffing presence					Author, year
	In homozygotes			In heterozygotes Interaction of homologs		
	P_1		P_2	Synapsis	Asynapsis	
64C	D. melanogaster, vg 6		D. melanogaster, Oregon			Ashburner, 1967, 1969 a,b
3B₃	D. virilis		D. texana			Poluektova, 1970 a,b
64C	D. melanogaster, vg 6.		D. melanogaster, In(3LR)65C			Ashburner, 1969
46A	D. melanogaster		D. simulans			Ashburner, 1969
BR₄	Chironomus pallidivittatus		Chironomus tentans			Beermann, 1961
22B	D. melanogaster, Pacific		D. melanogaster, Oregon			Ashburner, 1969
63E	D. melanogaster		D. simulans			Ashburner, 1969
2B₁₃₋₁₇	D. melanogaster, hfw		D. melanogaster, FM6			Rayle, 1967
12b₃	Sciara ocellaris		Sciara ocellaris			Perondini, Dessen, 1969

Fig. 47. Puffing in the homologous loci of *Drosophila*, *Chironomus* and *Sciara* chromosomes. (After Korochkin and Belyaeva, 1972)

to the chromomer size. There are oocytes with reduced loops on both homologous chromosomes. Therefore, in cells of heterozygous individuals, variable suppression of the development of the giant granular loop was obtained.

1.3.3.6 The Differential Activity of the Nucleolar Organizer

According to modern ideas (see review of Kiknadze and Belyaeva, 1967), the nucleolus is a specific active region of the chromosome where RNA synthesis and the first steps of ribosome formation take place. In this chapter we will discuss only those cases of nucleolus heteromorphism where the differences in activity between homologous regions cannot be explained by structural mutations (i.e., deletion and duplication).

It is well known (Heitz, 1931) that in organisms with a pair of nucleoli, one is bigger then the other. These differences are not stable and within the same tissue there is a possibility of wide variation in the sizes of nucleoli. Apparently, the formation of different-sized nucleoli is caused by the asynchrony of condensation of the organizers in telophase. By this time the differences in nucleus size can be noticed (Belyaeva, 1965).

Heteromorphism of the homologous nucleolar organizers was demonstrated in most papers where the morphology of the satellite chromosomes was studied. In any tissue, all possible combinations in pairs of morphological types of satellite chromosomes were obtained (Gindilis, 1967; John and Lewis, 1968; Deryagin and Iordansky, 1970). Data demonstrating the morphological variability of the nucleolar zones of chromosomes are in good agreement with the idea of a competitive relationship between nucleolar organizers (McClintock, 1934). The extreme expression of the competitive relationship between nucleolar regions (leading to the total inactivation of some nucleoli) was obtained in hybrids. For the first time, Navashin (1934) demonstrated examples of secondary constriction (the pulling in of the satellites) by some parental chromosomes in the cells of *Crepis* hybrids. Similar results were obtained in other hybrids (Evans and Jenkins, 1960; Keep, 1962). The effects of inhibiting the activity of some nucleolar organizers by others was observed in animal cells as well, i.e., in somatic hybrids of mouse and humans. Human chromosomes in these cells are not able to produce rRNA (Eliceiri and Green, 1969).

This phenomenon was distinctly demonstrated for interspecies hybrids of *Xenopus laevis* × *X. mülleri*. Spacers separating the 28S and 18S regions are different for these species in the 40S precursors of rRNA. DNA-RNA hybridization was used for identification of rRNA specificity.

It was demonstrated that there is preferential transcription of *X. laevis* rRNA in hybrids. Up to the swimming tadpole stage, the rRNA of *X. mülleri* is totally or almost totally repressed regardless of the direction of crosses (Fig. 48). After this stage a very low level of rRNA from *X. mülleri* is obtained. The population of hybrid frogs is heterogenous for this trait. Some individuals demonstrate high synthesis of *X. mülleri* rRNA, whereas others do not. It is possible that the embryo populations are heterogeneous also. However, an analysis of the tadpole individuals (or embryos) was not done.

X. laevis rRNA is able to repress *X. mülleri* rRNA activity as well as the maternal cytoplasm. This was demonstrated in experiments using *X. laevis* normal females and females with a deletion of the nucleolar organizer (1^--strain). When *X. laevis* rDNA is present, this inhibition is permanent. Cytoplasmic repression, on the other hand, is reversible and, as a rule, only temporary (Honjo and Reeder, 1973). It should be noticed that in this case the differential inactivation is specifically restricted to the nucleolar organizer region, because many enzymes which are controlled by paternal as well as maternal genes were active (Wall and Blackler, 1974).

All the above examples illustrate the reality of differential inhibition of transcription in one homologous loci (and even one homologous chromosome or the whole paternal genome). This proves the presence of true allelic exclusion, although its molecular mechanism is still unknown.

It is possible, then, to conclude that a whole spectrum of differential activity of the paternal alleles is possible: from complete inactivation to equal activity of

Fig. 48. Synthesis of *mulleri* rRNA during early development in *Xenopus* hybrids. Nuclei were isolated from embryos at various stages and analyzed for their ability to synthesize rRNA as described in the text. Embryos from the same mating have the same symbol. *Open symbols, dotted line*, embryos from the cross $l/l^- \times m/m$. *Closed symbols, solid line*, embryos from the cross $l/l \times m/m$. (After Honjo and Reeder, 1973)

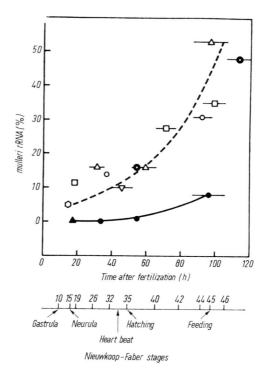

both genomes. In ontogenesis it is possible to see differences in the initiation of paternal and maternal gene action. The possibility is not excluded that the asynchrony of parental gene action during individual development is the first step from an evolutionary point of view, on the way to total inactivation of one genome. It is possible that this asynchrony of action is dispersed on the sex chromosomes as well as on the autosomes (Korochkin and Belyaeva, 1972) (Fig. 49).

What is the relationship between differential activity of the homologous loci and the events of allelic exclusion?

Bennett and Owen (1966) suggested that the phenomenon of functional haploidy is analogous to X chromosome inactivation in mammalian females. Those cases which have served as a basis for the definition of "allelic exclusion" (the characteristics of Ig synthesis, synthesis of the myeloma proteins and some isozymes) cannot be attributed, strictly speaking, to the events that are defined by this term. There are additional difficulties with the concept of allelic exclusion conception caused by the complicated construction of loci that are responsible for Ig synthesis (Mage et al., 1977).

Only strong proof of a correlation between the inhibition of transcription in one homologous locus (i.e., the heteromorphism of puffings) and the absence of the product related by this gene (i.e., either an isozyme or Ig) will be a sufficient basis for the use of the term allelic exclusion. Satisfactory investigations are possible only when the classic genetic (the establishment of the character of genetic control and gene localization) and biochemical (the control of characteristic ex-

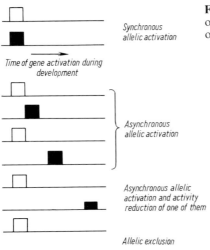

Fig. 49. Hypothetic scheme of the allelic exclusion origin on the basis of the asynchronous activation of homologous genes

pression) methods are combined. The detailed analysis of molecular events leading to the differentiation of antibodies-producing cells, performed in the recent times, leads to the conclusion that the allelic exclusion is realized at the level of a transcription.

In the process of the differentiation of lymphocytes the V- and C-genes (coding for variable and constant chains of polypeptides in antibody molecule) change their spatial positions and connect together. These genes are localizing in one chromosome but the prolonged region of DNA to be inserted between them. At the beginning stage of antibodies synthesis the excision of this internal DNA takes place with the following formation of one transcriptional unit containing both V- and C-genes.

The evidence favoring this supposition has been obtained in experiments with embryonic and differentiated cells. Thus, the transcripts of V- and C-genes were hybrizided with different fragments isolated from mice embryos. However, if DNA was isolated from differentiated myeloma tissue (i.e., such tissue which produced immunoglobuline very intensely) the V- and C-genes transcripts were hybridized with the same clearly characterized DNA fraction (Mozumi and Tonegawa, 1976; Brondz and Rokhlin, 1978; Rabbits, 1978; Monjo and Kataoka, 1978). Some authors suppose that an immunoglobulin heavy chain gene is formed by at least two recombinational events: (1) V_H-gene segment-J_H-segment (associating with C_H-genes) joining and (2) C_H-gene of corresponding switching type.

The synthesis of IgM, IgA or other types of immunoglobulins depends upon DNA rearrangement of the second type (Davis et al., 1980). Apparently the mechanism of allelic exclusion resulted from the following events: (1) On one of homologous chromosomes the joining of V- and C-genes did not arise and as a result this chromosome did not produce IgmRNA. (2) Another homologous chromosome was characterized by the corresponding rearrangement of DNA on the region of V- and C-genes. V- and C-genes of this chromosome produced IgmRNA (Monjo and Kataoka, 1978).

1.4 Nuclear-Cytoplasmic Interactions. The Basis for the Regulation of Gene Activity

1.4.1 Segregation of Ooplasm

As suggested by Morgan (1937) differential activity of nuclei is caused by differences in cytoplasmic environment; therefore, the first step of differentiation depends upon the nature of the cytoplasm in various parts of the egg (Lehman, 1957; Briggs and King, 1959; Nikitina, 1964). Cytoplasm heterogeneity first appears during oogenesis as a result of the process designated ooplasmic segregation. The essence of this process is the formation of egg polarity and the animal–vegetative (and in many animals, dorsal–ventral as well) gradient distribution of active biological compounds such as RNA's, proteins, etc. The animal part of the egg is characterized by a higher level of metabolism (Brachet, 1960, 1961; Raven, 1964).

These gradients play an important role in morphogenesis. The distortion of RNA metabolism with respect to the animal–vegetative gradient in the egg will depress embryonic development. Changing the primary polar gradient of RNA in amphibians by centrifugation prevents head development. Sometimes centrifugation increases RNA synthesis in some embryonic structures. As a result, individuals with two heads are formed. Similarly, sometimes the differentiation of three systems at the axis organs was observed (Brachet, 1960).

The formation of egg polarity is caused by the position of the egg in the ovary and by physiological peculiarities of the interaction of the egg with the blood system and the nurse cells which surround the egg. The animal and vegetative poles of frog oocytes get different blood supplies: the animal pole is surrounded by arterial vessels which carry blood rich in oxygen, whereas the vegetative pole is supplied by venous vessels. Naturally, these differences could affect the regional nature of metabolism which possibly precedes the development of the gradient (Child, 1948). Nurse cells which nourish the insect oocyte during maturation are regularly located near one pole. In turn, this distribution could affect the nature of the biochemical organization of the oocyte. The transport of various compounds from nurse cells into the oocyte was convincingly demonstrated. There are special structures which make such transport possible. For instance, in *Drosophila*, circular canals that connected the oocyte with the nurse cells have been described (Koch et al., 1967). These canals are filled with mitochondria which are moving through them into the oocyte. These electron microscopic observations were confirmed by observations in the light microscope (Hsu, 1952).

The transport into the oocyte of RNA synthesized by nurse cells has been demonstrated (Fig. 50). There is very little endogenous synthesis of RNA's in such oocytes. Radioautography was used to demonstrate that RNA is synthesized in the nurse cells and then transferred into the oocyte (Bier, 1963).

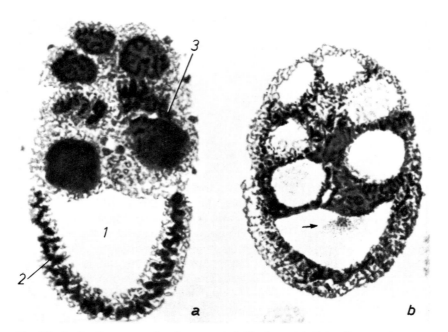

Fig. 50 a, b. Radioautograph of oocyte *1*, follicular epithelial cells *2*, and nurse cells *3* in *Musca domestica*. **a** *Musca domestica* follicule incubated 30 min with H³-cytidin. **b** The same after 5 h incubation. It is noticeable that labeled RNA migrated into oocyte from the nurse cell (indicated by an *arrow*). The heavy labeled RNA is obtained in cytoplasm of the nurse cells instead of in nuclei, where it was located before. (After Bier, 1963)

In many cases, these nurse cells have polyploid nuclei and synthesize large amounts of RNA's. In oocytes with metabolically inactive nuclei, the amplification of rDNA does not occur and rRNA are synthesized by trophic cells (Bier, 1965, 1967). In some beetles, for instance *Dytiscus marginalis*, active RNA synthesis is going on in trophic cells as well as in oocytes. Oocytes contain extrachromosomal DNA that is represented by Djiardine bodies (Bier et al., 1967).

In addition, it has been demonstrated that RNA synthesis in the maturing oocytes of mammals is under the control of the surrounding follicle cells. RNA synthesis in various stages of oocyte maturation was changed (Fig. 51) and was terminated in large oocytes. It was found that the anthrum fluid, which is produced by follicular cells, contains compounds that inhibit DNA dependent RNA-polymerase activity in vitro (Moore et al., 1974). A proteinous factor was extracted from embryonic vesicles of amphibian oocytes. This factor inhibits transcription of the ribosomal genes (Crippa, 1970; Crippa et al., 1972).

Apparently, the differences in various parts of the cytoplasm are caused by the maternal genotype (Child, 1948; Neyfach, 1962; Raven, 1964). These differences are due to the action of certain gene groups and to the synthesis of various RNA fractions in the oocytes. These RNA's are stable in *Xenopus laevis* and remained in the cytoplasm for a long time during embryogenesis. Approximately 90% of this

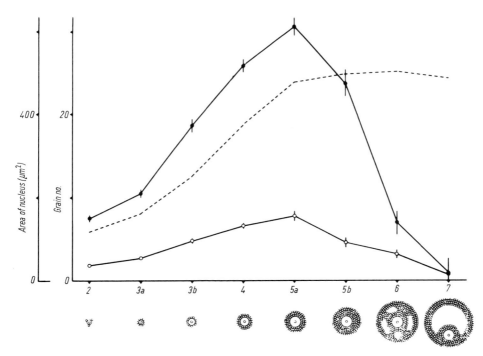

Fig. 51. [³H] uridine incorporation into the mouse oocyte at different stages of follicle growth expressed as the mean number of grains over the nucleus. *Open circles*, grains over the nucleolus; *closed circles*, grains over the whole of the nucleus. *Vertical lines* represent the standard error of the mean. The *broken line* depicts the change in area of the oocyte nucleus during growth. (After Moore et al., 1974)

RNA is rRNA (Davidson et al., 1964, 1966; Crippa et al., 1967; Davidson, 1972). Only 3% of the total genetic information is transcribed in the lampbrush stage in *X. laevis*. The amount of functional DNA in this case has been estimated by Davidson to be equal to 8.2×10^5 genes. Each of these genes can potentially code for a protein molecule as large as the β-chain at hemoglobin (Davidson et al., 1966; Crippa et al., 1967). Davidson has suggested a number of stages of RNA synthesis according to the stages of oocyte development (Table 23; Davidson et al., 1964). A large amount of the long-living mRNA is synthesized off the lampbrush chromosomes. These mRNA's can remain functional in the embryo until the late blastula stage. This was demonstrated by DNA-RNA hybridization (Davidson, 1972).

In *Xenopus*, from early embryogenesis until the 500–2000 cell stage (early blastula), the store of mRNA that was synthesized on the lampbrush chromosomes did not change. However, 28% of this store disappeared during the 1.5 h–2.0 h that separate the early and late blastual stages (7 and 9 respectively) (Crippa et al., 1967).

In the sea urchin *Paracentrotus lividus*, mRNA's which are associated with ribosomes are synthesized. These mRNA's are not immediately active and cannot synthesize proteins. The treatment of such ribosomes with trypsin stimulated the

Table 23. RNA synthesis in *Xenopus* oocytes. (After Davidson et al., 1964)

Stage	State of nuclear apparatus	Diameter of oocyte (mm)	µg RNA[a]	Relative specific activity of RNA 3–7 days after labeling[b]	Average specific activity (cpm/µg RNA) of RNA after labeling with 1 mC uridine-H^3 in one representative experiment[c]				
					1 day	3 days	14 days	42 days	75 days
6	Lampbrush retracted	1.0–1.2	2.6	16	14	10	30	53	60
5	Late lampbrush	0.6–0.8	2.8	77	40	65	100	38	57
4	Lampbrush maximal	0.5–0.6	1.9	100	37	81	108	32	34
3	Beginning lampbrush	0.4–0.5	0.7	49					
Total cpm in the RNA of an average Stage 4 + an average Stage 5 + an average Stage 6 oöcyte					226	417	441	382	392

[a] Values based on 20–30 determinations for each stage of oocyte
[b] Values based on 10–12 determinations for each stage of oocyte. Value for Stage 4 set at 100
[c] Each lot of toads tends to label at a different over-all rate, in our experience, and it is consequently difficult to pool absolute specific activity data. However, relative synthetic activities of the diferent stages of oöcyte agree well in different experiments. The animals were labeled with 3, 0.33 mC injections of uridine-H^3 (sp. ac. 3.73 C/mM) given on the 3rd, 4th, and 5th days following stimulation with bovine pituitary extract. The animals were caused to ovulate again at 30 days, to remove previously ripened Stage 6 oöcytes and thus prevent dilution of the newly maturing Stage 6 oöcyte population

mRNA for protein synthesis by derepressing these RNA's (Monroy et al., 1965). Apparently, there are two mechanisms which activate protein synthesis in sea urchins after fertilization: (1) release of mRNA from complexes with informosome, and (2) increasing of the T_2 elongation factor whose concentration is very low in eggs before fertilization (Monroy, 1970).

There is reason to suppose that during oocyte maturation, some regulatory proteins have been synthesized. These proteins play the important role in oogenesis and embryogenesis. Among them, the factor (FSM) which stimulates maturation should be mentioned. If a mixture of karyoplasm and cytoplasm from oocytes at the stage of the destruction of embryonic vesicles is injected into nonmature oocytes, the latter are induced to undergo normal maturation (Detlaff et al., 1964). It was shown that the main active component in the cytoplasm is not the gonadotropic hormone but compounds which are synthesized in the oocyte. However, these are in the cytoplasm of oocytes before the destruction of the embryonic vesicles. The oocyte recipients reached metaphase at the second division of maturation and were capable of being activated, when they were injected with cytoplasm of maturating oocytes (Masui and Markert, 1971; Masui, 1972). At the same time, various inhibitors are synthesized in the oocyte as well. It was found that the injection of cytoplasm from maturating oocytes into the blastomeres of dividing eggs stopped the cleavage of these blastomeres, whereas the intact blastomeres continue normal cleavage. The injection of cleaving embryos or activated egg cytoplasm does not suppress cleavage. This factor obtained from the cytoplasm of the maturating oocytes and which suppresses cytotomia and mitosis was named the cytostatic factor (Masui and Markert, 1971).

Table 24. Expression of Pu gene in F_1 of inter- and intraspecies crosses. (After Sokolov, 1959)

Cross, P	Number of individuals		Total number	Individuals Pu, %
	Normal	Pu		
♀ D. virilis $\frac{Pu}{Pu}$ × ♂ D. virilis $\frac{+}{+}$	10	539	549	98.2
♀ D. virilis $\frac{+}{+}$ × ♂ D. virilis $\frac{Pu}{Pu}$	7	539	546	98.7
♀ D. virilis $\frac{Pu}{Pu}$ × ♂ D. lummei $\frac{+}{+}$	216	420	636	66.1
♀ D. lummei $\frac{+}{+}$ × ♂ D. virilis $\frac{Pu}{Pu}$	210	2	212	0.9

There is evidence that during egg maturation, embryonic inductors are also synthesized. These compounds stimulate the first organogenetic events and the differentiation of the axial organ system (neural tube, somites and chord). Some inductors are inhibited. In the mature egg these inhibitors form distribution gradients. As a result, the inhibitors predominate in the vegetative part of the egg which later will form endoderm and mesoderm (Tiedemann, 1970).

Some isozymes that play an important role in embryogenesis are stored during oogenesis. It was suggested that in some animals these compounds are synthesized in the nurse cells and then transported into the oocyte (see previous chapter). Therefore, "the beginning of individual development, or more exactly its conditional beginning, should be placed somewhere before the mature egg" (Astaurov, 1974).

Sokolov used methods of classical genetic to investigate nuclear-cytoplasmic interactions in relation to the maternal effect caused by cytoplasmic compounds that are stored in the oocyte. He found that in developing interspecies hybrids between *D. virilis* and *D. lummei*, mitosis depends upon the direction of the crosses. If the hybrid received cytoplasm from *D. lummei*, development was normal. Cytoplasm from *D. virilis* caused the development of abnormal embryos due to abnormal mitosis. Apparently the compounds of the ooplasm of the mother have an important role in the regulation of mitotic cell division (Sokolov, 1959).

The interspecies plasm contains products that can temporarily change the dominance of paternal traits. For instance, trait dominance was changed in the F_1 in heterozygotes for the dominant mutation Puffed (rough facets, peculiar bubbles on eyes and altered eye size) when *D. virilis* individuals were crossed with *D. lummei* (Table 24). Therefore, the Pu allele which is obtained from *D. virilis* females is dominant in 66.1% of the hybrids, i.e., less than in the interspecies crosses. The wild-type allele of this gene in *D. lummei* is hypermorphic to the Pu allele of *D. virilis*. This mutation is dominant only in 0.9% of the cases when the hybrid receives the Pu allele from *D. virilis* males.

The time factor is important in this case. It was demonstrated that the plasm of one species gradually changes its properties under the influence of the nucleus

Table 25. Segregation of Pu in F_2 of inter- and intraspecies crosses. (After Sokolov, 1959)

Cross, P	Number of F_2 individuals		Total number	Individuals Pu, %
	Normal	Pu		
♀ D. virilis $\dfrac{Pu}{Pu}$ × ♂ D. virilis $\dfrac{+}{+}$	156	404	560	72.2
♀ D. lummei $\dfrac{+}{+}$ × ♂ D. virilis $\dfrac{Pu}{Pu}$	181	480	661	72.6

of another species. This conclusion followed from data obtained in experiments where the inheritance of the Pu trait in the F_2 was investigated. In both interspecies and intraspecies crosses this trait appeared as a dominant character in the F_2 (Table 25).

Sokolov wrote, "The comparison of data demonstrates that plasm properties reflect the nature of the nucleus which was in contact with it. These properties change during ontogenesis due to the interaction of allelomorphic and nonallelomorphic genes." This situation is depicted schematically in Fig. 52. Naturally, gene mutations which control the synthesis of compounds important for morphogenetic processes during ontogenesis will affect embryonic development in very early stages. Addition of these compounds into mutant eggs could normalize development. The recessive mutation "0" in *Amblystoma mexicanum* causes slow embryonic growth and reduced ability of amputated extremities to regenerate. Heterozygotes (+ /0) at this locus do not differ from wild-type individuals. At the same time, the fertilization of 0/0 eggs with + / + or + /0 sperm results in abnormal embryo development which terminates at the gastrula stage because the blastopore does not close.

In rare cases, when the blastopore does close, resulting embryos do not form a nerve plate. In a number of experiments, eggs which were obtained from ♀ 0/0- × ♂ 0/0 or 0/+ individuals were injected with cytoplasm or nucleoplasm from + / + oocytes. The injection of nucleoplasm normalized embryonic development.

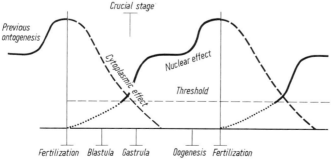

Fig. 52. The general scheme of nucleus-cytoplasm interaction during succeeding generations. (After Sokolov, 1959)

Table 26. Lethal effect of the dor mutation on embryonic development in *Drosophila melanogaster*. (After Garen und Gehring, 1972)

Genotypes of parents	Genotypes of fertilized eggs	Developmental phenotypes
dor/dor ♀ × dor/Y ♂	dor/dor ♀	Embryonic lethal
	dor/Y ♂	Embryonic lethal
dor$^+$/dor ♀ × dor/Y ♂	dor$^+$/dor ♀	Adults formed
	dor/dor ♀	Adults formed
	dor$^+$/Y ♂	Adults formed
	dor/Y ♂	Adults formed
dor/dor ♀ × dor$^+$/Y ♂	dor$^+$/dor ♀	Adults formed
	dor/Y ♂	Embryonic lethal

Normal development of one blastomere was induced when nucleoplasm was injected into one blastomere at the two blastomere stage. The "normal" cytoplasm normalized development to the neurula stage only, even though the amount of cytoplasm added was high (1% of the total volume). It was suggested that in this case there is a "maternal effect" as a result of the storage of products at the normal "0" allele in the oocyte nucleus. These compounds could be RNA's (Briggs and Kassens, 1966).

A similar explanation has been offered for the deep orange mutation in *D. melanogaster*. This mutation is on the X chromosome, is lethal in the homozygous condition during early stages of embryogenesis and demonstrates a maternal effect (Table 26).

The injection of the normal cytoplasm into deep orange embryos at the syncytialic preblastoderm stage normalizes the processes of ontogenesis. After injection, 50% of the dor/dor embryos reach late stages at development. It was suggested that the dor$^+$ allele controls the synthesis at pteridines which are necessary not only for pigment formation but also serve as cofactors for the hydroxylation of folic acid components. Injection of cytoplasm into dor/dor eggs resulted in pteridines penetration (Garen and Gehring, 1972).

The embryonic cells of dor mutants demonstrate a number of morphological defects when cultured in vitro: syncytium formation of muscle cells, spherical cell formation, little cell proliferation, etc. These defects disappear when extracts of nonfertilized wild-type eggs are added to these cultures. Heat treatment (80 °C for 10 min) inactivated this extract. The embryos from ♀ dor/dor × ♂ dor$^+$/y crosses were used as donors at different stages of development. Extracts from embryos at the blastoderm stage did not affect the abnormalities, whereas extracts from stages later than gastrulation were effective. Apparently, some components which are responsible for reparation are produced by the paternal dor$^+$ allele at the period between the 3rd and 5th h after fertilization. It is possible that these compounds are temperature-sensitive proteins (Kuroda, 1974/75).

The presence of some specific cytoplasmic fraction in certain regions of the egg could determine the development of certain parts of the embryo. In insects, for instance, the polar plasm contains polar granules which determine germinal cell formation (Hagner, 1911; Jazdowska-Zagrodzinska, 1966). These polar granules are associated with a group of ribosomes and contain RNA which goes into the cytoplasm during polar cell formation. This RNA seems to play a significant role in the process of embryonic cell determination (Mahovald, 1971, 1977). Ultraviolet treatment before blastoderm formation of the polar segment of *Drosophila* egg cytoplasm results in the absence of genitals in developing flies, although these flies are absolutely normal for other characteristics. The same result was obtained when the polar granules were displaced by centrifugation (Jazdowska-Zagrodzinska, 1966).

The posterior portion of polar plasm from wild-type *D. melanogaster* eggs in the early cleavage stage was injected into the anterior of mwhe embryos at the same developmental stage. Cells with polar granules and rounded nuclei which were similar to normal polar cells and are the predecessors of the embryonic cells were obtained in this area following the injection. These cells were introduced into the posterior region of ywsn[3] *D. melanogaster* embryos at the same stage to determine if these cells function normally. Subsequently, the adult female flies were crossed with ywsn[3] males. Progeny with the ywsn[3] and wild-type phenotypes were obtained (Illmensee and Mahovald, 1974).

The injection of polar (but not anterior) plasm from unirradiated *D. melanogaster* eggs into ultraviolet-treated eggs prevents the development of sterility. Moreover, some homozygotes for the lethal rudimentary mutation (embryo perished during development) can develop normally until imago if they are injected with cytoplasm from unfertilized wild-type eggs (1.5% of their volume) at the preblastoderm stage. Similar effects were obtained when eggs were injected with 0.01 µg of pyrimidine nucleotides. These experiments suggest that the *rudimentary* mutation leads to an insufficient amount of pyrimidine in the embryo (Okada et al., 1974a, b).

A hypothesis which explained the nature of polar plasm was suggested by Brown and Blackler (1972). According to this hypothesis, the early segregation of germinal cells is explained by the episomic inheritance of amplified rDNA which is an important compound of embryonic (polar) plasm. Cells which contained such extra copies of rDNA develop into primordial germ cells. It is natural that the removal of polar plasms induces the abnormal segregation of germ cells and sterility in developed animals. The episomic mechanism of heredity suggests that extra copies of rDNA are released during meiosis from the nucleus into the cytoplasm. This rDNA is stored in the polar plasm. It is transported first into the cytoplasm and then into the nucleus of future germinal cells. This hypothesis was not proved experimentally. However, it is possible to say that during ooplasmic segregation the store of genetic information, which determines embryo development over a long period of time, is accumulated as a result of the activity of certain segments of the whole genome.

The actual biochemical heterogeneity of cytoplasm has a differential effect on nuclear activity. This hypothesis was investigated using nuclear transplantation which was developed by Briggs and King (1959) and later improved by Gurdon (Gurdon and Laskey, 1970; Laskey and Gurdon, 1970).

1.4.2 Control of Nuclear Function by Cytoplasmic Factors

In Gurdon's experiments (Gurdon, 1970, 1971) transplantation of amphibian nuclei convincingly demonstrated that factors in the cytoplasm regulate nuclear function. For instance, autoradiography showed that the neurula stage nucleus which synthesizes all types of RNA's will terminate this synthesis 1 h after its transplantation into egg cytoplasms. RNA and DNA synthesis rates were determined with the aid of labeled precursors (Table 27).

As seen in this table, the activity of the transplanted nucleus corresponded to the condition of the cytoplasm into which this nucleus is transplanted. The polyacrylamide gel of the RNA's synthesized by this nucleus demonstrates that this RNA synthesis is controlled by the surrounding cytoplasm. During development of the embryo from the egg with the transplanted nucleus, the transplanted nucleus starts to synthesize the high molecular weight nonribosomal RNA's in the middle of the cleavage process, as in normal development. Then these nuclei synthesize tRNA's at the end of cleavage and, during gastrulation, the synthesis of rRNA is obtained. Since RNA synthesis is first suppressed in the transplanted nucleus, but later resumed, it is possible to suppose the presence of some component in the cytoplasm which controls the synthesis of each RNA type independently (Fig. 53). When the nucleus of the amphibian *Discoglossus pictus* is transplanted into anucleolate *Xenopus* eggs, the resulting embryos from these eggs develop normally until the late blastula stage. They do not, however, gastrulate. These embryos were able to synthesize the high molecular weight RNA's and tRNA's but not rRNA.

The original method for cultivating isolated *Xenopus laevis* cells was found by Shiokawa and co-workers (Shiokawa et al., 1967). They dissociated cells of the blastula (B-cells) and late neurula (N-cells) stages. The cultivated B-cells synthesize tRNA, but are not able to synthesize rRNA (the ratio of rRNA/tRNA is 0.5–0.8). N-cells synthesize both types of RNA's and the ratio of rRNA to tRNA is more than 5.0. This ratio is reduced to 2.0 if N-cells are cultivated together with B-cells

Table 27. The cytoplasmic control of chromosome activity in transplanted nuclei. (After Gurdon, 1970)

Chromosomal activity and cell-type				Chromosomal activity of transplanted nuclei
Donor nucleus		Host cell cytoplasm		
RNA^{++} DNA^- DNA^{--}	Neurula Adult brain	RNA^{--} DNA^{++}	Activated unfertilized egg	RNA^{--} DNA^+ DNA^{++}
RNA^{--} $Mitosis^{++}$ DNA^{++}	Mid-blastula	RNA^{++} $Mitosis^{--}$ DNA^{--}	Growing oocyte	RNA^{++} $Mitosis^{--}$ DNA^{--}
RNA^+ Condensed chromosomes^{--}	Adult brain	RNA^- Condensed chromosomes^{++}	Maturing oocyte	RNA^{--} Condensed chromosomes^{++}

$++, +, -, --$, indicate the rate or frequency of synthesis

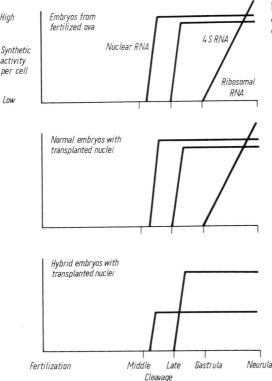

Fig. 53. The cytoplasmatic control of the nuclear RNA synthesis. (After Gurdon, 1971)

or B-cell extract. This changing rRNA/tRNA ratio is caused by the increase of tRNA synthesis rate and not because of a decrease in rRNA synthesis. Washing the N-Cell cultures from B-cell extracts returned the normal level of rRNA synthesis. This observation suggests that in the B-cell extract there is a factor which selectively suppressed the synthesis of rRNA. The concentration of this factor is reduced during cleavage (or, as an alternative, it is inactivated). This induces the activity of loci that synthesize rRNA during the gastrula stage. DNA and protein synthesis is not sensitive to this hypothetical factor (Yamana and Shiokawa, 1966, 1975). A thermostable cytoplasmic compound which stimulates DNA and RNA synthesis in hepatic cell nuclei of mice has been found (Thompson and McCarthy, 1968). However, this observation has not yet been repeated.

The induction of tRNA gene transcription by existing tRNA in ova has not been found in fish. The beginning of tRNA synthesis does not require preliminary activation of heterologous RNA synthesis (Solovijeva et al., 1973). These data suggest that in the ova cytoplasm there are a number of compounds which control nuclear activity and which affect independently different genes or groups of genes (Gurdon, 1964; Gurdon and Woodland, 1970; Laskey and Gurdon, 1970). The noncoordinated stimulation of tRNA synthesis was demonstrated for eukaryotic somatic cells also (Willis et al., 1974).

Table 28. Mean numbers of grains per nuclear area for a minimum of 100 nuclei ($\pm$ s.e.m.). (After Berstein and Mukherjee, 1972)

Developmental stage	Embryo control	Number of embryo nuclei in heterokaryons	Number of A9 nuclei in in heterokaryons
2-cell	38.2 $\pm$ 2.9	84.5 $\pm$ 5.3	18.2 $\pm$ 1.1
4-cell	383.8 $\pm$ 36.5	718.0 $\pm$ 32.7	189.7 $\pm$ 9.0
Ratio of 4-cell: 2-cell	10.0:1 [a]	8.5:1 [a]	10.4:1 [a]

[a] If the 95% confidence limits of the means are calculated, the ratio of 8.5:1 may range from a minimum of 6.9:1 to a maximum of 10.6:1. Similarly, the ratio of 10.4:1 may range from 8.4:1 to 13.0:1. Therefore, within the 95% confidence limits, no two ratios are statistically different from each other

The regulation of activity of the genetic apparatus by morphogenetic substances of the cytoplasm is especially noticeable in mosaic eggs. It is known that the elimination of the polar part of egg in *Dentalium* before its first division induces development of larva without an apical tuft and posttrochalic region. If the polar region is eliminated before the second division, the apical corolla is present but the posttrochalic one is reduced. Centrifugation does not change the result of this experiment. Therefore, from these experiments, the conclusion is that morphogenetic factors are located in the cortical layer of the polar zone (Wilson, 1940; Moore, 1963; Verdonk, 1968). These compounds affected RNA synthesis in the nuclei of developed embryos.

The removal of the third polar lobe which is absolutely necessary for normal differentiation of *Ilyanassa obsoletta* eggs (which develop similarly to *Dentalium*) leads to a reduction of RNA synthesis during cleavage and gastrulation. It was suggested that the cytoplasm of this polar lobe affects the character of gene transcription and embryonic differentiation (Collier, 1966, 1977). Apparently, the control of DNA and RNA synthesis in differentiating systems of mosaic eggs depends upon cytoplasmic factors which possibly begin their action early in mosaic eggs.

Similar events occur during the hybridization of somatic cells (Harris, 1970, 1973).

Controls of RNA synthesis during early embryogenesis in mammals is similar to that in amphibians, as was demonstrated in hybrids of mice blastomeres and fibroblasts of the A-9 strain. Nuclei of A-9 cells, which are characterized by intense RNA synthesis, lose this ability in heterokaryons from the two-cell blastomere stage and incorporate the essential amount of H^3-uridine after fusion with blastomeres of the four-cell stage (Table 28). These observations led to the conclusion that the blastomere cytoplasm of the two-cell stage contains factors which can suppress RNA synthesis. By the four-cell stage these factors are absent or they are in very low concentration. The addition of nuclei and cytoplasm of A-9 cells to the two- and four-cell stages stimulate RNA synthesis in "embryonic" nuclei of the heterokaryon (Table 29). The amount of RNA synthesized in the two-cell blastomeres and in the heterokaryon "two-cell stage/A-9" is small compared with the four-cell blastomeres and the "four-cell stage/A-9" heterokaryons. The incorporation of label in the acid soluble pool was identical in these experiments. Therefore,

Table 29. Synthesis of RNA and DNA in heterokaryons. (After Harris, 1973)

	RNA	DNA
Cell type		
HeLa	+	+
Rabbit macrophage	+	0
Rat lymphocyte	+	0
Hen erythrocyte	0	0
Cell combination in heterokaryon		
HeLa-HeLa	+ +	+ +
HeLa-rabbit macrophage	+ +	+ +
HeLa-rat lymphocyte	+ +	+ +
HeLa-hen erythrocyte	+ +	+ +
Rabbit macrophage-rabbit macrophage	+ +	0 0
Rabbit macrophage-rat lymphocyte	+ +	0 0
Rabbit macrophage-hen erythrocyte	+ +	0 0

0, No synthesis in any nuclei; 00, no synthesis in any nuclei of either type; +, synthesis in some or all nuclei; + +, synthesis in some or all nuclei of both types

in spite of the differences in mouse and *Xenopus* morphogenesis, the control of nuclear RNA synthesis during development in both of these animals, apparently, is similar (Bernstein and Mukherjee, 1972).

Experiments were done where HeLa cells, which are able to synthesize a lot of DNA and RNA, were fused with chicken erythrocytes, the nuclei of which are inactivated and do not synthesize RNA or DNA. The incubation of such heterokaryons in a medium containing H^3-uridine demonstrated that the synthesis of RNA and DNA occurs not only in the HeLa nuclei but also in the erythrocyte nuclei (Fig. 54; Johnson and Harris, 1969; Harris, 1973). Table 29 represents the results of various experiments in which the nuclear activity in heterokaryons was analyzed. These data show that (1) if either parental cell was able to synthesize RNA, RNA synthesis will occur in both nuclei in heterokaryons; (2) if either parental cell was able to synthesize DNA, DNA synthesis will occur in both nuclei of the heterokaryon; (3) if neither parental cell could synthesize DNA, the heterokaryon is not able to synthesize DNA (Harris, 1973).

Darlington (1977) obtained hybrids of somatic cells which were isolated from hepatic and other cell cultures of mice (i.e., of hepatic and nonhepatic origin). The hepatic cells HePa la and HH synthesize albumin, transferrin and α-fetoprotein. Most somatic hybrids between these cells and L-cells did not demonstrate albumin and α-fetoprotein synthesis, but continued to synthesize transferrin. This type of expression was similar for hybrids with one and two doses of the hepatic genes. In hybrid cells between hepatic cells and embryonic hepatic cells, 3 of 17 cells synthesized embryonic α-fetoprotein. The investigators suggested that albumin and α-fetoprotein production in the hybrid cells depends upon the epigenetic status of the nonhepatic parent.

Fig. 54. Autoradiograph of heterokaryon incubated 20 min with H³-uridine. Both the mouse nucleus and the erythrocyte nuclei are very heavily labeled, and there is also substantial cytoplasmic labeling. The labeling is increased with the increasing of the nuclear size. (After Harris, 1973)

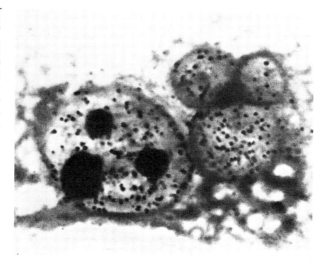

Interesting data have been obtained during hybridization of haploid mice spermatids with proliferating HeLa cells. Heterokaryons which contain spermatid nuclei were obtained when polyethylen-glycol was used as a fusion agent rather than inactivated Sendar virus. The intensity of DNA synthesis was similar in spermatid and HeLa nuclei. The synthesis of DNA in HeLa nuclei started earlier than in the spermatid nuclei. The results of observations show that the reactivation and reparation of haploid spermatid genomes is possible when they are hybridized with proliferating cells (Elsevier and Ruddle, 1976).

The investigation of the early cell hybrids between human and mouse cells (Rechsteiner and Parsons, 1976) revealed that human and mouse chromosomes are often separate from each other inside metaphase and interphase nuclei. Although this segregation disappears during successive divisions, the sectoral type of nuclei was obtained for a number of mitotic divisions.

Nuclear activation in heterokaryons differs from that of nuclei which are transplanted into eggs. After transplantation into the egg, total dedifferentiation of the nuclei was obtained; these nuclei began to be capable of determining the development of the whole organism. When these nuclei were implanted into other somatic cells, nuclear reactivation was restricted in certain functions. For instance, the nuclei of chicken erythrocytes in heterokaryons with myoblast cells of rats began to synthesize RNA. Various chicken antigenes appeared in the heterokaryon phenotype. However, the chicken myosin was never synthesized. In the rat myoblast–chicken myoblast heterokaryon, myosin is produced from both genomes (Ringertz et al., 1971, 1972; Ringertz, 1974). Moreover, the specific stability of differentiation of nuclei from specialized cells was shown for nuclei from Harding-Passey's melanoma. When these nuclei were transplanted, using microsurgery, into muscular cells, they retained their ability to perform specific functions (the control of the melanosomes synthesis) under entirely new conditions. In general, data demonstrate that tissue and cellular differentiation is determined and supported by nuclear-cytoplasmic interactions, similar to the cases of early embryogenesis (i.e., experiments with transplanted nuclei). During cell development, the nuclear-cytoplasmic system is formed. All the components of this system are rigidly coordinated with each other. Apparently, this coordination prevents the transformation and metaplasia processes.

It was suggested that in heterokaryons there are two types of substances used in differentiation: inhibitors and activators of differentiation. The following facts corroborated the presence of inhibitory components: (1) Pigment inhibition was obtained after the fusion of pigment cells and fibroblasts (Davidson, 1972). (2) The synthesis of S-100 protein and glycerol phosphate dehydrogenase was reduced or totally absent in hybrid cells of rat glia and mouse fibroblasts. The S-100 protein is characteristic of glial cells, and glycerol phosphate dehydrogenase lost its ability to be induced by hydrocortisone (Minna et al., 1972). (3) The synthesis of growth hormone was inhibited in somatic hybrids of pituitary cells and mouse fibroblasts (Sonnenschein et al., 1971). (4) The syntheses of hepatic ADH and aldolase B were suppressed. Activity levels of tyrosine amino transferase and alanine amino transferase were reduced. Moreover, the inducibility of these enzymes disappeared and the glycogen store was depleted (Croce et al., 1974; Davidson, 1974a, b; Weiss, 1974).

The presence of substances which activated differentiation can be assumed on the basis of experiments with somatic hybrids of hamster ovary cells and human fibroblasts. In this case, three new types of esterase are synthesized in the heterokaryons. They are similar to esterases from the hamster ovary which, however, are not synthesized in these cells during in vitro cell cultivation. The synthesis of these esterases depends upon the presence of a specific human chromosome which apparently controls the synthesis of an activating substance (Fa Ten Kao and Puck, 1972; Davidson, 1974a, b).

1.4.3 The Nature of Cytoplasmic Factors Controlling Nuclear Function

What is the nature of the cytoplasmic regulatory factors and in what period of ontogenesis are they synthesized? There are numerous observations which lead to the suggestion that the cytoplasmic substances which control nuclear activity are proteinaceous in nature. Gurdon (1971) performed a series of experiments where highly purified DNA was injected into oocytes. If was noted that such DNA can serve as a template for RNA transcription. Therefore, it was suggested that the cell cytoplasm contains all the compounds necessary for the determination of the traits associated with the transplanted nucleus. In delicate experiments Arms (1968) demonstrated that labeled cytoplasmic proteins migrate into the transplanted nuclei. It was also shown that the labeled proteins accumulate in the nuclei and that this occurs concurrently with the nuclear morphological changes which reflect the transition of this nucleus into a new functional status (Arms, 1968) (Fig. 55). It was later (Gurdon, 1971) demonstrated that histones (mol.wt. 10,000–20,000) as well as

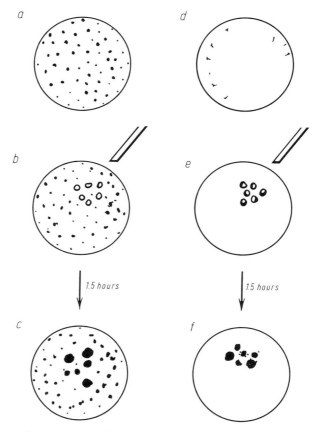

Fig. 55a–f. Experiment demonstrating the accumulation of the cytoplasmic protein in the nuclei of adult animal brain cells which were transplanted into nonfertilized ova and induced for DNA synthesis. Puromycine is an effective inhibitor of the protein synthesis. **a** The nonfertilized ova labeled during oogenesis through the intraperitonal injection of H³-amino acids. **b** The transplantation of brain cell nuclei and puromycin injection. **c** The labeled protein is concentrated in the most swollen cells. **d** The nonfertilized, nonlabeled ova. **e** The transplantation of the brain cell nuclei and the injection of H³-thymidine. **f** The labeling appears to be heaviest in the most swollen nuclei. (After Gurdon, 1971)

acidic proteins (mol.wt. 62,500) penetrate into the transplanted nucleus during the time that it is changing activity. Reinjection of various fractions of the labeled proteins into amphibian oocytes showed that these proteins could be divided into three groups, according to their behavior: (1) proteins which concentrate mainly in the nucleus, (2) proteins which predominate in the cytoplasm, and (3) proteins which are evenly distributed in the cell (Bonner, 1974).

Gurdon (1967) specified the stage of egg development when the cytoplasm becomes inducible for DNA synthesis. For this purpose he injected nuclei isolated from *Xenopus laevis* brain into eggs of these animals at various periods after egg activation. The incorporation of H³-thymidine into the transplanted nuclei was

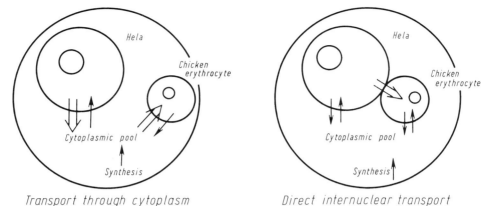

Fig. 56. Nuclear-cytoplasmic interactions in a heterokaryon. (After Ringertz et al., 1972)

studied. Maturation of oocytes was stimulated by hypophyseal hormones. It was observed that the factor which induced DNA synthesis is absent in the nucleus and cytoplasm of the growing oocyte, but appears here several hours after the injection of the pituitary hormones. This factor remains in the egg cytoplasm for at least 1 h after fertilization or activation. Similarly, Skoblina (1974) investigated in detail the behavior of sperm nuclei which were injected into either maturing oocytes of *Bufo viridis Laur*, or into oocytes which had no germinal vesicle. She concluded that DNA synthesis in the sperm nuclei starts at the final stages of maturation. A "DNA synthesis factor" (apparently DNA-polymerase; Gurdon and Speight, 1969; Crippa and Lo Scavo, 1972) appears (or is activated) in the amphibian oocyte cytoplasm shortly before the end of the maturation process. During oocyte development in *Xenopus*, the distortion of the germinal vesicle and nuclear maturation are induced by cytoplasmic factors which are produced or demonstrates its activity independently on the nucleus (Balakier and Czolowska, 1977).

In chicken erythrocyte – HeLa cell heterokaryons, growth and changes of erythrocyte nuclear morphology depend upon the accumulation of HeLa cell cytoplasmic proteins in the nucleus. These proteins can migrate from the cytoplasm of the HeLa cells or can go directly from the HeLa nucleus into the nearby erythrocyte nucleus (Fig. 56). There is a characteristic negative correlation between the number of erythrocyte nuclei and the amount of cytoplasmic proteins incorporated into each erythrocyte nucleus. The RNA synthesis in these nuclei is also reduced proportionately. Therefore in this case nuclear reactivation is affected by cytoplasmic proteins (Ringertz et al., 1972).

It is interesting that many primary and secondary embryonic inductors which determined the process of cell differentiation have an proteinaceous nature. For instance, the factor which induces the formation of mesodermal derivatives during embryogenesis (mesodermal inductor) has a molecular weight of about 25,000 (Tiedemann, 1966, 1967). Nerve growth factor (NGF), which controls nerve tissue growth, has a molecular weight of 20,000–40,000 (Levi-Montalcini and Angeletti, 1965). Epithelial growth factor (EGF) has a molecular weight of about 74,000. This

Fig. 57. The inductive action of purified neural (E 335/12) and mesodermal (E 351/5) inducing fractions either alone or in different combinations ranging from the ratio 1:1 to 50:1; *n*, number of cases (Tiedemann, 1967)

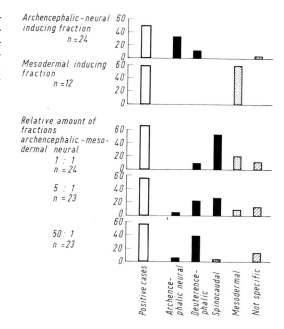

growth factor consists of two subunits (the molecular weight of each is 30,000). These subunits demonstrate esterase activity. Two smaller subunits (mol.wt. 6500 each) have EGF activity (Rutter et al., 1973). Tissue-specific regulators of mitosis (inhibitors of cell cleavage), known as chalones, are glycoproteins with a molecular weight of about 25,000 (Bullough et al., 1967; Bullough and Laurence, 1968; Bullough, 1969). These factors penetrate the cytoplasm of competent cells and cause an increase in the synthesis of various RNA's in these cells.

Especially interesting are experiments that analyze the molecular mechanisms of primary embryonic inductors (see Saxen and Toivonen, 1963; Ignatijeva, 1967, 1971; Toivonen et al., 1976). Tiedemann (1966, 1967) purified two protein compounds from chicken tissues. One of these induces the formation of neural structures (brain, eye, lens, acoustic vesicle, etc.). The other induces formation of mesodermal structures (blood island, mesothelium, muscle tissue, etc.). Tiedemann introduced various concentrations of these proteins into the blastocele of developing embryos. The results of these model experiments are presented in Fig. 57. In experiments with fluorescent antiserums, it was shown that the inducing agent is transported into the nucleus and induces a change in the functional status of this organelle.

According to Tiedemann (1970), protein inductors are present in the unfertilized egg in an inactive form. The discovery of the inhibitor pointed to a complex mechanism of regulation of inductor action. The basic experiments which led to the separation of the inductor from the inhibitor are as follows. Frozen and defrosted chicken embryos were homogenized in two volumes of 0.1 M NaCl and centrifuged for 2 h at 105,000 g. The supernatant, after precipitation with ethanol,

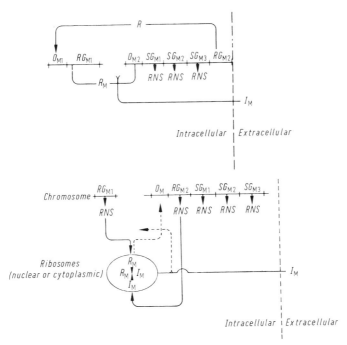

Fig. 58. Models for control of gene activity during primary embryonic induction. (After Tiedemann, 1967)

demonstrated only a weak inducing activity. After treatment of the supernatant with phenol at 60 °C, the proteins extracted from the phenol layer could induce trunk and tail structure formation. Apparently, the supernatant contains the induceable factor in a nonactive complex which dissociates during phenol treatment at 60 °C. The inhibitor is detectable in the water layer. If the components of the water layer combine with the induceable protein in the presence of 6M urea and this mixture is then dialyzed against water, the induceable factor is once again inactivated.

Discovery of inhibitors allows an explanation of the phenomenon of "self-differentiation" on the molecular level.

It was shown long ago that ectodermal differentiation at the gastrula stage can be induced by various exogenous compounds. Among these compounds are alkaline urea, thiocyanate, and acids. The common factor is that all of these are toxic to ectoderm. The differentiation of ectodermal explants can be induced by changing the ionic composition of the medium (Holtfreter, 1948). Self-differentiation of the gastrula ectoderm can be caused by a nonactive inducing factor in the ectoderm (Holtfreter, 1948). Apparently, changes in cell permeability are induced by changing the ionic parameters of the medium. This leads to the departure of the inhibitor from the cells. In addition, inductor factors which, along with inhibitors, are attached to membrane structures could be released from these structures. A direct effect on the interaction of proteins and nucleic acids in the chromatin is less likely (Tiedemann, 1966, 1967).

The distribution in the developing embryo of inducing agents and the inhibitors which may be connected to them is heterogeneous. The factor which induces the formation of mesodermal structures was found in an active form in the future endoderm of the gastrula as well as in the mesoderm. The active factor which induces the formation of neural structures, however, was found only in the future mesoderm. Tiedemann suggested that the inducing factors are prelocated inside the embryo before gastrulation. This could possibly occur in the egg before fertilization.

After being released from the inhibitors, the inductors organized the differential expression of various genes in the specializing cell system. This takes place in a number of stages.

Immediately after inductor action the cell is sensitive to the inhibition of RNA synthesis. This means that the embryonic inductor itself does not directly affect gene expression. It was suggested that inductor can act in two ways: (1) the inductor determines the utilization of a succession of metabolic reactions which later cause the activation of certain genes, or (2) the inductor releases a specific mRNA which is synthesized in a competent cell during the early stages of development, but which is stored in an inactive form (Ansevin, 1965). The presence of such mRNA has been indicated in biochemical investigations (Brown, 1964).

Tiedemann (1967) supposed two possible models for the control of gene activity by inductor action (Fig. 58). According to the first model, the inductor is bound to the repressor and this inhibits transcription; according to the second model, the inductor interacts with factors which inhibit the processes of translation in the competent tissue.

Therefore, beginning very early (oogenesis) and continuing throughout the different stages of embryonic differentiation, there are a number of essential cytoplasmic factors. Some of these are proteinaceous in nature. They synthesize and accumulate in an inactive form long before their utilization as regulators of differential gene activity.

1.4.4 Molecular Mechanisms for the Interaction of Regulatory Proteins with Chromosomal DNA

Regulatory proteins affect the functional status of the nucleus. This can occur through the control of the transcriptional activity of the chromatin. Chromatin and DNA extracted from *Rana pipiens* embryos at various developmental stages demonstrate (without additional RNA–polymerase) comparatively low transcriptional activity. This activity decreases from the gastrula stage to the stage of actively feeding larvae. The addition of RNA-polymerase to the chromatin preparation leads to increased transcription (Table 30).

Differences were found in in vitro and in vivo transcriptional activity of chromatin from various parts of the embryo (Flickinger et al., 1965). These differences can be explained by differences in permeability of animal and vegetative halves of the embryo. These differences might also be explained by a reduction of the amount of nuclear RNA-polymerase in the chromatin preparation from late

Table 30. Template activity of chromatin from different regions of the developing frog embryos. (After Flickinger et al., 1965)

Stage and region	μg Protein/DNA	Cpm
Blastula, stage 9		
Chromatin of animal halves	3.00	6,150; 5,460
Chromatin of vegetal halves	1.10	6,275; 6,340
Gastrula, stage 11		
Chromatin of dorsal halves	3.16	15,404
Chromatin of ventral halves	1.14	8,930; 10,266
Late neurula, stages 15–16		
Chromatin of embryonic axis	3.28	20,172; 22,138
Chromatin of bellies	1.02	16,961; 18,979
Chromatin of bellies minus $MnCl_2$ and RNA polymerase	1.02	79
Tailbud, stage 18		
Chromatin of embryonic axis	3.33	25,884; 26,914
Chromatin of bellies	1.52	37,468; 38,018
Chromatin of axis, minus $MnCl_2$ and RNA polymerase	3.33	47

stage embryos (Dingman and Sporn, 1964; Flickinger et al., 1967). Characteristically, the ratio protein/DNA is similar in chromatin preparations from *Rana pipiens* embryos at various stages of development (Flickinger et al., 1965; Table 30). The histone/DNA ratio in chicken embryos remains constant in all tissues during all stages of development. At the same time, the content of nonbasic proteins and the RNA/DNA ratio in chromatin preparations are higher in actively synthesizing embryonic tissues than in mature and nonactive tissues (Dingman and Sporn, 1964). Apparently, during embryogenesis, the increased transcriptional activity of chromatin from various parts of the embryo is independent of the amounts of protein which exist in this embryo. This could reflect either some conformational change in the DNA molecule or a different distribution of the same amount of protein in chromatin from different cells (and at different development stages).

Questions were asked related to the structural–chemical basis for this stability, regardless of the internal properties of the nuclear system or its interaction with the cytoplasm. There seem to be two mechanisms which do not exclude each other but are functionally linked: supercoil versus nonsupercoil formation and repression–derepression of various DNA segments by either proteins or other kinds of compounds. In some organisms, the differentiation of blastomeres of primary germinal and somatic cells occurs together with the differentiation of the nucleus. In *Ascaris* and most copepods, this differentiation manifested itself in the nucleus of one blastomere. Here, the nucleus either: (1) does not undergo a typical interphase stage; (2) goes through mitosis slowly; (3) repeats itself only in one sister component and retains its individuality through a number of cell generations; or (4) its sister compounds are characterized by an instability of the chromonema (diminution of chromatin in *Ascaris*; Beermann, 1959; Prokofieva-Belgovskaja, 1960).

Prokofieva-Belgovskaja designed a schematic representation of the mechanisms of differentiation of nuclear characteristics in somatic cells:

General changes in the cyclic properties of the nucleus

1. Differentiation of the mitotic rate (blastomeres)
2. Changes in the total mitotic cycle differentially in various tissues
   - endomitosis
   - amitosis
   - polytenia
   - meiosis
3. Differentiation of the cyclic conditions of chromosomes (super coiling) in the resting cells
   a) chromosomes in prophase
   b) chromosomes in metaphase
   c) typical resting nuclei
   d) lampbrush chromosomes
4. Differentiation of physicochemical properties: DNA content, relative iso-electric point of chromosomes (depending upon the degree of polyploidy, polyteny, cell age, cell type etc.).

Differential changes in the cyclic properties of the chromosomal complexes, separate chromosomes or chromosomal segments

1. Heteropycnosis of the sex chromosomes
   a) differentially in males and females
   b) differentially in various tissues.
2. Differential reduction of various chromosomal segment activities: heteropycnosis of heterochromatic and linked euchromatic segments (differentially in various tissues and various developmental stages)
3. Differential activation of various chromosomal segments: transition to resting condition differentially in various cells in various developmental stages
4. Differential reproduction
   a) separate chromosomes
   b) chromonemes in various chromosomal segments
5. Chromosomal elimination
   a) whole chromosomal complex
   b) separate chromosomes
   c) part of chromosomes differentially in various tissues
6. Genetic mechanisms
   a) somatic crossing over
   b) somatic mutation
   c) genome segregaration
   d) breakage and chromosome mutation.

Particular segments of chromosomes lose their ability to remain unsupercoiled in interphase as cell development occurs.

These segments remain in the nucleus as very small granules which contain DNA. In these granules the activity of the segment which synthesizes nucleoli is increased (Prokofieva-Belgovskaja, 1960, 1963; Belyaeva, 1965).

The mechanism of "repression" and "derepression" of various genes is linked to an interaction of chromosomal DNA with proteins (Moore, 1963; Elgin and Bonner, 1973). An inhibitory affect on embryogenes was obtained for polyamines, cadaverine and spermidine. The injection of proteins causes "blastula arrest" or the termination of development in eggs of *Rana pipiens* at the late blastula stage (stage 9). These proteins are injected into the fertilized egg. The injection of globulin plus albumins inhibits development 98%–100% of the time; the injection of histones in 59% and nucleoproteins only 9% of the time. It was suggested that these proteins affect the nuclear system because transplantation of nuclei from eggs blocked at the blastula stage into anucleated egg results in the same affect. Nuclei that have been through five generations give the same results: the termination of development at the blastula stage. Noticeable differences in karyotype morphology were not seen between control and experimental embryos. Apparently, the termination of development is caused by many different abnormalities of several levels (Markert and Ursprung, 1963; Ursprung, 1964, 1965; Ursprung and Smith, 1965).

In 1950, Stedman and Stedman suggested that "the basic proteins of the nucleus are inhibitors for genes; protamines and histones suppress the activity of groups of special genes" (Stedman and Stedman, 1950). A significant number of publications which appeared in connection with the biochemical investigations of Huang and Bonner have dealt with the effects of protamines and histones on embryogenesis. These authors demonstrated that the separation of pea germ DNA from histones induces an increase in the RNA in cell-free systems (from Bonner, 1967).

Preparation	Incorporation of nucleotides (mmol/10 min/mg DNA)
Nonpurified chromatin	1,175
Purified chromatin[a]	1,175
Deproteinated DNA with exogenous polymerase	90,000

[a] was obtained from nonpurified chromatin by centrifugation in the density gradient

Histones were considered as DNA repressors (Huang and Bonner, 1962; Bonner, 1967). Histones, which are in chromatin, are organized into special structures known as nucleosomes. Each nucleosome is made of a DNA segment of 200 nucleotide pairs complexed with octamers of histone proteins (Kornberg, 1974, 1977; Noll, 1977; Felsenfeld, 1978). The octamers contain two copies of the lysine-rich histones, H2A and H2B, and arginine-rich histones, H3 and H4. H1 is associated with the nucleosome but is not one of the structural components. Data on the composition of histones isolated from third-instar larvae and adult flies of *D. melanogaster*, *D. hydei*, and *D. virilis*, created special interest. These species demonstrate significant differences in the amounts of the two arginine-rich histones during development. The amount of H4 is higher and H3 lower in larva than in the imago.

Differences in the amount of postsynthetic modifications (acetylation and phosphorylation) were also observed at various stages of development. The imago demonstrated a higher level of such modification. There are interspecies differences in the distribution of histones for both developmental stages. The acetylation rate of H3 may be exceptional in *D. virilis* because it is at a relatively low level at every stage of development (Holmgren et al., 1976).

Franke et al. (1976) studied transcriptionally active chromatin from lampbrush chromosomes and nucleoli of oocytes of amphibians and *Acetabularia* nuclei and noted that the nucleosomes which are visible in the control preparations are not detectable in transcriptionally active chromatin. This was also seen in spacer regions of the transcriptionally active chromatin. DNA of nucleoli is packed into nucleosomes when the main part of the ribosomal genes is not active.

In some papers describing similar systems, however, the association of nucleosomes with both transcriptionally active and nonactive chromatin was seen (Reeves and Jones, 1976; Reeves, 1977). Some researchers suggested that the processes of selective transcription in differentiating cells could be associated not only with specific amounts of histones but also with methylation and acetylation of histones bound to genes (Kischer et al., 1966).

Acetylation of histones reduces their ability to inhibit RNA synthesis in vitro (Allfrey et al., 1964). The acetylation rate and concentration of acetyl-group histones are higher in transcriptionally active chromatin. Deacetylation of histones, at the same time, does not slow down. The rate of histone acetylation increases in types of cell where RNA synthesis is stimulated, i.e., spleen cells treated with erythropoetin, liver cells under the influence of cortisol, milk gland cells treated with insulin, etc. (Allfrey et al., 1972). Inhibition of nuclear activity is accompanied by the deacetylation of histones.

Phosphorylation of histones is another mechanism for the neutralization of basic proteins. The phosphorylation reaction occurs after the completion of the histone synthesis. It is reversible. Phosphorylation of the lysine- and arginine-rich histones precedes an increase in RNA synthesis. This was found in transforming lymphocytes, in the regenerating liver and pancreas, etc. Histones and protamines are phosphorylated at the final stages of spermatogenesis (see review of Allfrey et al., 1964, 1972). Specific kinases which participated in the transport of phosphorus groups from ATP to histones are stimulated by cyclic AMP (Allfrey et al., 1972).

Histone phosphorylation can occur at two levels: (1) molecular, which includes the nucleofilaments and their supercoil formation, and (2) microscopic, which included chromosomal condensation. The specific phosphorylation of up to three serines in the H1 histone is correlated with interphase microscopic changes in chromatin organization and therefore affects the molecular level of organization. The phosphorylation of additional serine and threonine residues in H1 and H3 histones is correlated with the mitotic condensation of chromatin (Curley et al., 1978). This condensation could be caused by changing the distance between adjacent DNA chains in the chromosome. This condensation may be due to H1 histone making transversal bridges between DNA molecules (Sluyser, 1977). H3 histone phosphorylation in the mitotic chromosome influences heterochromatin condensation. Histone H2a phosphorylation is very important for interphase organization of heterochromatin and is different from H3 and H1 histones phosphorylation in that it occurs at all stages of the cell cycle.

Therefore, various histone fractions have different specializations in chromatin structure. The phosphorylation of H1 histone determines the molecular level of chromatin organization; the phosphorylation of H2a histone determines the heterochromatization of chromatin and the superphosphorylation of H1 histone, and normal phosphorylation of H3 histone affects the microscopic level of organization (Curley et al., 1978).

The methylation of histones, in contrast to acetylation, is a relatively late event in the cell cycle which does not demonstrate any correlation with DNA, RNA, histones or other proteins (Inoue and Fujimoto, 1969). However, this reaction is related with changes that occur in the nucleus before mitosis, when chromatin condensation occurs and the nucleus begins to be less synthetically active (Tidwell et al., 1968).

Later, two subfractions of histones (Bush, 1965) were described and their biological activity was investigated. It was demonstrated that the treatment of anuran (*Discoglossus pictus*) eggs with arginine- and lysine-rich fractions of histones inhibits development. The same effect was obtained for *Pleurodeles waltlii* eggs when they were treated with lysine-rich histones. The degree of the effect depends upon the time when eggs are treated (Brachet, 1964; Molinaro and Cusimano-Carollo, 1966; Sherbet, 1968). In general, the lysine-rich histones are more effective.

Embryos of various species demonstrate significant H^3-arginine and H^3-histidine incorporation and accumulation of histones at the late blastula stage. Therefore, it was suggested that the nuclei of embryos at early stages of development are functionally inert. This was shown by investigations of androgenetic hybrids in which developmental abnormalities do not occur until the gastrula stage. These experiments allowed the conclusion that at that time, most of the structural genes are "closed" by histones (Moore, 1963). What is the mechanism of inhibition of DNA transcription in vitro by histones? According to one hypothesis, histones, which are assembled with DNA, prevent RNA-polymerase motion during transcription (Shin and Bonner, 1970; Koslow and Georgiev, 1971; Pospelov, 1973; Pospelov and Pupishev, 1973). It is possible that this "obstacle" is simply mechanical in character and is the result of the position of histones in the minor groove of the double helix of DNA, along which the RNA polymerase is moving (Reich, 1964).

Another hypothesis suggests that the structure of DNA is altered as a result of histone–DNA association, and supercoiling of DNA occurs (Richards and Pardon, 1970). It was suggested that RNA-polymerase is not able to transcribe those regions which are supercoiled (Johns, 1972).

Pospelov (1973) developed the idea that histones play an important role in the configuration transition of DNA from the A-form (bases are at an angle of $20°$ to the axis of the spiral, the distance between them is ~ 2.5 Å and 11 nucleotides are in one coil) to the B-form (the bases planes are almost perpendicular to the long axis of the helix, the distance between nucleotides is 3.4 Å and 10 nucleotides are in one coil). It is possible that DNA-polymerase uses only the B-form for replication, whereas RNA-polymerase is active on DNA in the A-form (Arnott et al., 1968). Histones and polylysine block B→A transition.

Some investigations supposed that the mechanism of action of various fractions of histones can differ, i.e., the lysine-rich histone I (f1) inhibits transcription when it is associated with DNA; the arginine-rich histone III (f3) inhibits RNA-polymerase activity when the two are associated (Spelsberg et al., 1969).

There are data, however, from *Pseudococcus obscurus* where (1) there are no qualitative differences in histone content between active and inactive segments of the chromosome; (2) there are no differences in the lysine/arginine ratio from these regions (Berlowitz, 1965a). The histone/DNA ratio is also the same in eu- and heterochromatic segments of chromosomes (Frenster, 1965). The histone composition of various tissues is similar for the same organism (Butler et al., 1968), in metaphase and interphase cells (Comings, 1967), in embryonal and adult tissues (Gineitis et al., 1971), and in organisms from different taxonomic groups (Fambrough et al., 1968). At the same time, specific histones were obtained from amphibian and bird erythrocytes (F_{2c} histone fraction; Edwards and Hnilica, 1968), echinoderms, and fish and mammal sperm (arginine-rich histones and protamines; Gineitis et al., 1970). The differences in the number of subfractions of lysine-rich histones was described for some organisms (Berdnikov et al., 1976).

The interaction of DNA and histones cannot explain all types of the regulation of differential genome transcription. A number of investigators have tried to explain the significance of the interaction of histones and nonhistone proteins in vari-

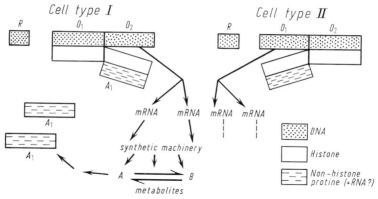

Fig. 59. Model of the genetic control of differentiation. *R*, regulator gene; *O* operator. (After Ursprung, 1964)

ous cells. It was suggested that histones play a passive role in the regulation of gene activity. The significance feature is the association of histones with acidic proteins. Ursprung has suggested that inactive genes are affected by DNA-histone associations. Nonhistone proteins can "take out" histones and thereby "open" some regions of the DNA for transcription (Fig. 59; Ursprung, 1964, 1965; Ursprung and Huang, 1967).

It was demonstrated that nonhistone proteins which assemble with DNA:
1. have a very high rate of synthesis and metabolism;
2. are present in high concentration on the chromatin of metabolically active tissues;
3. are preferably located on those parts of the chromatin which demonstrate especially high levels of RNA synthesis;
4. are phosphorylated during the gene activation period;
5. their synthesis is sensitive to hormone action (Allfrey et al., 1972).

It was demonstrated also that the acidic proteins stimulate synthesis in isolated chromatin fractions and increase the transcriptional activity of the DNA-histone complex (Frenster, 1965; Teng et al., 1970). These proteins promote the transcription of free DNA (Allfrey et al., 1972), and they determine transcriptional specificity. The accumulation of acidic nucleus proteins with a molecular weight of 20,000–40,000 occurs during puff formation in *Drosophila* salivary gland chromosomes (Berendes, 1972). An important role in this process is played by phosphoproteins which demonstrate a high level of tissue specificity (Allfrey et al., 1972; Rickwood et al., 1972; Fig. 60).

The nuclear phosphoproteins meet most requirements of regulators of gene activity:
1. they are located on active segments of chromatin;
2. their distribution in various cell types is specific;
3. their activity (the phosphorylation of hydroxyl groups of serine residues) reflects the intensity of RNA synthesis in this tissue;
4. they are species-specific;

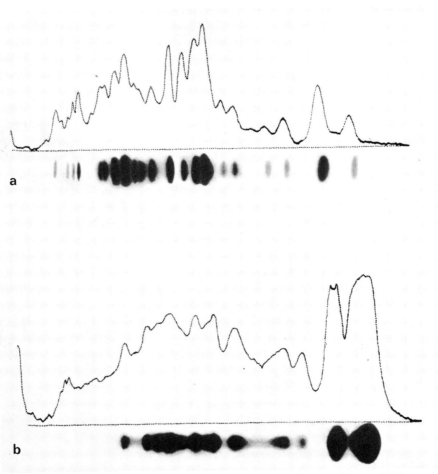

Fig. 60 a, b. Electrophoretic separation of acidic proteins from rat liver and kidney nuclei. The proteins were stained with Amido Black 10 B. A densitometer tracing is aligned over the corresponding banding pattern. (After Allfrey et al., 1972)

5. the transcription of a DNA/phosphoprotein complex is greater than that of DNA alone (Allfrey et al., 1972). However, in *D. hydei*, a correlation between the induction of puffing and chromosomal protein phosphorylation was not found (Ziegler and Emmerich, 1973).

Chromatin extracted from various differentiated cells demonstrates transcriptional activities characteristic for these cells from which the chromatin originates (Fig. 61; Paul and Gilmour, 1968). For instance, the globin mRNA is transcribed only by chromatin extracted from erythropoietic tissue (Gilmour and Paul, 1973). This transcriptional specificity is due to the protein part of the chromatin because deproteinization affects the synthesis of various RNA's. The experiments dealing with RNA synthesis in chromosomal preparations under the control of exogenous RNA-polymerase yielded the same conclusion. In these experiments, special treat-

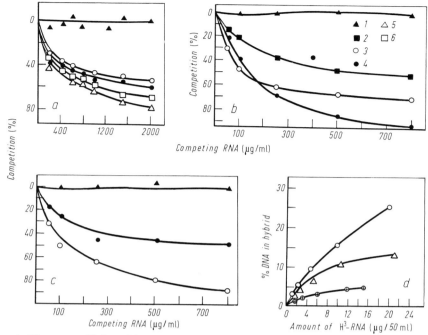

Fig. 61 a–d. Tissue specificity of isolated nuclei and chromatin template activity. **a** RNA synthesized by the isolated thymus cell nuclei was hybridized with DNA in the presence of nonlabeled competide RNA of other origin. The thymus RNA is the more effective competitor than RNA obtained from heterologous tissues. **b** RNA extracted from liver is the most effective competitor in the hybridization reaction between DNA and RNA synthesized by liver chromatin in in vitro condition. **c** Experiment as above with RNA from kidney. **d** Kinetics of hybridization to calf thymus DNA of H^3RNA's made with the following primers; whole calf thymus chromatin (● —● —● —); deproteinized chromatin (○ —○ —○ —); dehistoned chromatin (△ —△ —△ —) and DNA. *1*, E. coli; *2*, brain; *3*, kidney; *4*, liver; *5*, spleen; *6*, thymus. (After Paul and Gilmour, 1968)

ment affects the deproteinization of chromosomes (Sederoff et al., 1973). In a number of cases, RNA preparations can also increase the rate of transcription. The possibility is not excluded that there is a specific, derepressing RNA which, similarly to the acidic proteins, releases specific DNA regions from histones and, as a result, determines the nature of local supercoiled DNA and the synthesis of specific mRNA's (Frenster, 1965; Barbett et al., 1974).

Treatment of nuclei from rat liver with deoxyribonuclease II (which reacted mainly with the transcriptionally active parts of chromatin) produces a chromatin fraction which is rich with the transcribing nucleotide sequences. The chromatin from this fraction is organized as subunits with 100 Å diameters which are located along the chromosome. Extracted subunits differ from nucleosomes of nonactive chromatin. They contain significantly more nonhistone proteins and are enriched with newly synthesized RNA (Gottesfeld and Butler, 1977).

Some nonhistone proteins that were synthesized in the S-phase in HeLa cells are able to activate the transcription of histone genes in the G_1-phase. During this activation in reconstructed chromatin from liver cells, the simultaneous activation of globin genes does not occur. Therefore, it was suggested that this effect is not associated with total chromatin derepression (Park et al., 1977; Jansing et al., 1977).

In *D. melanogaster*, the specificity of the distribution of various proteins along the chromosomes has been demonstrated (Hill and Watt, 1977). The nonhistone protein D_1 was specifically located in a specific region of chromomeres which contains DNA enriched with AT-pairs. Another nonhistone protein, D_2, is located mainly in transcriptionally active chromosomal segments and mainly in nuclei and puffs (Alfagerne and Cohen, 1975/1976).

The exact DNA-protein interactions but not the synthesis of "early RNA" (see Chap. 1.1,2) affects the process of determination in cells (specialization during development). Results obtained in our laboratory corroborated this suggestion. The genitals imaginal discs obtained from various stages of male larvae of *D. melanogaster* were transplanted into female larvae which are ready to pupate. This experiment was done to determine the period when the determination of cells occur for the presumptive synthesis of esterase-6. An analysis of esterase activity in transplants demonstrated that cells in which the esterase-6 should be synthesized in the future appear 8–10 h after the beginning of the third larval stage.

The readiness of cells from the genital imaginal discs to synthesize esterase-6 reflects the transition of cells from one level of determination to another more specific level. It is possible that this transition occurs during one cell cycle because the readiness for differentiation is seen in cells within 4–7 h, i.e., in the time period in which only one mitotic cycle can take place (Nöthiger, 1972). The relationship between the determination of the Est-6 gene action and its transcription and translation was studied with the aid of inhibitors of protein synthesis (cycloheximide) and RNA synthesis (actinomycin D). The injection of these inhibitors suppresses esterase-6 activity in the male genitals during the first 6 h after hatching. These results suggest that in this period the beginning of transcription and translation of the Est-6 gene takes place. Therefore, a significant interval was found between the appearance of cells which are able to synthesize the esterase-6 and the time of expression of the structural gene which controls this enzyme. On the basis of observed data, the suggestion was made that cell determination is not caused by the accumulation of the "long-living" mRNA or by the synthesis of "early RNA". Rather, determination depends upon changes in the configuration of the DNA of the chromosomes. In other words, the DNA-protein complexes which are formed as a result of nuclear-cytoplasmic interactions initiate the cell to choose one from many possible modes of development.

In this case, the intensive RNA synthesis during pregastrulation and gastrulation probably does not have any relation with determination and embryonic inductor formation because these are already in the egg. This RNA synthesis apparently affects the status of the embryonic layers, which is defined as "competency" – the physiological ability to react with inductor. This early RNA may code for the number of cell receptors which recognize subsequent protein inductors and which affect their transmission into the reacting system. Although this is only a hypothesis, it shows direction of further experimentation designed to answer specific questions related to mechanisms of development.

Part 2
Gene Interaction at the Tissue Level

2.1 Functional-Genetic Mosaicism in Cell Population

2.1.1 Determination and Differentiation – the Properties of the Cell Populations

One of the important results of nuclear-cytoplasmic interactions is the realization of the complex processes which caused cell determination – the main event in ontogenesis. In discussing cell determination in his classic work in experimental embryology, Spemann (1938) wrote, "Usually we follow the definition given many years ago by K. Gaider, according to whom determination is the causal conditionality of the future fate of the embryonic parts. This causal conditionality is occurring or has occurred." Moreover, Filatov (1934) defined the process of determination as "the effect of some parts of the developing organism on other parts which, as a result, go through part of their development under specific conditions." Hadorn (1965) characterized determination as a "process, the result of which allows a competent cell to choose one from many developmental pathways."

The sequence of events during such a choice are demonstrated schematically in Fig. 62. In the first step, which is designated maturation (MI), cell competence is established: the epidermis (E) is able to react to the induction which begins nervous system (N) formation. Determination I (DeI) leads to the separation of the E and N systems. Next, maturation II (MII) occurs, which results in the acquisition of competence which will allow the formation of more specialized components, etc.

Tissue and cellular differentiation are the external expressions of determination which is manifested as the synthesis of specific RNA's and proteins. These RNA's and proteins, in turn, are responsible for the formation of the morphological and functional peculiarities of the specialized cells. The determination process is the result of interactions of regulatory proteins and DNA. Such interactions determine differential gene activities. An important question, however, is whether the steps of determination and differentiation are dependent upon events which occurred only inside the cell. Such steps may, on the other hand, require not only the inducing stimulus, but also some of the intercellular interactions described below. There is reason to believe that "the ability of a previously differentiated tissue to undergo differentiation ... depends upon the presence of a certain minimal mass which is necessary for future morphogenesis" (Lopashov, 1935; Saxen and Toivonen, 1963). Interesting data related to this problem were obtained by Grobstein and Tseng Mi-pai. Grobstein found that transplantation into the eye chamber of pieces of the presumptive neural field of a mouse embryo at the stage of the primitive streak caused teratoid growth of the nerve tissue. The nerve tissue in the implantations developed more poorly if the embryonic pieces were first cultivated on glass. However, even after preliminary cultivation, the embryonic pieces that were

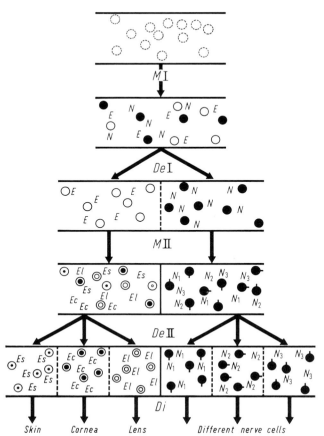

Fig. 62. Tentative scheme of the sequential processes of maturation, determination, and differentiation occurring in the ectoderm of an amphibian embryo. (After Hadorn, 1965)

isolated later from the head process stage are differentiated in a neutral direction in the anterior chamber. The effect was dependent upon the size of the transplanted pieces. Pieces smaller than 0.1×0.1 mm are weakly differentiated, even if several of these pieces are transplanted in close contact. By the late medullar plate stage, however, fragmentation of the presumptive nerve tissue cannot avert their differentiation (Grobstein, 1955).

Tseng Mi-pai investigated the influence of the mass on neural differentiation with experiments on salamander embryos (*Cynops orientalis*). She cultivated various numbers (1, 2, 5, and 10) of pieces of presumptive neuroectoderm from the early gastrula stage. These pieces were treated with calcium-free Goltfreter's solution. The data obtained are in Table 31. There was a significant amount of nerve tissue formation (86% and 88%) when the explants contained one or two pieces (respectively) of ectoderm. While histological differentiation proceeded normally, there was no noticeable regionalization. The frequency of nerve tissue formation was higher in group III (91%). Moreover, regionalization of neural structures was evident. Differentiation of neural structures was not better in group IV, where the explant contained ten pieces. Necrotization was, however, obtained. These data in-

Table 31. Dependence of neural differentiation upon number of ectodermal pieces. (After Tseng Mi-pai, 1960)

Type of differentiation	Group			
	I	II	III	IV
	Number of ectodermal pieces			
	1	2	5	10
Atypical epithelium	1	1 12%		
Transitional type	6 14%	2	2	9%
Localization was not identified				
Segments containing nerve cells	31	14	6	14
Whole explant is formed by the neural tissue	6 89%	2 86%		35%
There is a cavity in the neural tissue	2	3	1	
Cerebrum			88%	11 91%
Epiphysis		86%		2
Olfactory bulb				9
Nerve tissue with expressed localization				
Without pigment layer and lens	2	2 14%	65%	
With pigment layer and without lens	2 11%	3+1	2+6	1
Without pigment layer and with lens	1	1		1
With pigment layer and lens			4	
Mesenchyma			2	
Total number of experiments	51	25	22	21

Percentages were calculated from the total number of slices used in series of experiments

dicate that there is an optimal size of neuroectoderm explant which will result in differentiation (Tseng Mi-pai, 1960).

These experiments demonstrate that by the early neurula cell differentiation and specificity are determined by the cells (which are functioning as independent units) rather than by intercellular factors. Therefore, cells in the neurula stage are "programed" for a specific development. Cells from the gastrula stage do not show this specialization where they are cultivated as a suspension of separate cells. Aggregates of such cells, however, demonstrate noticeable specialization (Elsdale and Jones, 1963; Jones and Elsdale, 1963). Wilde (1961) investigated cell differentiation during cultivation of both separate cells and pieces of tissue of the embryo *Ambystoma maculatum*. Differentiation was not observed within 6 days in cultures which contained only one cell. Such cultures also had no mitotic divisions. Differentiation did not occur in any culture consisting of less than 16 cells when a standard medium was used. If this medium was replaced by a solution of intercellular compounds, however, cell aggregates of 2–15 cells began differentiation (as measured by the formation of pigmental, muscular, and nerve cells). In both media, cultures consisting of more than 15 cells demonstrated cell differentiation. The number of cultivated cells undergoing differentiation in the intercellular solution medium, however, was higher than in the synthetic standard medium.

Lopashov attached great importance to the interaction of embryonic cells with intercellular medium. According to his view, these interactions could have an important effect on primary cell differentiation. Specifically, the internal medium of the blastocele and gastrocele – into which a surface layer of compact cells turns – could contain certain proteinaceous compounds. In the presence of these compounds the cell surfaces do not compact. Therefore, these cells retain their ability for inductional interactions (Lopashov, 1965, 1977).

A similar phenomenon is apparently obtained in a later period of embryogenesis, i.e., during the development of kidney tubules in mice. Auerbach (1964) used the heterogenic induction – the pectoral part of the spinal cord – for the stimulation of development of the metanephrogenic mesenchyme in the kidney tubules. The factor which induced tubule formation was shown to have the following properties:
1. the factor goes through a filter with a 0.1 μ pore diameter,
2. after the initial contact with the system, 30 h are necessary for a positive response,
3. a minimal volume of inducting tissue is necessary.

Results of these experiments could be represented as follows:

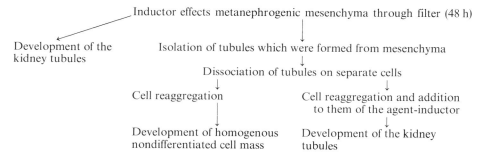

The data obtained demonstrated that in the early stages of induction, development in a certain direction is stable at the tissue level (the isolated pieces of mesenchyme developed into kidney tubules after contact with inductor), but is labile at the cell level (the disaggregation of cells prevented their differentiation). More prolonged contact of the reacting system with the inductor leads to the stabilization of determination on the cell level also. The previously dissociated cells, after reaggregation, are restored to the level of differentiation characteristic for the tissue.

Recently, the molecular mechanisms of the intercellular interactions have begun to be better understood. The synthesis of rRNA in the cells of the sea urchin embryo is initiated in the mid-gastrula stage. The disaggregation and following reaggregation of embryonic cells in this period do not disturb the rRNA synthesis in the reaggregated cells. In the sea urchin cells which were dissociated in the mesenchyme stage and then reaggregated in sea water, the synthesis of rRNA is significantly inhibited compared with normal embryos. The synthesis of 28 rRNA is very low. Apparently, normal cell interactions are important for the start of rRNA synthesis in embryonic development between the mesenchymal-blastula and gastrula stages. Such cell interactions, however, are not necessary to maintain rRNA synthesis once this synthesis has begun (Giudice et al., 1967).

The next steps in the interaction of the inductor with a competent tissue can be distinguished as follows:

1. The activation period – when intercellular interactions occur. These interactions are accompanied by the interchanging of biologically active macromolecules that penetrate into the cells of the reacting system and interact with cytoplasmic elements. Transcription, apparently, does not change during this period.

2. The period of stabilization of determination at the tissue level (sometimes called maturation). This stabilization occurs with the interaction of the protein regulators with chromosomal DNA (see Part 1) and with the intensification of transcription.

3. The period of stabilization and determination at the cell level, called programing by Elsdale and Jones (1963). In this period, the functional condition of the chromosomes is stabilized on the molecular level. This stabilization sets the development of the cell in a certain direction. The ability of cells to maintain this determination after several cell generations has been termed epigenomic heredity by Olenov (1967) and Vachtin (1974).

The certain lability of determination allowed the transdetermination.

2.1.2 The Phenomenon of Transdetermination in Cell Populations

Determination can be considered a multistepped process which is closely associated with maturation and differentiation. Maturation, in turn, involves a number of processes which result in the appearance of competence. There are two types of determination: (1) determination that directly leads to differentiation (for instance, the development of specialized nerve cells), and (2) determination which leads to the repetition of cell conditions that, however, are not differentiated (for instance, cambial cells in the epidermis of *Stratum cylindricum*, perithelium, cryptic epithelium of intestine, etc.). It was demonstrated with repeated transplantations of determined parts of *Drosophila* imaginal discs that these transplants retain their condition of determination and will differentiate into the tissues expected (Fig. 63; Hadorn, 1965). This is known as autotypic determination. Parallel to this is allotypic determination, or transdetermination. This event is rare and is seen, for instance, when antennal cells develop into the blastema of the legs (Hadorn, 1965; Kauffman, 1973; Gehring, 1976 b; Ouweneel, 1976).

There is a certain probability of development in a direction of any condition of determination. In other words, there is a frequency of transdetermination. For example, cells of the genital discs, with a given probability, are transdetermined into cells of the head or legs. The secondary transdetermination of this allotype leads to the appearance of the wing. There was never, however, a direct transformation of genital to wing cells. In general, the tendency of transdetermination can be schematized as follows:

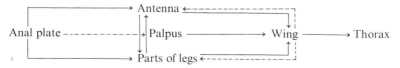

(*Solid lines*, more common events; *broken lines*, less common events.)

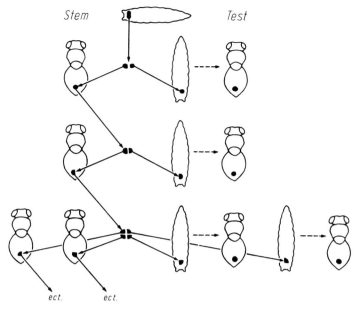

Fig. 63. Method used for permanent cultures in vivo. Stem line fragments grow but remain undifferentiated; test fragments pass metamorphosis and develop into adult structures. (After Hadorn, 1965)

Transdetermination often leads to the appearance of a phenocopy of a known mutation. Moreover, there are many so-called homeotic mutations in *Drosophila* (first described by Astaurov in 1927), which cause allotypic organ formation. Aristopedia is one example of such a mutation (leg instead of antennal structures are formed).

The mechanism of transdetermination is not clear. It is, however, known that only cultures with a high rate of proliferation will undergo transdetermination. There is, in fact, a positive correlation between the rate of proliferation and the frequency of transdetermination. Some hypotheses for this were experimentally examined (Hadorn, 1967; Gehring and Nöthiger, 1973):

1. The migration of transdetermined cells was rejected in experiments with tissue transplants from ebony to yellow mutants of *Drosophila*. The transplanted structures formed were always black, regardless of their transdetermination.

2. There may be cell heterogeneity in the transplanted tissue. Therefore, for instance, the genital disc may contain several cells which can form legs; normally, however, the proliferation of these cells is suppressed. Somatic crossing-over was used to investigate this hypothesis. Imaginal discs for transplantation were obtained from larvae which were heterozygous for both the yellow and singed mutations. Somatic crossing-over was induced by X rays. If this crossing-over takes place between the centromere and the marker genes, then two cells with different genotypes will be formed – double mutant homozygous and double wild-type homozygous. Using these cells as markers, it was shown that it is unlikely that trans-

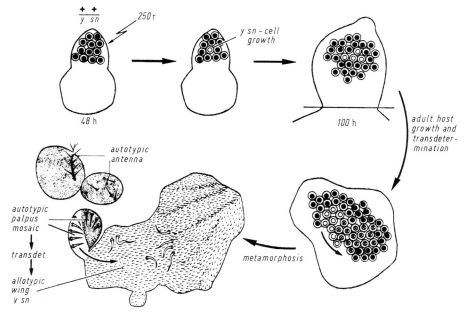

Fig. 64. Transdetermination in a clone. Cells with *black nuclei* are of the $++/y$ sn geno-type. Cells with *light nuclei* are hemi or homozygous for y sn. A *cross* indicates the transde-termined offspring within the y sn clone. As an example, a differentiated test implant is shown with auto- and allotypic structures. (After Hadorn, 1967)

determination is caused by the selective proliferation of some cells which carry a different program (Fig. 64).

3. Transdetermination may be the result of somatic mutation. The frequency of obtained transdeterminations, however, is several orders higher than the frequency of somatic mutations (although this frequency is not known exactly in cultures of proliferating tissues).

From the above, then, it appears more likely that transdetermination affected the cell's heredity by causing a change at the functional level (Hadorn, 1967; Gehring, 1968).

It was suggested that cells comprising the *Drosophila* imaginal discs are able to retain their determined condition for hundreds of generations. Similar systems are known in vertebrates, such as the cambial tissues – e.g., *Stratum germinativum* of the epidermis. There are effectors of determination (E) and carriers of determination (C) in cells. Effectors correspond to genotropic substance (Waddington, 1964), and their nature is unknown. They could be proteins with a low molecular weight, various small molecules, ions, hormones, and either cytoplasmic or nuclear components. The genotropic effectors could activate, repress or derepress various genes (Fig. 65a). The products which control the specificity of differentiation are designated D. When determined cells divide, the carriers of determination are distributed between the daughter cells and, as a result, the cellular concentration decreases. The condition of determination is stable if the carriers are synthesized at

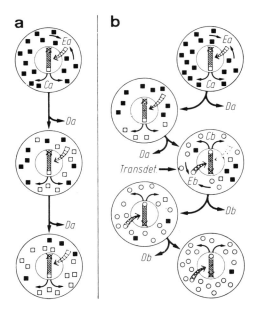

Fig. 65 a, b. Relation between rate of cell division and rate of synthesis of carriers of determination. **a** Unchanged cell heredity. **b** Dynamics leading to transdetermination; *Ea, b*, genotropic effectors; *Ca, b*, carriers of determination; *Da, b*, specific differentiation. *Black squares*, pre-existing carriers for **a**; *light squares*, newly synthesized carriers for **a**; *Circles*, carriers for the transdetermined state **b**. Within the nucleus two genes or gene teams are shown either active or repressed *(cross)*. (After Hadorn, 1967)

a rate which compensates for the reduction in concentration. Thus, there is a steady state between the rate of carrier dilution and synthesis of new determination factors.

If this steady state is changed as a result of increased cell proliferation, the changed concentration of carriers of determination could affect the genotropic effectors in such a way that they will repress some genes and activate others. At the same time, new carriers of determination could be synthesized which replace the previous carriers (Fig. 65b).

The discussion above supposed that determination and transdetermination occurred not only on the level of the separate cells, but also on the level of all populations which formed as the result of the proliferation of cells.

2.1.3 The Heterogeneity of Cell Populations

Every cell population is heterogeneous because the individual cells are in various functional conditions, in different stages of the cell cycle, in different stages of differentiation or degeneration, etc. The most important for development are those types of heterogeneity which are inherited genetically – i.e., by somatic crossing-over, somatic mutations, the position effect, by differential gene activity (allelic and nonallelic), etc. Such events can result in the mosaic construction of cell groups. Although related morphologically and functionally, these cell groups are not identical in their functional genomic organization.

All cases of mosaicism can be divided into two large groups if the classification of Russell (1964) is slightly modified: (1) structural–genetical mosaicism where the gene action in either group of cells is affected by the structural changes in the genome, and (2) functional–genetical mosaicism which results from differential inactivation at homologous loci.

2.1.3.1 Structural–Genetical Mosaicism

The structural–genetical chimerism in mammals is caused by the participation of a polar body in embryo formation or by the fusion of two different zygotes. In the first case, the mosaics could have a different ploidy of cells: 1n/2n; 2n/2n; 2n/3n. Fertilization conditions would determine which of these exist. Diploid-triploid mosaics were found in man (Böök and Santesson, 1961) and the cat (Chu et al., 1964). Mosaics are known in honeybees which are the result of the peculiar fusion of pronuclei. Two variants of such fusion have been proposed: (1) the division of the maternal nucleus before fertilization and the fusion of the sperm nucleus with one of these, or (2) the penetration of two sperm nuclei into the egg. One sperm nucleus fuses with the maternal nucleus while the other sperm nucleus replicates without fusion (Stern, 1968). Rare cases of double fertilization have been described for *D. melanogaster* (Stern and Sekiguti, 1931) and *Bombyx mori* (Goldschmidt and Katsuki, 1931). Finally, gynandromorphism is known in *Drosophila melanogaster* and other insects, which is caused by the elimination of the unstable X_R (ring X) chromosome during the first mitosis in an $X_N X_R$ strain (Morgan and Bridges, 1919; Stern, 1968).

Hotta and Benzer (1972) capitalized on this effect for the mapping of certain morphogenetic traits in *D. melanogaster*. It is known that the unstable ring X chromosome $In(1) W^{vc}(X_R)$ is eliminated in the first cell division. As a result, two cell clones are formed – one with one X chromosome ($X_N O$) and the other with two ($X_N X_R$). In the $X_N X_R$ portions of the embryo, where the X_R chromosome carries the wild-type alleles, recessive mutations are suppressed. In the $X_N O$ portions of the embryo the recessive alleles were expressed phenotypically. (Benzer used genes which controlled eye and cuticle color, bristle shape, and some behavior traits.) The distribution of the mutant and normal fields in the adult fly varied and was dependent upon the distribution of the cells in the forming blastoderm. This distribution is determined in the very early stages of development. The first ten nuclear divisions in *Drosophila* occur without cytokinesis. The resulting nuclei migrate to the egg surface where they undergo three more divisions. Cell membranes are formed around the nuclei and the blastoderm is thus formed. As mentioned above, two cell clones – $X_N X_R$ and $X_N O$ – are formed after the first cleavage. Each clone, under ideal conditions, should occupy approximately half of the blastoderm. However, if there is random mixing of nuclei after the first division, then the borders between normal and mutant clones should also be random. The probability that the mutant–normal clone border occurs between two cells in the blastoderm that will eventually develop into two different organs is small. On the basis of the deviation of the observed mutant frequency from the ideal mutant frequency of 0.5, a preliminary map of embryonic development was made. Thus, the method used here is in certain ways similar to the classic method of genetic mapping which uses recombinant frequencies. The difference between the two methods is that the

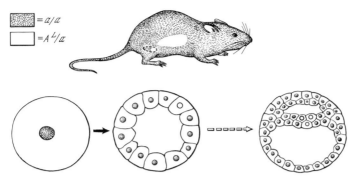

Fig. 66. *Above*, a mouse homozygous for not-agouti (*a/a*), mosaic by having an agouti-like (*A^L/a*) patch. The ovary, too, is mosaic, as indicated by the outline above the hind leg. *Below*, the fertilized egg and two later embryonic stages in cross sections. It is assumed that a mutation from *a* to *A^L* occurred in one nucleus of the blastocyst *(middle)* and that descendants of this cell became incorporated in the embryonic disc *(right)*. (After Stern, 1968)

morphogenetic map has two dimensions. Benzer's work is an excellent example of the successful use of genetic methods to solve problems in experimental embryology.

Mosaicism is very often caused by somatic mutation. The same types of mutation that occur in germ cells may also occur in somatic cells (Shapiro, 1966). The frequency of mutation in vitro in somatic cells is similar to the frequency in vivo in germ cells. Cell clones can result from the division of somatic cell mutants. These mutants may somehow change the direction of development. The mutants will also change the observed phenotype (Searle, 1968).

Mosaicism in female mice homozygous for the recessive nonagouti color gene (aa) is a well-analyzed example of such mutations (Bhat, 1949). The mosaic individual has an agouti spot which is caused by the somatic mutation a→A^L (Fig. 66).

Table 32. Autosomal germinal (G) and mixed (M = germinal-somatic) mosaics in mice. (After Grüneberg, 1966)

Gene	Type of mutation	Type of mosaic	Sex	Strain origin	Ratio of normal: mutant gametes
Agouti	A →a	G	?	Het.	6(?): 2
Albino	+ →c⁻	M	♂	+/c^{ch}	58:17 [a]
Varitint-waddler	+ →Va	G	?♀	?Het.	24: 2
Agouti	a →A^w	G	?♀	DBA	18: 3
Agouti	a →A^w	M	♀	Het.	61: 2
Agouti	a →A^w	M	♂	C58	21: 3
Tail-short	+ →Ts	G	♂	BALB/c	120: 8
Trembler	+ →Tr	G	?	?Het.	5: 4
Crooked-tail	+ →Cd	G	?♀	A	6: 2
Twirler	+ →Tw	G	?	Het.	14: 2
Lurcher	+ →Lc	G	♂	Het.	59: 3
Rib fusions	+ →Rf	G	?	129	29: 2

[a] The mosaic transmitted 58+, 71 c^{ch} and 17 c⁻ gametes

The progeny of mutant cells is incorporated into a hair follicle in a certain body region. It is of interest that the somatic mutation which occurred during early embryogenesis can affect cells that are predecessors for both germ and somatic cells (Russell, 1964). Several examples of such somatic mutations are known for mice (Table 32). Similar mutations are known for *Drosophila* (Altenburg and Browning, 1961).

Distributions of human embryonic cells leading to distinct mosaicism in complexes of sex chromosomes have been described (Fig. 67; Klinger and Schwarzacher, 1962). Mosaics for differing karyotypes have also been described for the black fox (Volobuev and Radjabli, 1974).

Another event that can result in mosaicism is somatic crossing-over. This event was first described for *Drosophila* (Stern, 1936). Figure 68 gives a schematic demonstration of schematic crossing-over (Stern, 1968). In this case, females which were heterozygous for the recessive yellow body and singed bristle genes (ysn/y$^+$sn$^+$) showed patches of yellow body, singed bristle phenotype on an otherwise black body, normal bristle fly.

Somatic crossing-over has also been described for mice (Carter, 1952; Grüneberg, 1966). W^v/+ heterozygotes had field of wild-type color and W^v/W^v color. Some gonad cells also had the W^v/W^v genotype (Carter, 1952). Obviously, somatic crossing-over occurred very early in embryogenesis and affected not only predecessors for other somatic cells, but also predecessors for germ cells.

Somatic crossing-over may also take place in cells which originate in germinative tissues. In one study, genes from the sixth linkage group were used as markers. These genes include Caracul (Ca, bristles are twisted and hairs are wavy) and belted (bt, white abdomen). The distance between Ca and bt is 12 map units. In progeny from the cross

$$\frac{Ca +}{+ bt} \times \frac{+ bt}{+ bt}$$

two males were found which were homozygous for the Ca allele. One of these was

$$\frac{Ca +}{Ca \, bt}$$

This genotype can be explained if the sequence of genes is

$$\frac{Ca + centromere}{+ \, bt \, centromere}$$

Then, somatic crossing-over between Ca and bt will produce

$$\frac{Ca + centromere}{Ca \, bt \, centromere} \quad \text{and} \quad \frac{+ \, bt \, centromere}{+ \, + \, centromere}$$

cells. Somatic crossing-over between the bt gene and the centromere will produce

$$\frac{Ca + centromere}{Ca + centromere} \quad \text{and} \quad \frac{+ \, bt \, centromere}{+ \, bt \, centromere}$$

cells (McNeil, 1957; Grüneberg, 1966).

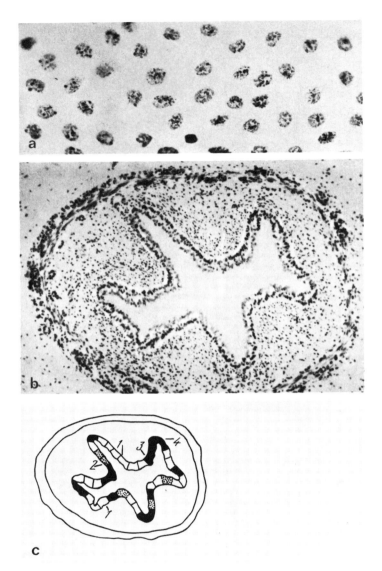

Fig. 67 a–c. Distribution of cells with sex chromatin in an XY/XXY mosaic human embryo. **a** A portion of the amnion epithelium. *Left half,* a sex chromatin positive area; *right half,* a negative area. **b** Transverse section of the esophagus. **c** Tracing of the section with information on the incidence of sex chromatin in the epithelial layer. *1* = 0–9%, *2* = 10–19%, *3* = 20–29%, *4* = >30%. (After Stern, 1968)

Another type of mosaicism is observed in position effect. Position effect describes a phenomenon of gene suppression due to cis position rearrangement of chromosomes (Baker, 1968). Suppression occurs when genes are translocated to heterochromatic sections of chromosomes.

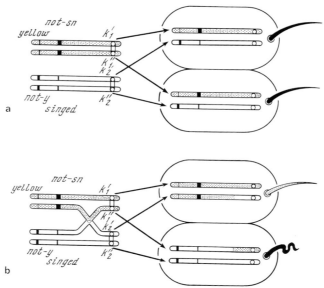

Fig. 68 a, b. Distribution of chromosomes heterozygous for two genes in the trans position. *Left*, the chromosomes are doubled. The sister kinetochores of the chromosomes are labeled K'_1 and K''_1 and K'_2 respectively. *Right*, the chromosomes are distributed to daughter cells. **a** No crossing over. The daughter cells are alike genetically, and they or their descendents may form not-yellow, not-singed bristles. **b** Crossing over between the locus of singed and the kinetochores k. The daughter cells differ genetically and may form a twin spot of yellow not-singed and not-yellow singed bristles. (After Stern, 1968)

An illustrative example of the position effect in *D. melanogaster* involves an X chromosome translocation with the normal white allele on chromosome IV. This translocation is designated $R(w^+)$. In $R(w^+)/(w)$ or $R(w^+)/R(w^+)$ the eyes are mosaics: part of the cells are wild-type and the rest of them are white (Fig. 69; Becker, 1966). This effect was not obtained for $R(w^+)/(w^+)$ or $R(w)/(w^+)$ flies. Therefore, the mosaic effect was shown only in the case when the translocation is heterozygous and flies are heterozygous for the mutant allele, or when both the translocation and mutation are homozygous.

Parental somatic cells affected by the position effect are inherited for a number of cell generations. These cells formed the region of phenotypically mutant tissue (Baker, 1968). It was suggested that the gene inactivation resulting from the mosaic-type position effect is associated with a delay in replication of the heterochromatic regions (Schultz, 1956; Lima-de-Faria, 1969; Gvozdev et al., 1973).

Mosaicism of position effect is not due to somatic mutation because:
1. The Y chromosome antagonizes the expression of the mosaicism in *D. melanogaster*;
2. The number of white cells is significantly increased in the mosaic eye at low temperatures. The temperature-sensitive period is pupation, when the pigment is stored, and not the larval period when the eye primordium is determined;

3. The corresponding mutations are not expressed phenotypically in the primordial cells in individuals with the position effect of the mosaic type;
4. Biochemical analysis of mosaic eyes demonstrated an excess of compounds from the pteridine pathway. This pathway plays an important role in eye pigment formation in wild-type individuals. Such an excess is not seen in mutant individuals (Baker, 1963).

Anaviev and Gvozdev (1974a) have studied the position effect of the mosaic type in *D. melanogaster* with the chromosomal mutation Dp (1:f)R and the translocation of the euchromatic region 1A3-4-3A1-2 of the X chromosome to the centromere heterochromatin. According to their data, there is a delay of replication of the translocated euchromatin. RNA synthesis here was reduced to 60% of normal. These data show that the position effect is accompanied by alterations at the transcriptional level. The resistance of the eye mosaic peculiarities to inhibitors of cell division and RNA synthesis added at the end of the first larval stage does not contradict the above statements (Baker, 1967; Perez-Davila and Baker, 1967).

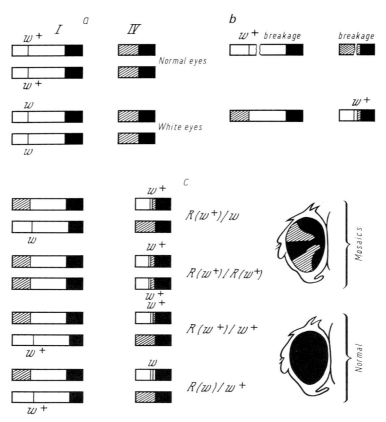

Fig. 69. a Eye phenotypes in flies with intact chromosomes 1 and 4. **b** The translocation induction between chromosomes 1 and 4 by Rentgen treatment. **c** The effect of translocations on eye phenotypes. (After Becker, 1966)

2.1.3.2 Functional–Genetical Mosaicism

The best known case of structural–genetical mosaicism is the Lyonization which was discussed in Chap. 1.3.3.2. This phenomenon is characterized by the formation of cell clones which differ in X chromosome activities.

Parallel to the cytological and genetic observations are biochemical observations of differential X chromosome functions. For instance, the G6PDH gene is on the X chromosome. This gene has two electrophoretically distinguishable alleles – fast (A) and slow (B). Davidson (1972) cultivated clones from separate fibroblast cells taken by biopsy from a G6PDH heterozygous (AB) human female. In more than 50 clones, when examined individually, only one G6PDH band was found. When the clones were mixed, two bands were observed. With the aid of biochemical and histochemical methods, two erythrocyte populations were established. The cells of one population demonstrated a high level of glucose-6-phosphate dehydrogenase activity while cells from the second population had almost no G6PDH activity (Beutler, 1964).

Cases of mosaic expression of pathological traits which are sex-linked are also known (Beutler, 1964). Individuals heterozygous for ectodermal displasy have patches of normal and abnormal skin. Also, heterozygous individuals have a spotty retina (Waardenburg and Van den Bosch, 1956). Similar observations were obtained in the case of choriodermy (Sorsby et al., 1952), color blindness (Beutler, 1964), sex-linked hypochromic anemia (Rundles and Falls, 1964), and hemophilia (Frota-Pessoa et al., 1960). Also, both normal and pseudodystrophic muscle fibers were obtained in humans heterozygous for the Dushen-Pearson's muscular dystrophy mutation (Pearson et al., 1963).

It has been demonstrated that lymphocytes do not differentiate if they are obtained from patients carrying the agammaglobulinemia mutation on the X chromosome. These cells have no γ-globulin synthesis. Two populations of cells were observed in lymphocyte cultures from a woman patient heterozygous for the agammaglobulinemia mutation. Cells from one of these populations were able to differentiate and synthesize γ-globulin. The second population of cells did not have this ability (Fudenburg and Hirschorn, 1964).

There are two types of gargoilism (abnormal mucopolysaccharide metabolism and accumulation): autosomal and sex-linked. Among clones which were begun from a heterozygote for sex-linked gargoilism, 72% were characterized by metachromatic staining with toluidine blue and by a high urea content. In the case of the recessive, autosomal form of this disease, inactivation of one homologous chromosome was not obtained (Danes and Bearn, 1966, 1967). Moreover, evidence has been obtained that individuals heterozygous for blood group determinants (Xg^{a}Xg) have two different erythrocyte populations. One female patient had one X chromosome with the Xg (+ve) allele and an allele for hypochromic anemia of the Cooly-Rundles-Falls type while the other X chromosome was marked with the X (-ve) allele and was wild-type for hypochromic anemia. Hence, this individual was a double heterozygote. Differential centrifugation was used to separate the erythrocytes into two groups: (1) normal Xg(-ve), and (2) abnormal, hypochromic microcytes, homozygous for Xg(+ve). The patient's mother was heterozygous for hypochromic anemia and homozygous for Xg^a. In this individual, the normal erythrocytes and abnormal microcytes were both Xg^a.

Separate cases of differential expression of traits which are controlled by autosomal genes are known. In general, however, homologous autosomes have a synchronized activity. An illustration of this is the immunological data dealing with the allotypic specificity of cells of an immuno-competent tissue which was discussed previously. The clonal principle of differentiation in cells that produce antibodies is generally recognized (Papermaster, 1967; Burnet, 1971).

Hair color mosaics in mice serve as another example of mosaic trait expression in autosomal heterozygotes. In such heterozygotes, normal colored spots alternated with mutant colored spots (Russell, 1964). The mutation variting-waddler (vo) is located on the XVI linkage group and, in homozygotes, demonstrates a white phenotype. Heterozygous individuals have a mosaic phenotype which is similar to the Lyonization effect. In many cases, the alteration of trait expression is due to the translocation of the effective autosomal part into the X chromosome (this event was first described for *D. melanogaster*). The suppression of activity of this autosomal segment was seen if this segment was incorporated into the heterochromatin of the X chromosome. If suppression occurred and the translocated segment contained a dominant allele, then the recessive phenotype was expressed (Ohno, 1971). When the dominant wild-type nonalbino allele (c$^+$) was translocated into the X chromosome (X^t) and the albino allele (c) was present on the intact autosome I, then X^{t}Y males and X^{t}O females demonstrated the wild-type phenotype. Heterozygous females (X^{t}X^n) are mosaics. The presence of two populations of somatic cells in such individuals was demonstrated by studying H^3-thymidine labeled chromosomes in prophase and metaphase. The X^n chromosome was heterochromatic in about half the cells and did not replicate its DNA until late in the S phase. The remaining cells had a heterochromatic X^t chromosome. This chromosome in these cells also replicated late (Ohno, 1967).

D. melanogaster heterozygous for an alcohol dehydrogenase deficiency (Oregon R × Adh-n$_1$ hybrids) demonstrated a peculiar mosaicism of enzyme activity in cells of the Malpighian tubules. Cells with a negative histochemical reaction for ADH were together with cells which had a positive reaction. This staining pattern is associated with a differential activity of homologous loci. The slightly reduced ADH activity in heterozygotes could have been caused by either lower enzyme activity in all of the cells or by the presence of cells with ADH deficiency (demonstrated with histochemical methods).

It seems that the uneven distribution of ADH in the Malpighian tubules of Oregon R × Adh-n$_1$ hybrids was already determined when the enzyme activity first began to increase (the end of the first and beginning of the second larval stage). Thus, this increasing enzyme activity is the first process which leads the biochemical mosaicism in third instar larvae (Korochkin et al., 1972b).

A similar situation which is determined, however, by the differences in nonallelic genes, was discovered in ♀*D. virilis* × ♂*D. texana* hybrids. Some of these hybrids were different in α- and β-esterase activities in the Malpighian tubules and salivary gland of third stage larvae. Most *D. virilis* strains demonstrated a low activity of thermostable β-esterase and a high activity of thermostable α-esterase. The content of β-esterase is high in *D. texana*, especially in the Malpighian tubules. Af-

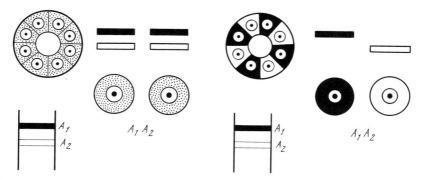

Fig. 70. Cellular (*left*) and tissue (*right*) levels of the expression of biochemical traits

ter a 2–3 min treatment of the hybrid larvae at 50 °C, histochemical methods revealed heterogeneity of esterase contents in the cells. In some of the cells the esterase activity was reduced (apparently, the β-esterases were in the majority). Other cells had a majority of α-esterases. Thus, the mosaicism in cell distribution was noticeable. This phenomenon was corroborated by the microelectrophoretic separation of esterase isozymes from separate cells. All hybrid organs were characterized by the heterogeneity of cells with respect to their esterase content. Some cells had the paternal esterase composition while other cells had the maternal esterase composition (Korochkin et al., 1973 b).

The heterogeneity of cell populations led to the realization of differential gene activity on the tissue level. The differences between cellular and tissue levels of differentiation can be illustrated when biochemical markers are used (Fig. 70). Suppose there are two electrophoretic variants of protein A (A_1-fast and A_2-slow) which are controlled by a pair of alleles. Codominant expression will explain the presence of both electrophoretic variants in heterozygous individuals. This occurs at the cellular level. There is, however, a second possibility. There are cells which can synthesize only the A_1 variant and cells which synthesize only the A_2 variant. The proliferation of these cells will lead to clones. Therefore, these traits are also codominant at the tissue level. However, in each separate cell there is only one allele.

The presence of complicated intercellular interactions, competition among cell clones, selective proliferation and elimination of certain clones allow us to suggest that in some cases the tissue level is the basis for dominant (or recessive) gene expression (Fig. 71). Genes will demonstrate dominant expression in tissues when clones contain only one active gene of the homologous locus. This gene will have a noticeable selective advantage and will predominate in the cell population. This possibility does not exclude genes with local action, i.e., as in the experiments with allophenic mice (Mintz, 1965; McLaren and Bowman, 1969). Two types of allophenic (chimeric) mice were created by artificially fusing blastomeres obtained from mouse lines which differed in the number of dominant and recessive traits. The mixture of normal and mutant cells formed tissues which could function normally when the product of the normal cell migrated into the mutant cells and, therefore, compensated for any deficiencies. This can also occur with genes with

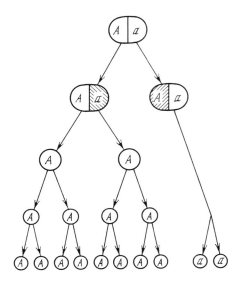

Fig. 71. Selection of cell clones with differential activity of homologous loci. (After Korochkin and Belyaeva, 1972)

a nonlocal type of action (e.g., the short ear gene which is active in cartilage). Such compensation is not possible for genes with a local action and, therefore, mixtures of these cells will form abnormal structures because of genetic mosaicism (i.e., the genes which are active in the hair matrix).

It is obvious that differential expression of nonallelic genes in cells which comprise embryonic primordium (see Part 1) significantly increases the heterogeneity of differentiating cells.

In general, the heterogeneity of cell populations (which form either organs or tissues) occurs early in embryogenesis as a result of differential gene expression (usually nonallelic, but sometimes allelic genes as well). Differentiation of cells depends not only upon the functional whole genome composition of the differentiating cell, but also upon the genomes of the surrounding cells. These interactions influence the character of cell differentiation and the growth and proliferation of cells.

2.2 Genetic Heterogeneity and the Establishment of Tissue Systems During Development

2.2.1 Interactions Between Cell Populations During Morphogenesis

The significance of trait expression at the tissue level during morphogenesis has already been demonstrated for the case of the early morphogenetic process of primary embryonic induction. The presumptive neuroectoderm is heterogeneous and is made of two cell populations. Both electron-dense and electron-transparent cells were obtained from this tissue from cynopus pyrrhogaster in the gastrula stage. The electron-dense cells degenerated while the electron-transparent cells developed into various mesodermal structures when explants of presumptive ectoderm were treated with mesodermal inductors. Dissociated cells of the presumptive ectoderm could be divided into two groups with the aid of electrophoresis: (1) cells which were able to react to a neuralizing inductor and which perished after the addition of mesodermalizing agent, and (2) cells with the opposite reaction to such treatment. It was demonstrated that the acquired competence of the dissociated cells of the presumptive ectoderm is correlated with the altered sensitivity to the inductor (see review Ignatijeva, 1967). It should be noted that there is a significant loss of cells during morphogenetic processes, and, therefore, a successive replacement of cell populations during ontogenesis. This process occurs in various tissues such as nerve (Hughes, 1968), muscle (Saunders et al., 1957), and others (Berdishev, 1968). Cell loss has also been described as part of the differentiation of insect imaginal discs (Friström, 1969; Michinoma and Kaji, 1970).

An especially illustrative example of the replacement of cell populations during ontogenesis is the changing hemoglobin spectrum. It was found that (at least in amphibians) the replacement of embryonic with adult hemoglobin was due to the replacement of a cell population specialized in embryonic hemoglobin (HbF) synthesis by a cell population specialized in the synthesis of adult hemoglobin (HbA). The HbF synthesizing cells are produced in the yolk sac while the HbA synthesizing cells are produced in the liver and red bone marrow (Maniatis and Ingram, 1971). This hemoglobin replacement is not the result of changing gene activities, as was once suggested. Immunofluorescent methods have conclusively shown that HbF and HbA cannot exist in a cell at the same time. However, others have obtained amphibian erythrocytes with both HbF and HbA (McLean and Jurd, 1971; Ben-bassat, 1974). Similar results have been obtained for mammals (for review see Medvedev, 1968; Sallei and Zuckerkandl, 1974).

The replacement of 1-Aph by p-Aph (isozymes of alkaline phosphatase) during insect development is also the result of processes at the tissue level. It is known that the larval hypoderm, for which 1-Aph is a biochemical marker, is completely replaced by the adult hypoderm during early pupae formation. Shortly after puparium formation, cells from the imaginal discs proliferate and form the adult hypoderm. Larval epidermal cells are shed into the body cavity where they are phagocyted first in the head region, then in the thorax (8 h after puparium formation), and finally in the abdomen. Thus, the disappearance of l-Aph can be explained by the disappearance of the tissues that produce it. The yellow body which contains p-Aph is formed from mid-intestinal epithelial cells of the larvae which invaginate into the intestine and which are formed as a result of proliferation of imaginal cells. Yellow body formation is concomitant with the appearance of p-Aph. Shortly after imago hatching, the yellow body is excreted from the digestive system as a meconium. Thus, p-Aph is absent on electrophoretic slabs made from adult flies (Schneiderman et al., 1966; Wallis and Fox, 1968).

An interesting case of differential expression of homologous sex-linked loci is obtainable in female hybrids between a horse and a donkey. The differences in gene expression are apparently dependent upon the selective proliferation of one cell clone. In this case, the product of the maternal (horse) allele for glucose-6-phosphate dehydrogenase was solely active in the blood, pancreas, brain, kidney, salivary glands, spleen, and lymphatic system. The paternal (donkey) allele was solely active in the liver. Both parental allele products were found in the lungs and thyroid gland (Hook and Brustman, 1971). This data might be explained by the pattern of proliferation of cell clones formed by the maternal and paternal cells.

Such replacement in cell populations affects the conditions and interactions of surrounding tissues. This, in turn, affects the formation of the organ, i.e., the establishment of its morphological, physiological, and biochemical traits. It was demonstrated that during liver differentiation there is a heterogeneous expression of biochemical traits in the cell populations of this tissue. This heterogeneity was demonstrated when cells from the human liver were cloned (Table 33; Kaighn and Prince, 1971).

In some cases, the differential expression of sex-linked homologous loci might be due to stochastic processes. It is possible that the expression of the parental alleles of glucose-6-phosphate dehydrogenase (Gpd) in hybrids of the black polar fox is an example of such a phenomenon. The investigation of glucose-6-phosphate dehydrogenase in the erythrocytes of the female hybrids demonstrates the significant variability of enzyme activity between the maternal heterozygotes and paternal hemizygotes. Individuals with 1:1 and 1:2 ratios of parental enzyme types occurred especially frequently. The frequency distributions in erythrocytes of female hybrid individuals with different quantitative expressions of the parental Gpd loci can be explained without assuming a selective advantage of any one cell population. Rather, this frequency can be based on the assumption that initial cell populations originate randomly (Serov and Zakijan, 1974; Serov et al., 1976).

The variability in cell composition of different cell populations is also seen in certain physiological differences between cells. Specifically, an analysis of cloned liver cells showed a wide variability in the amount of albumin in these cells. This variability is dependent upon certain physiological conditions (Bernhard et al., 1973). The development of methods to distinguish between genetic and physiological variability is essential to evaluate the role of gene interaction in the formation of observed heterogeneities.

Table 33. The heterogeneity of liver cells and cells from various clones in their biochemical properties. (After Kaighn and Prince, 1971)[a]

Antigen	Liver clones[b] Fibroblastic Fetal				Infant (7 Wk.)				Adult		Cell lines Epithelial Liver	Kidney	Carcinoma		Fibroblastic Skin		Lung
	C1	C2	C3	C4	3	4	6	7	20	21	CG[c]	HEK	HeLa	KB	S[d]	S/M[e]	W138
Fibrinogen	+	−	+	+	+	+	+	+	−	−	+	+	−	−	−	−	−
Transferrin	+	−	+	+	+	−	+	+	−	−	+	+	−	−	−	−	−
Albumin	−	−	−	−	−	−	−	−	+	−	−	−	−	−	−	−	−
Prealbumin	NT	−	+	−	+	+	+	+	−	−	−	−	−	−	−	−	−
α$_1$-Antitrypsin	−	−	−	−	−	−	+	−	−	−	−	−	−	−	−	−	−
α$_2$-Haptoglobin	−	−	+	−	−	−	+	+	−	−	−	−	−	−	−	−	−
β$_2$-Glycoprotein-I	−	−	−	−	−	+	+	+	−	−	−	−	−	−	−	−	−
α$_1$-Acid Glycoprotein	−	−	−	−	−	−	−	−	−	−	−	−	−	−	−	+	+
α$_2$-Macroglobulin	−	−	+	+	+	+	+	+	+	+	+	−	−	−	+	+	+
β$_1$-Lipoprotein	+	−	+	+	+	+	+	+	−	−	+	−	−	−	+	+	+
C′3-Complement	−	−	+	+	+	+	+	+	−	−	+	−	−	−	+	+	+

[a] Confluent Petri-dish cultures were washed three times with F12 medium (minus leucine) and 4 ml of F12 containing [14C] leucine (5 μCi/ml, 300 Ci/mol) were added. After 24 h of incubation, the medium was removed, dialyzed versus GC buffer containing 1 mM L-leucine, lyophylized, and dissolved in 0.1–0.2 ml of water. Sample wells of Ouchterlony plates were filled with 5 μl of appropriately diluted carrier human serum followed by three 5-μl applications of the sample. Monospecific antisera versus human plasma proteins were added to adjacent wells. When precipitin patterns had developed, the plates were washed for three days in PBS, 1 day in distilled water, dried and exposed to Kodak Royal Pan film for 2 weeks. Autoradiographs coincident with carrier patterns were scored as +, definitely labeled or −, unlabeled (PBS, phosphate-buffered saline)

[b] Sub-clones 3, 4, 6, and 7 of HFL2 were isolated from a 7-week-old infant with congenital brain deformity at autopsy; sub-clones 20 and 21 of HL8 from adult stabbing victim; C1–C4 from HLX

[c] Chang liver cells

[d] Skin fibroblasts

[e] Skin and muscle fibroblasts

Blank spaces indicate not tested because of unavailability of antisera at the time

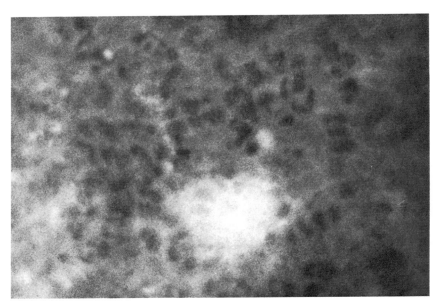

Fig. 72. Immunohistochemical reaction for aldolase C in liver of the 18-day-old rat embryo. In the *middle of field*, there is a group of fluorescent cells which contained aldolase C. Enzyme was not obtained in the surrounded cells (ocular × 7; objective × 40, preparation by Bochkarev)

Interesting data have been obtained in the investigation of tumor cell populations. It was shown that the karyotypes of cloned populations are variable. This variability is dependent upon three things: (1) the interclonal differences caused by the karyotypic differences among the ancestors of the cloned cells; (2) the intraclonal variability of the tumor cells; and (3) the variability of the ancestor cells. The differences among cells were found to play an important role. For instance, clones of mouse rabdomiosarcome virus were obtained which demonstrated karyotypic variability after 21 to 26 days. This variability was significantly higher than the chromosomal heterogeneity of the original tumor. Poorer cultivation conditions led to an increase in the heterogeneity of cloned populations (Vachtin and Borchsenius, 1969; Vachtin, 1972). Similar data were obtained by Ronichevskaya (1975).

Variations in aldolase C liver cell during rat ontogenesis have been investigated in our laboratory with the aid of the Coons immunohistochemical method (Botchkarev, unpublished data). Most cells from 12-day-old rat embryos contain high concentrations of aldolase C. There is, however, a decrease in the activity of this enzyme in liver cells on the 16th day. In 18-day-old embryos, high aldolase C activity was found in only a few groups of cells (10–25 cells in each group). Active aldolase C was not found in the cells surrounding these groups of cells with high activity (Fig. 72). Groups of cells with high aldolase C activity are very rare in adult animals. Such groups contain two–five cells. It seems possible, then, that during rat ontogenesis, there is a replacement of cell populations. Also, the remaining groups of cells with high aldolase C activity may be susceptible to hepatoma. The

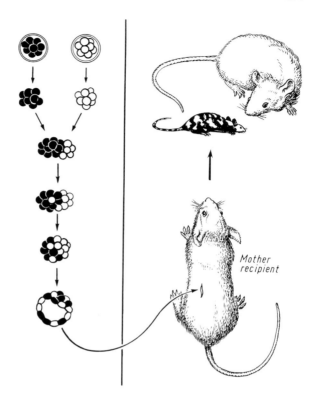

Fig. 73. Obtaining allophenic mice with striped coloration. (After McLaren, 1973)

Mother recipient

heightened ability of such cells to incorporate carcinogenic viruses or to integrate such viruses into the host's genome may be the cause of tumor degeneration. Embryonic proteins were found in these tumor cells. This may indicate that tumors have developed from the embryonic cells (Abelev, 1971; Doyle and Mitchell, 1975; Sato et al., 1975). Differences between animals with high and low abilities to develop cancer can be determined by differences in the number of microclones of "embryonic" cells which can initiate to form tumors. It is also possible that the number of such microclones in the adult animal is genetically determined.

The character of clones during normal tissue differentiation was demonstrated by model experiments with allophenic (chimeric) mice. These mice were genetic mosaics created from combinations of blastomeric cells from embryos with different genotypes (Fig. 73). Using such chimeric mice, it is possible both to determine the distribution of the mosaic spots and to predict the behavior of cell clones originated from a certain tissue or organ during ontogenesis. The probability of cells of both genotypes being present in the same primordium depends upon the number of cells that were precursors for this primordium. It is obvious that if only one cell is the ancestor for a primordium, then no mosaics will be obtained. If two cells are the ancestors for the primordium, then half the resulting individuals will demonstrate mosaicism. Similarly, if, for example, the investigated primordium does not show mosaicism in 1/16 of the chimeric individuals, then the conclusion can be made that this primordium originated from five cells. These five cells are representative of both genotypes in various proportions in different individuals. This conclusion follows from the fact that 1/16 is the probability of the random combination of five cells of the same genotype in the same primordium $[1/16 = (2)(\frac{1}{2})^5]$. This equation is true if the precursor cells proliferate at the same rate and if there is no selection for certain cells (McLaren, 1972). Numerous experiments allowed the formation of a conclusion about the number of cells and corresponding clones which compose an organ. For example, the system of mouse pigment cells is developed from 34 genetically determined primordial melanoblasts. Hair follicles are composed of 150–200 clones which originate from somites. Photoreceptor cells of the retina are formed from 20 clones. Reproductive cells have two–nine ancestral cells. A number of other cells and tissues in chimeric mice are composed of a certain number of clones (at least two). These include the liver, kidney, brain, erythrocytes, γ-globulin producing cells and others (Mintz, 1971). Also, while the separate somite is not a clone, it originates from several cells. Each myotome which is the precursor for an eye muscle is also not a clone but, rather, is formed from two–five precursor cells (Gearhart and Mintz, 1972). Proximal kidney tubules are derived from four–five primary cells (Tettenborn et al., 1971). Study of nerve tissue histogenesis in chimeric mice has demonstrated that the Purkinje cells of the cerebellum are formed from many cells (Dewey et al., 1976). The process of determination in these cells begins when there is a large enough number of cells in the primordium.

A typical example of such events concerns the determination of the cell initiators in the retina. Chimeric mice were used. These mice were obtained from the fusion of blastomeres with a genotype for normal vision $(+/+)$ and a genotype for degenerative retina (rd/rd). Mice with the rd/rd genotype have retinal cells which differentiate during embryogenesis but degenerate during the postnatal period. The localization of degenerated areas was used to perform clonal mapping (Fig. 74). The results of this investigation demonstrated that the retina of each eye is derived from ten cell clones (Mintz and Sanyal, 1970; Mintz, 1971, 1974).

Morphogenesis of chimeric animals, however, is apparently more sophisticated and is accompanied by various cell mass migrations and competition of cell clones. Tissue-specific proliferation may also take place. For example C3H-type cells predominate in the liver of the allophenic C3H↔AKR mice. This predominance is due to C3H hepatomes while AKR-type cells form the blood cells. AKR mice have a high incidence of leukemia. As a result, the chimeric mice also have AKR leukemia (Mintz, 1972). Selection against these cells was demonstrated – especially in the blood primordium of mosaic mice – with the aid of the microinjection of teratocarcinoma cells which were marked by a hypoxanthin phosphoribosyltransferase (HPRT) deficiency (Dewey et al., 1977). Two fibroblast populations – one with active HPRT and one deficient in this enzyme – were observed in patients heterozygous for HPRT deficiency (Migeon et al., 1968). These same patients, however, have normal erythrocytes (Nyhan et al., 1970).

Significant variability in the proportions of C3H and C57BL/6 cells in allophenic mice was observed during the investigation of the clonal growth model of the differentiation of the spinal column and skull parts. The cells of C57BL/6 mice predominated in the skull bones while C3H cells comprised most of the spine – especially the lumbar-sacral regions (Moore and Mintz, 1972).

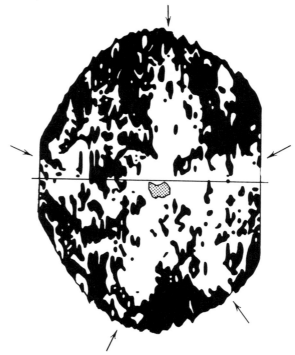

Fig. 74. Clonal origins of the photoreceptor cell layer of the mouse retina. The drawing is a two-dimensional reconstruction from serial sections of the retina of an allophenic mouse with approximately equal numbers of normal (+/+) and retinal degeneration strain (*rd/rd*) cells. The development, followed by death, of the *rd/rd* cells leaves null areas (shown in *black*) on which clonal mapping is based. A stellate pattern (*arrows* point to five +/+ components) suggests that the visual cells of each eye are derived from only ten clonal initiator cells that proliferate radially from a small circle. (After Mintz, 1971)

Cell selection was demonstrated for hair development in +/+↔Gs/Gs chimeric mice. Gs is the sex-linked semidominant mutation which affects the morphology and color of the hair. Phenotypic similarity of the hair of allophenic and regular Gs/+ heterozygous mice was obtained. This observation suggests that only one X chromosome is active in hair cells (Dunn, 1972).

Some investigators, however, have found differences in expression of morphological traits between chimeric Ta/Ta↔+/+ and Ta/+ heterozygous females (McLaren et al., 1973). (Ta is the semidominant mutation which causes the absence of hair on certain body parts, the modification of agouti color, respiratory defects, and a reduction in the aperture of the eyelid). Nesbitt (1974) has suggested that the differences in pigment distribution between heterozygous and chimeric mice are explained by a limited coherent clonal growth in developing mouse embryos.

A cytogenic analysis of mitosis in AKR↔CBA/H-T6 chimerics illustrated the possible role of the rate of cell division and also the rate of selective mortality during morphogenesis (Barnes et al., 1974).

A mathematical theory of clonal interactions during growth was formulated by Lewis and his co-workers (Lewis, 1973; Lewis et al., 1973).

Drosophila chimers have been obtained by transplanting the nuclei from one type of embryos into a phenotypically different embryos (Okada et al., 1974a, b).

Interesting data have been obtained from investigations of the genetic mechanisms of pigment formation in mice. It is known that mice homozygous for the recessive agouti allele have black hair. The dominant allele, on the other hand, affects yellow pigment formation on each hair. The dominant phenotype is brown. Genes that determine pigment synthesis are active in the melanocyte. The agouti properties, however, are dependent upon the hair follicles (Silvers and Russell, 1955). It was shown that the pigmentation pattern depends upon the interaction of two cell populations. Pigment synthesis depends only on the melanocyte genotype, whereas pigment distribution in the tissue depends on the surrounding cells, too. This was noted when the chimers of two mouse lines had different pigment synthesis and distributions (McLaren, 1973).

Approximate numbers of initiator cells were established for many organs of *D. melanogaster* with the aid of somatic crossing-over which was induced by X-irradiation. Antennae develop from 9 initiator cells (Postlethwait and Schneiderman, 1971). For tergites, the minimal number of initiator cells is 8 (Garcia-Bellido and Merriam, 1971). For legs 20 cells are necessary (Bryant and Schneiderman, 1969). Either 10 (Bryant, 1970) or 12 cells (Garcia-Bellido and Merriam, 1969) are necessary for wings. The eye-antennae discs require 23 cells (Garcia-Bellido and Merriam, 1969), and the mesothorax needs 8 or 16 cells (Stern, 1968).

The clonal character of cell differentiation has been demonstrated for other insects as well (Table 34).

Table 34. Initial cell numbers and clonal patterns in insects. (After Postlethwait and Schneiderman, 1971)

Order, Species		Imaginal structures					
		Leg	Antenna	Wing	Tergite	Eye	Genitalia
Diptera							
D. melanogaster	Initial cell number	20; 2	9	20; 12	12	20	2
(fruit fly)	Clonal pattern	Ls[a]	Ls	Ls	Ts[b]	S[c]	—
Hymenoptera							
Habrobracon	Initial cell number	2	2	2	—	2	—
juglandis (wasp)	Clonal pattern	Ls	Ls			S	—
Apis mellifera	Initial cell number	2	—	—	—	8	—
(honey bee)	Clonal pattern	Ls	Ls			S	—
Lepidoptera							
Golias sp.	Initial cell number	—	—	5	—	—	2
(butterfly)	Clonal pattern	—	—	Ls	—	—	—
Eacles imperialis	Initial cell number	—	2	—	—	—	—
(moth)	Clonal pattern	—	Ls	—	—	—	—

[a] Longitudial strips; [b] transeeted strips; [c] strips

It has been suggested that the realization of the inherited program in insects is accompanied by the successive formation of compartments and subcompartments: regions which are formed by polyclones (Garcia-Bellido et al., 1973, 1976; Crick and Lawrence, 1975). This process begins in the blastoderm stage (approximately 4000 cells) when the borders of compartments begin to be discernable. These compartments are the initiators of cell systems which develop in different directions.

The analysis of compartmentalization in *Drosophila* has most commonly been done in genetic mosaics which resulted from the induction of somatic crossing-over. The Minute mutation has been useful in such experiments. The M^+/M^+ cell has a greater growth rate than the M/M^+ cell (Garcia-Bellido et al., 1976).

On the basis of such experiments, the specific sequence of compartment formation in the wing imaginal disc has been determined:
1. anterior–posterior between 3 and 10 h of development
2. dorsal–ventral at about 30–40 h of development
3. wing-thorax at about 30–40 h
4. scutum–scutellum and postscutum–postscutellum at about 70 h
5. proximal–distal wing at about 96 h (Lawrence and Morata, 1977).

An important discovery was finding the compartment for aldehydeoxidase in the imaginal discs of *Drosophila* and other *Diptera* (Janning, 1973, 1974; Gehring, 1976a; Kuhn and Cunningham, 1977; Sprey, 1977). The analysis of enzyme distribution in the wing imaginal discs of *D. melanogaster* third instar larvae showed zones which possibly included dorsal and ventral spaces of the wing and the proximal and distal wing. When homozygous, the engrailed (en) mutation changes the rearrangement of aldehydeoxidase (Kuhn and Cunningham, 1977).

Homeotic mutations, which cause the replacement of one body part with another (see Ouweneel, 1976) often transform whole compartments. This means that the activity of the homeotic genes is restricted by the specific polyclones, i.e., by the cell groups that compose the compartment. The activity of the homeotic genes determines the appearance of certain aspects of determination. It is feasible to suppose that the compartment is the unit of determination. The final determination of cells which are ancestors of the compartment is the result of a number of events. Each determinational event may be accompanied by the activation of gene selectors. Some of these events are common for various polyclones, but the combination of active and nonactive gene selectors is unique for each compartment (Morata and Lawrence, 1977). Kauffman et al. (1978) have postulated a single biochemical system containing two components with concentrations $X(r,t)$ and $Y(r,t)$ at position r and time t which are being synthesized and destroyed at rates $F(x,y)$ and $G(x,y)$. Both components are diffusing throughout the tissues. Differences among compartments were explained with the presence of an "on–off" system which could control developmental specificity. This model can be used to explain determination and transdetermination.

The tissue level of control of biochemical events, on the other hand, is due to the regulation of cell numbers in the tissue. For example, it is known that triploid chickens have erythrocytes which are one and a half times larger than those in diploid chickens. The RNA and DNA content of the triploid erythrocyte is also higher than that in the diploid erythrocytes. The concentration of hemoglobin in the blood, however, is the same in triploids and diploids because the triploids have a reduced number of erythrocytes. Thus, there is a homeostatic mechanism for maintaining hemoglobin concentration at a constant level, regardless of triploidy and erythrocyte sizes (Abdel-Hameed, 1972).

Apparently, the same mechanism is involved in the regulation of blood hemoglobin and muscle enzyme concentrations in the tetraploid amphibian *Odontophrynus americanus*. These parameters are similar to those of the diploid *O. cultripes*. It was initially thought that the extra genomes in the tetraploid are inactivated. This, however, has been shown not to be the case (Beçak and Pueyo, 1970; Beçak and Goissis, 1971).

2.2.2 The Genetic Basis of Inductive Interactions of Embryonic Layers During Early Embryogenesis

The interaction of cell systems which are differentiating in different directions can be considered as a tissue level of inherited information. Together with the purely physical interactions of the embryonic layers, there is an interchange of macromolecules between these layers. These molecules diffuse from one primordium to another and cause specific effects (RNA and protein synthesis, regional differentiation, etc.). The essence of embryonic induction consists in such mutual complex spatial–temporal interactions of inductor tissue (for instance, chordomesoderm) and competent tissue (for instance, ectoderm). This induction causes a specific regional effect. Disturbances of morphogenetic processes could therefore be caused by mutations in either inducing systems or competent tissues. For example, mutations are known in mice which cause a disturbance in neuroembryogenesis by affecting chordomesoderm differentiation. There are primary defects in neuroectoderm as well. There is a T locus on the ninth linkage group. Of the numerous alleles which comprise this locus, several are lethal (Dobrovolskaia-Zavadskaia and Kobozieff, 1927; Dunn and Gluecksohn-Waelsch, 1952; Bennett, 1975). Mice with certain of these alleles have a greatly reduced tail or none at all. The viability of animals with different genotypes of this locus are given in Table 35. The T/T homozygotes die in the uterus on the 10th or 11th day of development. The t^o/t^o homozygotes die earlier than this (on the 5th–6th day of embryogenesis – i.e., shortly after embryo implantation). The t^1/t^1 homozygote even dies before implantation. The t^{12}/t^{12} homozygous embryos terminate their development on the 4th day when it is an undifferentiated embryonic primordium which is surrounded by the trophoblast (Smith, 1956). At the same time, t^o/t^1 heterozygous embryos are viable and develop into mice with normal tails. Nevertheless, some of these individuals have head abnormalities which are lethal.

The frequency of T locus mutations is high: approximately 11 of 3500 gametes were mutant here (Hadorn, 1961). These mutations were found in both laboratory strains and natural populations. Some workers suggest the T locus is a series of pseudoalleles (Lewis, 1951). A detailed investigation of these mutations has shown an interrelationship between mice notochord and the presumptive nerve plate. This interrelationship is similar to the interaction of corresponding embryonic layers in

Table 35. Phenotypic effect of various alleles of T locus in mouse. (After Dunn and Glueksohn-Waelsh, 1952)

Genotype	Viability	Phenotype
$T/+$	Viable	Short tail, brachiuria
$t^{na}/+$	Viable	Normal
T/t^n	Viable	Tailless, anuria
T/T	Lethal	Died, 11th day of embryogenesis
t^0/t^0	Lethal	Died, 6th day of embryogenesis
t^1/t^1	Lethal	Died, 5th day of embryogenesis
t^4/t^4	Lethal	Died, 8th day of embryogenesis
t^9/t^9	Lethal	Died, 9th day of embryogenesis
t^{12}/t^{12}	Lethal	Died, 4th day of embryogenesis
t^3/t^3 (t^7, t^8, t^{11})	Viable	Normal
t^0/t^1	Semilethal	Normal

[a] A lethal mutation

the amphibian. It was demonstrated that there is an abnormal development of inducing tissue (chordomesoderm) in mice with T group mutations. The T/T embryos do not differ from T/+ and +/+ embryos prior to notochord formation. However, by the 4–8 somite stage (8.5th–9th day of pregnancy), the following changes were noted:

1. the appearance of double and single vesicles near the body medial line;
2. abnormal neural tube development;
3. somite expression is lower than in normal embryos.

The vesicles in 1. above have disappeared in stage 10–12 somites (9th day of pregnancy). The abnormalities of notochord and development of the caudal part of the neural tube are more noticeable: these structures failed to open and on the 11th day the embryo died. Grüneberg (1958) pointed out that it was impossible to identify the notochord in T/T embryos. In the rare cases when it was identifiable, the notochord was represented by a plate of degenerative cells which were not separate from nerve tissue.

The chord in T/+ embryos fails to reach the caudal region (neural tube and/or caudal part of the archenteron). As in normal mice, the chord is completely separated from the neural plate (or neural tube) and from the closely associated mesoderm. The growing end of the chord in T/+ embryos, however, was longer and bulkier than normal; the base, which is linked with the neural tube, is wider than normal; the ventral zone is sharper (Grüneberg, 1958).

Ephrussi has investigated tissue culture cells from T/T embryos. These cells were isolated 6–12 h before embryonic death. The cells had high viability and had been differentiated into fibroblasts, myoblasts, and undetermined epithelium. All these cells proliferated rapidly and it is possible that after a long period of time, they were similar to normal cells. The cells from the lethal embryos also demonstrated a normal ability to differentiate. Cartilage formation was obtained even from cells at the end of the embryo where embryogenesis appears especially abnormal. This was done in vivo. Therefore, the lethal effect of the mutant genes is not manifested in vitro. It was suggested that embryonic mortality is caused by some toxic effect of maternal factors. Cultivation in a number of different plasmas, however, demonstrated that maternal plasma does not in any way affect "lethal" or normal tissues. Neither did uterine extract have any influence. Apparently, the abnormalities of embryo development are caused by disorders in the interaction between the inductor and the competent tissue (Ephrussi, 1935).

It has been suggested that in some cases the T locus mutations are manifest as distortions of the surface properties of the chord and, as a result, the chord tends to remain in the T/t and T/+ primordial tissue for nervous system formation. Structures close to the chord (neural tube and intestine) did not demonstrate any abnormalities (except where caused by the behavior of the chord). It cannot be excluded that the first manifestation of T-mutations occurs in primitive streak formation. The serious distortions of chord development are not seen until a later stage. This effect is more obvious in the case of the Sd (short-tail) mutation in both homozygous and heterozygous conditions (Grüneberg, 1958). This mutation is located on the fifth chromosome (Dunn et al., 1940). Abnormalities of the tail, kidney, and skeleton, and transformation of the neural plate into the neural tube are characteristics of this mutation. The regressive processes are accompanied by nuclear picnosis and by degenerative changes of blood vessels. The Sd/Sd homozygous mice died by 24 h after birth. In Sd/+ and especially in Sd/Sd embryos, the chordal growth zone was significantly reduced and disintegration was observed. There is reason to believe that the Sd mutation begins to exert its effect during the primitive streak stage – before the segregation of the chordal primordium. Abnormalities in chordal segregation from the mesoderm demonstrated that the disorders are not restricted by the chord but, instead, are mostly the result of changing behavior of the near axial mesodermal cells. The distortion of neuropore development, spina bifida formation and the appearance of the abnormalities can be explained with the scheme of Fig. 75 (Grüneberg, 1958).

Comparison of T and Sd mutants shows that in Sd/+ and Sd/Sd embryos, cell picnosis occurs significantly later than the structural abnormalities in chord development (Grüneberg, 1958).

Some mutants demonstrated a disbalance between nerve system differentiation and mesodermal cell elements. In t^9/t^9 homozygotes, for instance, the nerve tissue is hypertrophied in early embryogenesis at the expense of mesoderm (Gluecksohn-Waelsch, 1955, 1965). Similar abnormalities were demonstrated for another mutant – kinky tail – which causes a loop-shaped tail and which is located on the ninth chromosome. Shortly after implantation (7th day after fertilization), the Ki/Ki embryos can be easily defined by the presence of an extra embryonic axis and hyperplasia of embryonic tissue. In later stages (8th day of embryogenesis), the duplication of embryonic axes and hyperplasia of neuroectoderm and mesenchyma are clearly noticeable. It is sometimes possible to see more than two embryonic axes. A characteristic of Ki/Ki homozygotes is the formation of a spherical mass of unorganized tissue (Bauchstück). Duplications of various organs are easily noticeable on the 9th–10th day of development. The head and neural fold distortions are also noticeable. The rate of Ki/Ki embryo development is not different from normal until the appearance of duplications. Once these appear, development slows down. Thus, the T/T, t^o/t^o and Ki/Ki mutants all have something common in embryogenesis; t^o/t^o embryos have serious distortions in mesodermal development; Ki/Ki mutants cause a disorder in the interaction between inductor and competent tissue.

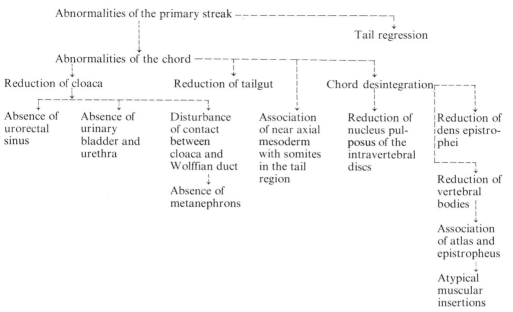

Fig. 75. Effect of abnormalities of the primary streak on embryogenesis. (After Grüneberg, 1958)

In all of these cases, the chordomesodermal primordium – the tissue of primary organization — is damaged. Therefore, only one short region of the ninth chromosome can control the development of the whole embryonic regulator system, i.e., the system of the primary inductor (Gluecksohn-Schoenheimer, 1949).

A similar case of control in mesodermal and neuroectodermal embryogenesis has been described for *D. melanogaster* (Poulson, 1968). In *Drosophila* embryos, a deficiency of a small section of the X chromosome in the Facet and Notch region increases the formation of neuroblast cells from abdominal and head ectoderm. This process occurs at the expense of the skin cells and other ectodermal derivatives. As a result, these embryos develop a nervous system which is three times larger than normal. On the other hand, the mesoderm remained underdeveloped and the muscles, heart, and fat body did not develop totally. Poulson demonstrated, also, that various Notch (N) factors which are lethal in homozygotes are associated with different length deficiencies (Fig. 76). The final result of development, however, is the same regardless of the number of missing bands (in the 261-38 deficiency, 45 bands are absent; in 264-19, only 1 band is absent). Deficiencies that did not include the Notch locus (3C7) had various effects on development (Table 36; Poulson, 1940). Poulson later investigated abnormalities in Notch mutants with dominant imaginal effects but recessive lethal effects (N^{55e11}, N^{264-40}, $N^{264-103}$, N^{60}, N^{60g11}). These mutants were tested in all hemizygous, homozygous, and heterozygous combinations with each other. Neuroblastic primordium abnormalities were identical in hemizygotes and heterozygotes. Complementation was not obtained in heterozygous combinations. Apparently, all five sites which are located along the X chromosome are important for normal embryogenesis (Poulson, 1968).

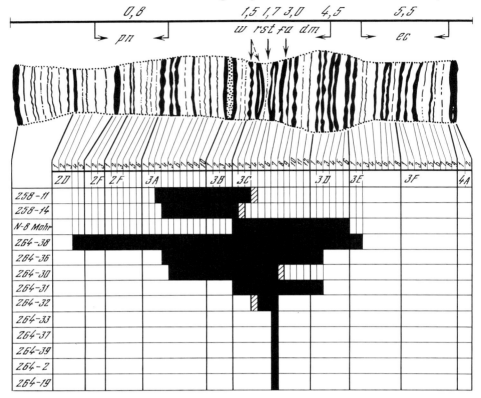

Fig. 76. Cytological and genetic map of left segment of X chromosome *D. melanogaster*. (After Hadorn, 1961)

Table 36. Summary of effects of the series of X chromosome deficiencies *D. melanogaster*. (After Poulson, 1940)

Deficiency	Size	Time of action	Effects
Nullo-X	Entire X	1st h	Abnormal distribution of nuclei; no blastoderm
Half-X	Half of X		
1st type	XR	1st–2nd h	Incomplete blastoderm
2nd type	XL	2nd–3rd h	Fails to separate germ layers; no differentiation
Notch-8	Includes bands 3 C 1–3 D 6	6th h on	Nervous system fails to separate from rest of ectoderm and hypertrophics. Mesodermal organs lacking. Fore-gut rudimentary; mid-gut incomplete
Scute-8	y Hw ac missing at tip of X	Late embryonic stages	Nearly complete larva; no air in tracheae; little or no muscular movement
White	Missing 1–16 bands	12th h on	Ectoderm ± normal, abnormal mesoderm and endoderm

The described cases all deal with abnormalities of neuroembryogenesis. These abnormalities involve underlying abnormalities in the development of chordomesoderm primordium (inductor) or disorders of its interaction with competent tissue. At the same time, there are mutations which affect primarily the competent tissues. The dominant mutation loop-tail, with incomplete penetrance (71%) in the A-Strong mouse line, belongs apparently to this group of mutations. Such mice have a delay in the transformation of the nerve plate into the neural tube. There are also defects in the development of various parts of the brain. However, chord and mesodermal structures develop normally. No deviations from normal were seen in the development of the optic, acoustic, or olfactory centers, or in the cranial and spinal ganglia. These observations suggest that this gene effect, on one hand, affects neural epithelium and, on the other hand, is very local (Stein and Rudin, 1953).

Similar distortions in neuroectoderm development were found in mouse embryos homozygous for the splotch (with spina bifida) mutation. These mice die in the uterus and are characterized by accelerated neuroepithelial growth. In this case, however, it is not clear if the effect of this gene is a primary or secondary character. The most intensive acceleration of nerve tissue growth occurs on the 13.5th day of embryogenesis. Maximal mitotic activity occurs on the 14.5th day. In the loop-tail mutation, neuroepithelial mitotic rates are maximal on the 9th–10th day of embryogenesis (Hsu and Van Dyke, 1948). Extraproliferation of nerve tissue during human embryogenesis has also been described (Patten, 1959; Diban, 1952). The causes, however, are not clear, although there is evidence they could have a primary character and are not associated with developmental abnormalities of chordomesoderm (Diban, 1952). Only recently, molecular biological and submicroscopical investigations of such events during early organogenesis were started. DNA was extracted from the spleen cells of mouse hybrids ($t^{12} \times$ C57BL) and from C57BL homozygotes. This DNA was hybridized with labeled rRNA extracted from the spleen of C57BL mice. The level of saturation of DNA from $+/t^{12}$ heterozygotes with RNA from $+/+$ animals was 30% lower than for saturation of DNA from $+/+$ mice. It was suggested that $+/t^{12}$ heterozygotes have a deletion in the loci which are responsible for rRNA synthesis – i.e., in the nucleolar organizer region (Klein and Raska, 1968). These data, however, have not been confirmed (Hillmann and Tasca, 1973; Erickson et al., 1974). At the same time, evidence of abnormal lipid production and utilization has been contained in the two-cell embryonic stage in t^{12}/t^{12} organisms. The appearance of cells with two nuclei could be a result of such abnormalities (Hillman and Hillman, 1975). Defects of aerobic metabolism were also noted. This could result in nonphysiological levels of intermediary metabolites. Cytoplasmic lipid filled cells. Mitochondria were often abnormal and contained crystals (Ginsberg and Hillman, 1975; Hillman, 1975). Similar defects of lipid metabolism and ultramicroscopic structure were obtained for t^0/t^0 homozygotes (Nadijcka and Hillman, 1975). In t^9/t^9 homozygotes which had reached further stages of development, abnormalities of intercellular interactions were demonstrated. These are apparently caused by changes in the pattern of surface-cell antigens (Spiegelman and Bennett, 1974; Bennett, 1975; Hillman, 1975). It was suggested that these morphological defects are the result of increased ATP concentrations and, hence, increased ATP metabolism. The t^{12}/t^{12} and t^{W32}/t^{W32}

embryos had an increased level of ATP metabolism until the termination of development. After this, the ATP usage dropped lower than controls and remained at this low level until its death (Hillman, 1975). It was also suggested that the doubling of Ki/Ki embryos could be caused by an abnormality in RNA distribution and by the formation of a double RNA gradient similar to that which was obtained during centrifugation of amphibian eggs. Cytochemical methods, however, did not demonstrate such a gradient (Gluecksohn-Schoenheimer, 1949).

Thus, at the present time a number of mutations are known which cause defects of early neuroembryogenesis on the levels of: (1) inductors, (2) competent tissue-neuroectoderm, and (3) interaction of inductor and competent tissue. However, the fine biochemical events at the cellular level and the resulting morphological expression at the tissue level are unknown.

2.2.3 Significant Time-Factors in the Genetic Regulation of Ontogenesis

As mentioned previously, during induction there is a spatial–temporal coordination of processes in both the induction and competent tissue systems. This coordination is the result of stabilizing selection (Smalgausen, 1938). Disturbances of this coordination modify the development of corresponding morphogenetic processes.

At present, many temporal patterns of genetic regulation of morphogenesis are known (Goldschmidt, 1961). For example, differences in behavior between strains of laboratory animals in some cases are the result of differences in the rates of neurochemical accumulation during development (King, 1967; Sviridov et al., 1971). The pattern of amphibian pigmentation is caused by the interaction of epidermis (which acts as an inductor) with tissue of the neural crest. The neural crest produces melanoblasts which migrate subepidermally under the influence of the inductor. There is a mutation (d) in *Ambystoma mexicanum* which, when homozygous, produces animals that are colorless (the so called white race of axolotl).

It appears in this case that the absence of color is due to the lack of coordination in time of the maturation of two interacting primordiums. It is not caused by the loss of inducing ability by the epidermis or by the inability of the melanoblasts to migrate or synthesize pigment. Regional pigmentation was demonstrated when pieces of the presumptive epidermis were transplanted from one colorless to another colorless embryos of axolotl when two individuals were different in age (Bogomolova and Korochkin, 1973). A change in rRNA synthesis is apparently the basis for the asynchronic differentiation, for the white race of axolotl was found to contain a deletion in the nucleolar organizer (Callan, 1966). It was suggested that this deletion results in a slow biochemical maturation of the presumptive epidermis in white animals. The synthesis of dehydrogenases occurs much later here than in the black race.

An example of a desintegration of such correlated systems is domestication. This was demonstrated by Smalgausen (1938) and then by Belyaev (1970, 1972). The correlated system has a number of imbalances. The functioning of the

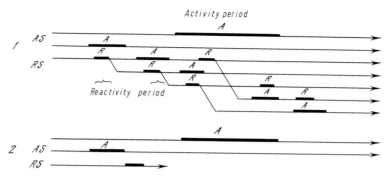

Fig. 77. Organ reduction during heterochrony in the inducible system. *1*, normal development of ancestral organ: *A*, activator; *R*, reactor; *AS*, activated system; *RS*, reacted system; *2*, organ rudimentation in descent as a result of heterochrony (delay of the reactor development) in the main inducible system. (After Smalgausen, 1938)

system, however, is not essential to life. The distribution of color spots in domestic animals (cow, dog, cat, etc.), for example, has an irregular character which cannot be found in wild animals. These animals have either one color or a regular alternation of colors (stripes or spots). It should be mentioned that the one-color type of coloration is established by a very complex genetic system. This system, however, can be out of balance. Belyaev pointed out a number of consequences of fox domestication: aberrant coloration and the taking-on of certain doglike characteristics. These do not occur in wild fox. Apparently, domestication in the fox leads to a number of morphological and physiological changes. These changes reflect a destabilization in the correlated system of ontogenesis (Smalgausen, 1938). Mutations which are phenotypically manifested during animal domestication acts at the level of the correlated interactions. (These mutations accumulated in natural populations, but it have not dispersed due to its low viability under the pressures of natural selection.) As a result of this mutation, previous patterns of correlation are broken up and new patterns established. The development of the crest and feathers on chicken legs and the fatty tail in sheep is principally the result of new correlations. The establishment of new traits (differentiation) is accompanied by the establishment of new correlations (integration). Smalgausen considered organ reduction as local disintegration of a correlated system and atavism as a local reintegration (Fig. 77; Smalgausen, 1938). These events are the consequence of displacing morphogenetic reactions in time.

What is the possible genetic basis for events which are characterized by the appearance of variability during morphogenesis? Apparently, the selection of mutants which regulate the spatial and especially the temporal nature of interactions of presumptive tissues can be such a mechanism. Suppose that there are two genes in a developing system, A_1 and A_2 (which can be allelic or nonallelic). These genes control the synthesis of a_1 and a_2 products. These products are responsible for morphogenetic reactions a_1^1 and a_2^2 respectively. If there is a multilevel control of trait manifestation, then the transcription of the A_1 and A_2 loci does not mean that the

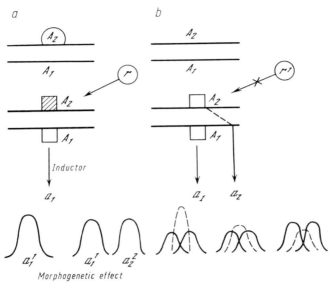

Fig. 78a, b. Hypothetical scheme of the effect of regulatory gene mutation on the morpho-genetic processes. **a** Normal morphogenesis; **b** variability of morphogenesis in mutant individual

traits will be expressed as part of the final phenotype. As discussed above, there are numerous genetic elements which can suppress trait expression at any step.

In addition, suppose that the morphogenetic reaction (Fig. 78a) which is controlled by the A_2 gene is not expressed as part of the phenotype because of its termination at either level of regulation: on the level of suppression of the a_2 synthesis or on the level of coordination of a_2 synthesis with time and the maturation of the reacting system. In this case, only the morphogenetic process associated with a_1 will occur.

Suppose that a mutation affects a regulator gene, r. This regulator matches the time of a_2 synthesis with the time of maturation of the system. As a result of this mutation, the A_2 gene activity is manifested in the phenotype (Fig. 78b). In this case, the a_2^2 event also occurs. If between the a_1^1 and a_2^2 reactions there is some interaction, then additional "intermediate" morphogenetic processes could also occur. Because numerous gene modifiers affect the expression of these reactions in the phenotype, the number of phenotypic variants can be almost endless (Korochkin, 1976).

The possibility that the r gene controls the level of a hormone in the developing organ should also be considered. In this case, an r gene mutation could play a significant role in hormonal balance. This balance is important for the regulation of certain traits. According to some (Salganik, 1968; Tomkins and Martin, 1970), hormones are genetic inductors. They have properties which allow them to effectively regulate differentiation and histogenesis (Mitskevich, 1966; Ohno and Lyon, 1970).

2.2.4 Interactions at the Cell and Tissue Levels in the Genetic Regulation of Ontogenesis

Although realization of inherited information at the tissue level is a principal new qualitative level of the regulatory process, it will be the result of interacting tissue primordia and cellular complexes. At the same time, however, there are periods during which the interacting primordia develop comparatively autonomously. For example, the autonomy of competent tissue maturation was described during primary embryonal induction (Holtfreter, 1948).

It is known that ectodermal cultures from salamanders can differentiate into neural and muscle tissues. For induction and histogenesis to occur, the tissues must be cultivated at least 30–60 h. Therefore, the neural, mesodermal, and ectodermal competencies of ectoderm begin to work successively and independently of surrounding tissue influences. Moreover, Flickinger (1952) has mentioned that the reacting tissue of the inducing system not only has an inherited mechanism to form various cell types, but also, a comparatively simple external stimulus (a pH change, for example) can initiate processes which are necessary for differentiation. In other words, the induction process itself cannot totally determine the specific nature of neural differentiation. Differentiation partially depends upon the metabolic processes that take place in the presumptive neuroectoderm (Flickinger, 1971).

Some degree of autonomy was described for the development of the inducing system also. Spinocaudal and mesodermal structures were induced when the dorsal lip of the amphibian blastopore was implanted into the blastocell. The embryonic tissue was obtained before invagination. The inducing effect changes once invagination has begun in that after invagination the upper region of the primary intestine cannot induce the development of the mesencephalic and spinocaudal structures; only prosencephalic structures were induced. If the inductor pieces are first cultivated in an explant and, after various time periods their effect is tested, it is found that the effect changes concordantly with in vivo tested inductors. Therefore, the changing of inducing effect of organizers during time is caused not only by invagination but also by its autonomic maturation (Saxen and Toivonen, 1963). Apparently, the neuralizing factors are formed later than mesodermalizing factors which are formed during oogenesis (Tiedemann, 1966).

Determination of competent tissues and inductors as well as differentiation occur at the same time as cell proliferation. This proliferation not only forms cell clones but also significantly regulates their interactions. Through proliferation, the cellular and tissue levels are related with each other. Actually, early embryogenesis is characterized by the formation and interaction of cell masses of certain volumes. These volumes are dependent upon different rates of cell division. In turn, DNA replication differs in various differentiating systems.

During the active synthesis of many specific RNA's there was an increase in the amount of early replicated DNA. Transcription of a small number of various RNA's, on the other hand, was accompanied by an increase in the amount of late replicated DNA (Table 37). Ectodermal cells from the early gastrula and cells of the dorsal axial region of tailbud have many chromosomal regions between telomeres which undergo DNA replication during the last quarter of the S phase.

Table 37. Comparison of metaphase plates from various parts of frog embryos at different developmental stages, characterized by the limit of terminal label (H^3-thymidine) or by heavy labeling of the chromosomal arms during the last quarter of S phase. (After Flickinger, 1971)

Development stage, embryo parts	Number of metaphase plates	Number of metaphase plates with limited terminal label	Number of metaphase plates with late heavy labeling	Percentages of metaphase plates with heavy labeling
Early gastrula, ectoderm	120	0	120	100
Neurula, neural plate-dorsal mesoderm	200	179	21	10
Neurula, ectoderm	127	82	45	35
Tailbud, dorsal axial ectoderm-mesoderm	200	46	154	77

The ends of many chromosomes begin to be replicated late in the early neurula. Conditions remain the same in the tailbud stage (Flickinger et al., 1967a).

Apparently, many DNA sequences along chromosomes are replicated during the last quarter of the 1.5 h S phase while in early gastrulation. During this period, most genes are still inactive. The future development of cells is not yet determined. Cells of the dorsal mesoderm (notochord and somites) and the ectoderm (neural tissue) are determined during gastrulation and early neurulation. There is late replicated DNA in the telomeres, but only a small part of the DNA between telomeres is replicated late in the 5 h–6 h S phase of the early neurula. Endodermal fate is determined during neurulation (Balinsky, 1965), but no differentiation has yet taken place. The notochord, neural tube, and somites are all differentiated in the tailbud stage (19th stage). In this region, numerous parts of late replicated DNA were found between telomeres. Thus, a lengthening of S phase and generation time between gastrulation and neurulation is accompanied by a transition from late to early S phase DNA replication in many DNA regions between telomeres. The telomeres, however, are always replicated late (Flickinger et al., 1967). The transition to early DNA replication between gastrulation and neurulation is correlated with an increase in the variety and amount of RNA (Brown and Littna, 1966; Denis, 1966).

Flickinger (1971) has suggested that the control of synthesis specificity depends more upon the number of cell divisions in a cell rather than upon the proliferation rate or length of the G_2 and S periods. In connection with this it is of interest to note Holtzer's ideas about the relationships between specific product synthesis, differentiation, and the number of cell generations. Holtzer distinguished between two mitotic types: proliferative and critical. The entry of a cell into a specific phase of differentiation depends upon the number of critical mitoses. Increasing the cell number without changing the phenotype is the result of proliferative mitoses. According to this hypothesis, two phases of differentiation can be distinguished. The first phase is the rapid transition of cells toward development in the early cleavage stages. In this respect, there are no significant differences between regulatory and

Fig. 79. A proposed cell lineage terminating in the chicken myoblast. Vertical arrows = proliferative cell cycles. Horizontal arrows = critical mitosis. (After Holtzer, 1970)

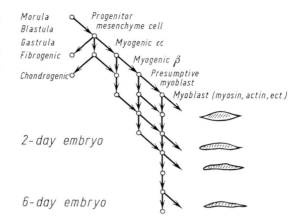

mosaic systems. The final result of early cleavage is the "definitive stem cell". This cell is able to produce daughter cells which can either add to the stem cell pool or represent the beginning of terminal differentiating cells. The second phase is the development of biological conditions which stimulate the growth and morphogenesis of systems derived from stable stem cells (Holtzer, 1970; Holtzer et al., 1972).

Mesenchyme cells go through four successive critical cell cycles (Fig. 79) which determine cell differentiation in the direction of myoblast development when the synthesis of specific muscular proteins start (Holtzer et al., 1972). It was suggested that every critical cycle derepressed part of the genome.

Alterations of DNA replication with bromodeoxyuridine (which randomly replaces thymidine with bromouracil) inhibits biochemical differentiation. For instance, the increase of creatine phosphokinase activity is reduced in differentiating muscles (Holtzer, 1970).

Various populations of neuroblasts have been established during nerve tissue development. These populations have a number of characteristics. The number of successive matrix cell divisions can be approximated beginning from the first presumptive cell:

$$Nn_{k+1} = [f(tn_1) + 1] [f(tn_2) + 1] \ldots [f(tn_k) + 1],$$

where Nn_{k+1} is the number of cells after n_{k+1} divisions;

$$f(n_{k+1}) = \frac{\text{proliferative pool}}{100} \text{ for } n_{k+1} \text{ cell division.}$$

In the tectum opticum of the chicken embryo, for example, 13 stem divisions have occured by the second day of incubation. This is a result of only eight–ten successive critical mitoses (i.e., asymmetrical and transformational mitoses) if one takes into account that neuroblasts are not formed until the 3rd–10th day of incubation.

The specificity of neuroblast differentiation which is derived from the matrix is determined by the number of previous mitoses. Thus, cells which have similar characteristics come from the same period of matrix multiplication. For example, long neurons are formed first, then neurons with short axons, and finally the small

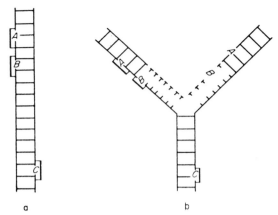

Fig. 80. Hypothetical scheme of the interaction of gene differential activity and their replication. (After Kaufman, 1967)

associative neurons are formed during embryogenesis of the central nervous system (Korochkin and Olenev, 1966; Olenev et al., 1968; Gratcheva, 1968).

It was demonstrated that two cycles of DNA synthesis and/or mitosis are necessary for dimethylsulfoxide initiation of hemoglobin synthesis in mouse erythrocytes infected with leukemia virus (McClintock and Papaconstantinou, 1974). Some workers have suggested that critical mitoses are necessary for the modification of nucleotides in the operator. This modification occurs with methylase enzymes. An $A\cdot T \rightleftarrows G\cdot C$ transition was assumed for operators during DNA replication. If cells have a specific methylase composition, this could influence operator function in a particular way and, as a result, differential genome activation could be achieved in various cell types (Holliday and Pugh, 1975). Kauffman has proposed a hypothesis where the rate of histone protein (which inhibits DNA transcription) synthesis falls behind the rate of DNA replication. As a result, there is a possibility of derepression of some loci in the daughter cells (Fig. 80; Kauffman, 1967). The data observed by DeRobertis and Gurdon (1977) do not prove Holtzer's idea. They investigated the expression of "silent" genes in nuclei transplanted from the kidney cells of *Xenopus laevis* into anucleated oocytes of the triton *Pleurodeles waltlii*. It was found that the oocyte cytoplasm contains factors which activate genes that were inactive during differentiation. Moreover, this gene activation was obtained without cell division.

A number of workers have demonstrated the role of cell division in the restriction of DNA template activity (Nakatsu et al., 1974). It was also suggested that the replicons in undifferentiated cells are not irreversibly inactivated. A few demonstrated the ability to synthesize RNA. All the DNA is replicated early and the absolute length of the S phase is relatively short. With cytodifferentiation, however, there is irreversible inactivation of many replicons. These begin to be replicated late and the length of the S phase becomes longer. Various tissues are characterized by a specific distribution of such late replicated regions of the genome (Fujita, 1974b).

Table 38. Duration of the cell cycle phases in the presumptive retina of the 10-day-old mouse embryo, in h. (After Konyukhov and Cazhina, 1970)

Genotype	$G_1 + M$	S	G_2	Total
$+/+$	2.5	6.5	1.0	10
or/or	8.1	5.7	1.2	15

An important role in these processes belongs to genes which control the length of the cell cycle. The activity of such genes affects crucial morphogenetic processes (Konyukhov and Sazhina, 1970). For example, in mouse embryos homozygous for the ocular retardation gene (or/or), the generation time (T) is 1.5 times greater. The G_1 and M phases in the retina primordium are three times greater than normal. It is possible that the length of T is determined by the length of G_1 but not M because the mitotic index in the or/or mutant is significantly lower than in normal embryos. Therefore, the or gene increases the length of the cell cycle and decreases the proliferative cell activity in the retinal primordium (Table 38). The localized result of the activity of this gene is a specific change in the structure of the retina primordium. In turn, this change can inhibit cell growth by influencing the biochemical processes which prepare the cells for DNA replication (Konyukhov, 1969, 1973).

Male rats carrying the H^{re} gene become sterile on the 90th day. It was suggested that this gene acts by activating adenylate cyclase to convert ATP to cyclic AMP (cAMP) in the Sertoli cells of the testis. The surplus of cAMP inhibits the mitotic divisions of the spermatogonia but stimulates meiosis in the remaining gametogenic cells. The final result will be a reduction in the number of sperm in the testis (Peeples and Ireland, 1974).

Gene interaction has an important role in cell division. For example, the effect of the microphtalmia (mi) gene is different from the or gene, but it exerts its action approximately the same time: 10 days after the beginning of pregnancy. This gene causes greater mitotic activity in the pigment layer of cells during eye formation. As a result, a thicker pigment layer is formed which distorts the eye morphogenesis, and the choroid fissura is not closed (Konyukhov, 1969). The effect of the or gene (reduced mitotic activity) was seen only in 12-day old embryos homozygous for or and mi (or/or, mi/mi). These double homozygotes also have a lower retinal growth and a thickening of part of the pigment layer compared with mi/mi embryos. Eyes of 20–30 day old or/or, mi/mi mice are more regularly shaped than in mi/mi mice and they have a better differentiated retina compared to or/or mice. Therefore, the structure of the eye in double homozygotes is partially normalized as a result of the mutual inhibition of the or and mi genes. The eye is, however, still abnormal (Konyuknov and Sazhina, 1966).

Genes that control growth and proliferation can be characterized by both temporal and organ specificity of action. For example, there are two alleles – bp^H (brachypodism-H) and cn (achondroplasia) – which cause abnormal collagen cell growth in mice. One of them – bp^H – is active during the embryonic stages of development whereas the other – cn – is active in the postembryonic period. In addition, the bp^H gene affects only the collagen-producing cells of the fibular bone primordium and does not affect the cells that produce collagen for the tibial (Konyukhov, 1969). Tibia growth, however, can be inhibited secondarily by inhibiting the transport of compounds from genetically damaged primordia of the thigh or peroneal bones or other embryonic tissues.

2.2.5 The Function of Genes in Terminal Cell Differentiation

Determined cell systems which have been formed during proliferation and complex morphogenetic events begin a four-phase differentiation process (Table 39; Wessels and Rutter, 1968):
1. predifferentiation, when an essential cell product is still absent and the cells are still actively dividing;
2. protodifferentiation, where a tissue-specific product has been synthesized for the first time and is in very small concentrations;
3. differentiation, when the cell undergoes its last cleavage and intensive synthesis of the tissue specific product takes place;
4. final phase differentiation, which results in partial or total genome inactivation and stabilization of mRNA (Rutter et al., 1968; Wessels and Rutter, 1968; Kafatos, 1972).
   Several tissue systems have been studied in detail and will be discussed below.
   A. Crystalline Lens. For this tissue, the various stages of differentiation are separated in space. Epithelial cells are in a stage of predifferentiation. Protodifferenti-

Table 39. General tendencies of the terminal differentiaton. (After Wessels and Rutter, 1968)

Predifferentiation	Protodifferentiation	Differentiation	Differentiated state
Cytological characteristics			
Cell cleavage, chromosomes are diffuse, basophylia	The first granules of the specific product, cell mitosis, rough endoplasmatic reticulum, basophylia	Last mitosis, big number of ribosomes, nucleus and nucleoli are large, rough endoplasmatic reticulum coloration turns on acidophylic	Acidophylia, number of ribosomes is decreased, nucleoli are small or absent, DNA is metabolically inactive, smooth endoplasmatic reticulum
Biochemical characteristics			
Specific synthesis is sensitive to DNA inhibitors, isozyme pattern is typical for nondifferentiated tissue	Appearance of specific product, sensitivity to actinomycine D	Changing of isozyme synthesis, beginning of intensive synthesis of specific product which is inhibited by actinimycine D	Isozyme pattern is characteristic for differentiated tissue, actinomycine stimulated specific synthesis, stabilization of RNA

ation in these cells is accompanied by the appearance of small amounts of α- and β-crystallines. Furthermore, the elongated cells are going through active differentiation and γ-crystalline synthesis begins. At this time, α-, β-, and γ-crystalline synthesis can be suppressed by actinomycin. The transition from LDH-5 to LDH-1 has already occurred. In the final differentiation stage in the fibroid cell zone, a stabilization of specific mRNA's can be correlated with a reduction of RNA synthesis. At this time, active crystalline and LDH-1 synthesis is taking place (Reeder and Bell, 1965; Reyer et al., 1966; Stewart and Papaconstantinou, 1966, 1967; Papaconstantinou, 1967). It has been suggested that crystalline synthesis is controlled by more than one gene (Zwaan, 1968) and that biochemical and morphogenetic processes are determined separately (Braverman and Katon, 1971)

Colchicine inhibition of morphological differentiation and cell division in the chicken embryo crystalline lens does not affect δ-crystalline synthesis and mRNA accumulation. The reduction of δ-crystalline synthesis rate during ontogenesis does not correlate with the reduction of cellular mRNA concentration (Beebe and Piatigorsky, 1977).

An attempt was made to clarify whether the selective increases of tissue-specific protein synthesis during cell specialization is a consequence of increased mRNA. Cultures of chicken crystalline lens epithelium were used to demonstrate that in vivo and in vitro differentiation of epithelial cells is accompanied by an accumulation of mRNA. A stimulation was observed between the 5th and 24th h of cultivation. The stimulation of protein synthesis can be explained by the increased synthesis and accumulation of mRNA (Milstone et al., 1976).

B. Pancreas. The role of intertissue and intercellular interactions in determination and differentiation were demonstrated using the pancreas as the model tissue. Mesoderm is necessary for the differentiation of pancreatic cells (in the predifferentiated condition) from the cell pool of the primary gut. This occurs on the 9th day of development in mice. Even very early primordia of this organ can differentiate in culture if the proper mesodermal elements are present. Mesodermal factors promote pancreatic acinar cell formation by increasing DNA synthesis and epithelial cell proliferation.

The dynamics of enzyme synthesis in differentiating rat exocrine cells is demonstrated in Fig. 81. A disproportionate increase in protein synthesis is seen. This increase is small during predifferentiation, but is increased more than 1000 times in protodifferentiation. Characteristically, a coordinated change in exocrine enzyme activity is absent during the transition from stage I to stage II. The differences in patterns of accumulation of various enzymes occur between the 15th and 19th days of embryogenesis. For example, trypsin and carboxypeptidase activities increase late compared with other enzymes, but lipase B activity could not be obtained until the 18th day. By this time, the activities of other enzymes are high. At the same time, however, the synthesis of a small group of enzymes is coordinated. This is true for lipase A and carboxypeptidase A, and trypsin and carboxypeptidase B – the activities of which change together (Grobstein, 1964; Wessels and Wilt, 1965; Rutter et al., 1968; Zwilling, 1968). It is clear that this process is controlled by a large number of coordinated genes. A scheme for the possible regulatory phases in the endocrine part of the pancreas is shown in Fig. 82 (Rutter et al., 1968; Wessels and Rutter, 1968).

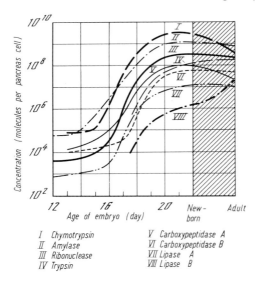

Fig. 81. Pattern of accumulation of the digestive enzymes is an index of exocrine-cell development. Extracts of pancreatic rudiments of various ages were prepared and assayed for enzyme content on the basis of specific catalytic activity. The midpoints between low and high levels differ for most of the proteins. Clearly they are not regulated as a single group. (After Wessels and Rutter, 1968)

I Chymotrypsin   V Carboxypeptidase A
II Amylase   VI Carboxypeptidase B
III Ribonuclease   VII Lipase A
IV Trypsin   VIII Lipase B

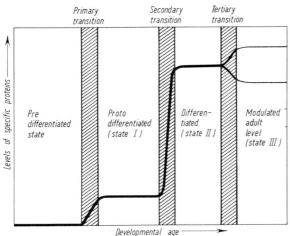

Fig. 82. Possible regulatory phases in cell differentiation are indicated by this schematic curve. The three stages of differentiation correspond to the numbered states associated with the exocrine enzyme levels. Each of them is introduced by a regulatory transition phase. (After Wessels and Rutter, 1968)

In the differentiation of the pancreas, the dynamics of mRNA synthesis correlate with the specific protein synthesis. A specific mRNA represents a small proportion of the total RNA pool of a rat embryo up to the 14th day. During the 14th–20th day of embryogenesis, the concentration of this specific mRNA increases. This increase is correlated with an increase in synthesis of specific organ proteins. At this time, the specific mRNA fraction represents approximately 2% of the total pancreatic RNA and this specific RNA is highly homogeneous. A second mRNA fraction is more heterogeneous but has an overlapping homology with the specific mRNA above. This heterogeneous fraction represents 0.03% of the total RNA pool of the pancreas (Harding et al., 1977).

Analogous tendencies were obtained for specific protein synthesis during the differentiation of other organs: cocoonase which is produced by specific cells of silkworm *Antherea poliphemus* (Kafatos, 1972), fibroine which is synthesized by the silk-producing glands of *Bombyx mori* (Jashiro et al., 1968), and ovalbumin which is produced by the chicken oviduct cells (Palmiter and Wrenn, 1971).

C. *Muscle*. Similar phases have been described for muscle cells during differentiation (Stockdale and Holtzer, 1961; Fieldman et al., 1964; Coleman and Coleman, 1968; Marchok and Wolff, 1968). During myogenesis, the replicating myoblasts withdraw from the mitotic cycle and fuse to form a multinucleated syncytium which is a myofibril. This process is accompanied by the synthesis of muscle structural proteins (Okazaki and Holtzer, 1966; Chi et al., 1975), and of membrane proteins associated with the contractile system (Prives and Paterson, 1974), and enzymes which are typically found in muscle (Easton and Reich, 1972). Undifferentiated myoblasts which remain in mitosis synthesize little myosin light chain. This synthesis increases significantly after cell fusion and the formation of the myotubules which will later differentiate into myofibrils. A poly-(A)-containing preparation of RNA was extracted from the dividing myoblasts and transcribed in a cell-free wheat germ extract. The RNA synthesized several polypeptides, of which myosin was one. The same RNA preparation from fused cells caused a higher rate of protein synthesis (Yablonka and Yaffe, 1977). Moalic and Padieu (1977) found that myoblast fusion in the newborn rat is accompanied (in 3-day-old cultures) by the synthesis of several poly-(A) RNA classes. There is a predominate class which sediments between 18S and 28S. This RNA class was identified as the one containing the mRNA for the heavy myosin chain.

Buckingham et al. (1977) demonstrated that a poly-(A)-containing 26S RNA which, according to its molecular weight was identified as myosin mRNA, was synthesized in undifferentiated myoblasts as well as differentiating myoblasts. This RNA was obtained from the heavy polysomal fraction of myotubules and a lighter (80S–120S) RNP fraction in myoblasts. The RNA, however, was not associated with ribosomes. According to Buckingham et al. (1976), poly-(A) RNA does not associate with polysomes and therefore is not long-living. This RNA assembles with ribosomes, forms polysomes and becomes stabilized before cell fusion when the cell withdraws from mitosis. Investigators have concluded that the mRNA for myosin and other muscular proteins preexist in myoblasts as a nontranslated fraction. This RNA accumulates during the stages which precede differentiation (Buckingham et al., 1977).

Paterson and Bishop, on the other hand, believe that the regulation of muscular proteins occurs at the level of transcription. They have examined in detail the population of mRNA molecules found in various myogenic stages. It was found that the prefusion cultures of myoblasts, fused myofibrillar cultures and cultures inhibited from fusing and undergoing myogenesis contain about 17,000 various mRNA sequences. About 20%–30% of the total mRNA was transcribed from the moderately repeated DNA fraction. The remainder of the RNA was transcribed from unique sequences. The 96 h myofibril culture contained 2500 sequences in high concentration and 6 sequences in very high concentration (each had about 15,000 copies per cell). It is possible that this mRNA is responsible for the synthesis of the specific muscular polypeptides, including myosin, actin, and tropomyosin.

D. *Erythropoiesis*. Before an erythrocyte is produced, a stem cell must divide and reduce in size and go through several stages: proerythroblast → basophilic erythroblast → polychromatophilic erythroblast → orthochromic erythroblast → reticulocyte → erythrocyte. The highest rate of hemoglobin synthesis occurs in the basophils. This rate is reduced in the polychromatophilic erythroblast. After this, the hemoglobin concentration reaches some critical value (13.5 µg per postmitotic cell); proliferation processes are terminated and cells begin terminal differentiation (Lajtha and Schofield, 1974). Although it is nonfunctional and shriveled, the nucleus remains in the fish, amphibian, and bird erythrocyte. Mammalian erythrocytes are without nuclei and their rate of production is about 20×10^{11} cells per day. The average life time for an erythrocyte in the peripheral blood system is 110 days. Thus, the average human produces 5×10^{15} erythrocytes.

Genetic factors have an important role in the regulation of ancestor cell proliferation. There are two mutations – W and sl – which cause the development of macrocyte anemia of mouse. Cultivation of ancestral cells from mice with macrocyte anemia in the spleen of healthy mice has shown that the W gene affects the number of cell colonies formed (there are less in anemic mice) while the sl gene affects the ability to maintain the growth of such colonies and does not affect the phenotypes of such cells (McCulloch et al., 1964). The stem cells rarely divided although they had a great potential for growth. No more than 32 cleavage divisions of the primary stem cells occur during embryo development. In postembryonic development there are no more than five or six such divisions. The unlimited ability of stem cells to grow is suppressed by specific inhibitors of mitosis (i.e., chalones) which circulate in this tissue. The reduction of inhibitor concentration causes an increased mitotic rate in the ancestral cells.

The differentiation of stem cells into erythroblasts is stimulated by the hormone erythropoietin which is produced in the kidney. A preliminary period of RNA synthesis is needed for the cell to be able to react to this hormone. At this time, the cell becomes sensitive to actinomycin. After this period, the cell can react to erythropoietin and becomes resistant to actinomycin. Apparently, by this time the specific cell receptor for erythropoietin has been formed (Baglioni, 1966). Once it acts on the cell, this hormone affects the genetic apparatus. Goldwasser (1966) has suggested two actions for erythropoietin: a repressor and a derepressor (Fig. 83). It should be noted that this hormone can change gene activity either directly or through the production of a mediator. Activation of the genetic system is expressed as an increase in RNA synthesis. This increase lasts a short time: before and during the stage of the basophilic erythroblast. Less than 300 molecules of globin mRNA are present in the proerythroidal nondifferentiated cell, whereas in the erythroidal cells the number of such molecules reaches 10^4.

The synthesis of mRNA terminates in the polychromatophilic erythroblast. Thus, no RNA is synthesized in reticulocytes and orthochromic erythroblasts. The mRNA synthesized in the basophilic erythroblast is stable and remains in the cell for long periods. This mRNA is active even in the reticulocyte stage. Protein synthesis starts after the termination of RNA synthesis in the polychromatophilic erythroblast. The rate of hemoglobin synthesis is maximal in polychromatophilic and orthochromic erythroblasts (Marx and Kovach, 1966). There is a correlation between the number of polysomes and the rate of hemoglobin synthesis (Table 40). This was demonstrated with the aid of electron microscopy and direct visualization of polysomes and separate ribosomes.

Fig. 83 a, b. Hypothetical interactions of erythropoietin with genes controlling the differentiation of erythrocyte precursors. **a** Erythropoietin *e* is a derepressor. Genes *J*, *K*, *L* which controlled the *j*, *k*, *l*, products synthesis were active before *e* action. Genes *F*, *G*, *H*, *I* are blocked by *u*, *v*, *w*, *x* repressors correspondingly. The erythropoietin repressor *x* complex derepressed gene *H*, etc. **b** Erythropoietin is a repressor. Genes *I*, *J*, *K*, *L* were active before its action. Gene *I* is inhibited by erythropoietin, etc. (After Goldwasser, 1966)

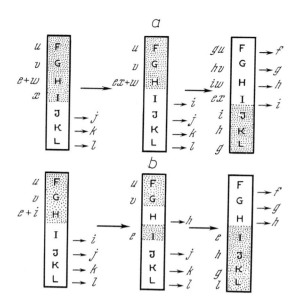

Table 40. The distribution of ribosomes as singles and polyribosomes during maturation of rabbit erythroid cells. (After Marx and Kovach, 1966)

Cell type	Total number ribosomes per standard area in 50 cells	% Ribosomes					
		Singles	Polyribosomes				
		1	2	3	4	5	6
Erythroblast	6730	1	11	22	43	16	7
Early reticulocyte	2775	22	25	29	18	4	3
Intermediate reticulocyte	1050	62	24	14	0	0	0
Late reticulocyte	350	58	39	3			

The total scheme of changes in RNA synthesis during erythropoiesis has been drawn up by Gazarjan (1972). In the first stage of erythroblast→reticulocyte transformation, the ribosomal genes and the loci upon which stable cytoplasmic RNA is synthesized are inactivated. The transcription of DNA regions that determine the structures of rapidly turned over RNA's still continues. In the next stage, some of these active regions are also inactivated (Fig. 84). This inactivation might be specific for histone F2C which is gradually replaced by histone F_1 at the terminal stages of erythropoiesis (Medvedev, 1972 a, b). Hemoglobin synthesis continues in the reticulocyte. The α-chain is translated in 140–150 s while the β-globin gene is translated in 200–230 s (Hunt et al., 1969). Globin mRNA has been extracted from erythrocytes and partially characterized. It is a 9–10 S RNA and is able to synthesize globin in a cell-free or Gurdon's system (Scherrer and Marcaud, 1968; Evans and Lingrel, 1969; Terada et al., 1971; Pemberton et al., 1972; Scherrer, 1973).

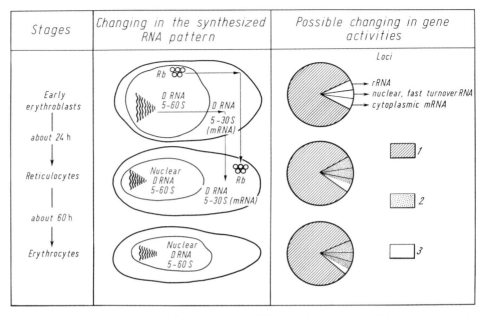

Fig. 84. The genetic control of the erythrocyte differentiation. *1*, loci which were repressed on stages before early erythroblast; *2*, loci which were repressed during differentiation erythroblasts into erythrocytes; *3*, active loci; *Rb*, ribosomes. (After Gazarjan, 1972)

The influence of actinomycin on the differentiation of erythroidal cells in the yolk sac of mouse embryos has been investigated. It was found that up to the 11th day of embryogenesis, a nonhemoglobin protein is synthesized. This phase is sensitive to actinomycin. Thus, a short living mRNA is responsible for the synthesis of this protein. After day 11, hemoglobin is synthesized. This synthesis is not sensitive to actinomycin treatment. The synthesis is, rather, associated with the stabilization of the mRNA template which controls hemoglobin molecule synthesis (Marx and Kovach, 1966; Wilt, 1967; Ryabov and Shostka, 1973).

The stability of globin mRNA is apparently correlated with the length of the poly-A tail. The reduction of tail length affects globin mRNA translation and termination of translation (Fantoni et al., 1976). The degradation of deadenylated mRNA depends upon the translation of the RNA and is accelerated by heme introduction (Marbaix et al., 1977). Translation is blocked in the early stages due to the absence of heme. The synthesis of heme depends upon the presence of the predecessor δ-aminolevulenic acid. Addition of this acid to cultures of early blastodiscs initiates hemoglobin synthesis, even if high concentrations of actinomycin are present. The synthesis of δ-aminolevulenic acid requires a specific synthetase. The template for this synthetase is transcribed before the first pair of somites is formed. The translation of this synthetase, however, does not begin immediately. It begins at the time when six somite pairs are formed. Before this point, synthesis can be blocked by actinomycin. After the six somite stage, hemoglobin synthesis begins regardless of the presence of high concentrations (100 mg/ml) of actinomycin, as long as tRNA's from late embryos are introduced into the medium containing, the blastodiscs. It was found that one of the minor isoacceptor fractions of alanine tRNA is responsible for this necessity. This fraction represents only 4% of the total tRNA fraction (Wainwright, 1971 a, b; Wainwright and Wainwright, 1972).

Heme will stimulate globin synthesis. It was found that an inhibitor of protein synthesis is formed in the reticulocyte lysate during the phosphorylation of proinhibitor. This phosphorylation is catalyzed by a cAMP-dependent protein kinase. Leucine prevents this transformation by blocking bond formation between cAMP and protein kinase. Thus, the protein kinase is inactivated by an allosteric mechanism (Datta et al., 1978).

The "hiding" of traits characteristic of erythroid differentiation (i.e., hemoglobin synthesis and the corresponding morphological changes) was obtained with dimethylsulfoxide (DMSO) treatment of somatic cell clone hybrids. It was found that treated hybrid cells synthesize the same amount of globin mRNA (9000 molecules/cell) as do parental cells. There is, however, no accumulation. Treatment of the hybrid cells with heme induces about 14% (1300 molecules) of the total globin synthesis. The treatment of such cells with heme and DMSO induces 25% (3000 molecules) of the globin mRNA synthesis. Also, differentiation was terminated. All the mRNA molecules are parts of polysomes and can therefore be translated. The incomplete differentiation of hybrid cells is due to a heme deficiency which restricts the expression of other coordinated functions of the erythroid cell (Harrison et al., 1977). An investigation of somatic hybrids of bone marrow and erythroleucosial cells has shown that the block of DMSO-induced accumulation of hemoglobin is connected with the presence of the X chromosome. This effect can be stopped by the addition of heme and δ-aminolevulinic acid, regardless of the increased concentration of globin mRNA in the cytoplasm. Apparently, the X chromosome acts at the level of inhibiting heme biosynthesis (Benoff-Rend et al., 1977). The interaction between cells and tissues is important in affecting the expression of biochemical traits. This is well illustrated with erythrocyte differentiation. Here, there is a replacement of fetal hemoglobin (Hb-F:$\alpha_2\gamma_2$) by adult hemoglobin (Hb:$\alpha_2\beta_2$) (Fig. 85). As previously mentioned, in amphibians this turnover is associated with

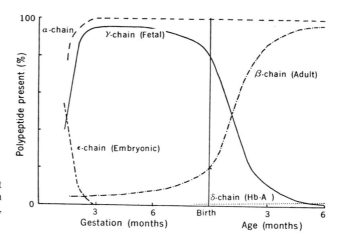

Fig. 85. The development of human hemoglobin chains. (After Kabat, 1972)

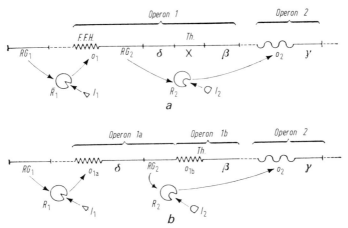

Fig. 86. Model of the regulation of hemoglobin chain synthesis. o, operator; RG, regulator-gene; R, repressor molecule synthesized under the control of RG; I, inducer; β, γ, δ, structural hemoglobin chain genes; X, unknown structural gene; $F.F.H.$, region of impact of mutation in familial fetal hemoglobinemia; $Th.$, possible regions of impact, discussed in the text, of the β-thalassemia mutation. (After Ingram, 1963)

the replacement of erythrocyte populations. Thus, there are two groups of stem cells in the primary erythroid cell pool. One of these groups (cells of the yolk sac) synthesizes only Hb-F while the other group (cells of the liver blood islets and bone marrow) synthesizes only adult hemoglobin (Maniatis and Ingram, 1971). The turnover of hemoglobin, however, is more complicated in birds and mammals (Niewisch et al., 1970; Vogel et al., 1970; Kabat, 1972; Ingram, 1972). In man and in the chicken (Fraser, 1964), erythrocytes were obtained which contained Hb-F and adult Hb simultaneously.

Therefore, two processes can be explained: (1) the formation of a heterogeneous erythroidal cell population according to the Hb type, and (2) the mechanism of rearrangement of cell composition during replacement of Hb-F by adult Hb.

Ingram (1963) and Baglioni (1964) have attempted to apply the Jacob and Monod scheme for gene regulation during erythropoesis. It was suggested that there is a gradual change in gene activities during this process. Thus, the transition from Hb-F to adult Hb can be explained at the cell level without considering the tissue level (Fig. 86).

This hypothesis has developed into a well-composed scheme of genetic regulation of hemoglobin synthesis by Zuckerkandl (1964) and Necheles et al. (1969). This scheme is based on the fact that the genes which control β and δ (adult Hb) and γ (Hb-F) chains are closely linked on the same chromosome. According to their hypothesis, a gene regulator (P_I) switches on operon I (with the β and δ genes). There is also a regulator P_{II} in operon I. This gene is stimulated by gene regulator P_I and turns off operon II which contains the γ-chain gene. The P_{II} product combines with an operon II structural gene product. This results in the turning-off of operon II.

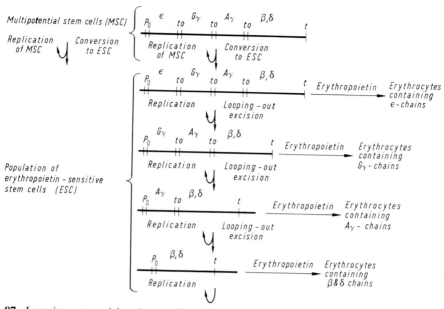

Fig. 87. Looping-out excision theory of erythropoiesis. Multipotential stem cells (*MSC*) exist in erythroid tissue and are converted occasionally into erythropoietin-sensitive stem cells (*ESC*) committed to erythroid differentiation. The ESC replicate rapidly and are converted by erythropoietin into terminally differentiating proerythroblasts. The proerythroblasts are precursors of the blood erythrocytes. The proposed chromosome structure for the human stem cells is indicated. *P* is the promoter locus and is the site of RNA polymerase attachment to the chromosome, whereas the t loci are the terminator sites at which transcription is terminated. The O sites are the operator loci and are derepressed in erythroid tissue. Only the promoter-proximal gene can be expressed because the more distal genes are separated from the promoter by a terminator. It is proposed that ESC occasionally undergo an intrachromosomal crossover event called looping-out excision which the promoter-proximal gene is excised from the chromosome as an acentric ring. The crossover occurs between the two homologous operator loci which are closest to the promoter locus and is analogous to the well-known excision of lysogenic bacteriophage λ DNA from the E. coli chromosome. The looping-out excision occurs independently on the two chromosomes in the diploid ESC. By this mechanism a maturing population of ESC is obtained. Although no information is available about its linkage, the ε gene is tentatively included in this model. (After Kabat, 1972)

The synthesis of Hb-F was noticeably inhibited and the synthesis of adult Hb higher (almost twice as high) in heterokaryons obtained from the fusion of frog and tadpole erythrocytes. These same changes in synthesis were observed when lysates of tadpole and frog erythrocytes were combined. This suggests that one (or more) regulating elements in these cells control the synthesis of the Hb fraction (Rosenberg, 1972).

An idea called the "looping-out excision theory of erythropoiesis" has been formulated by Kabat (1972). This theory takes into account the multipotential stem cells in the erythroidal tissue. These cells are sensitive to erythropoietin and can develop into erythrocytes (Till et al., 1964). The erythropoietin-sensitive cells rapidly multiply and are stimulated by erythropoietin to transform into terminally differentiated proerythroblasts. The hypothetical composition of human chromosomes of stem cells is present in Fig. 87.

It was suggested that crossing over occurs in erythropoietin sensitive cells. This is the loop-out excision stage. As a result, the proximal promotor gene is "cut out" of the chromosome as an acentric ring. Crossing-over occurs between two homologous operators which are close to the promotor. This event is analogous to removing the λ lysogen from the *E. coli* chromosome. The removal of the DNA loop occurs independently in both chromosomes in the diploid cell. This mechanism affects the maturation of the population of erythropoietin-sensitive cells and the heterogeneity of the population. As a result, various subpopulations are formed. The cells which synthesize the β and γ chains are selected and controlled by erythropoietin.

This hypothesis is a compromise between the "cellular" and "clonal" hypotheses. If this hypothesis is true then the erythroidal cell heterogeneity results from the nonsimultaneous looping-out excision in the cells. Moreover, in this hypothesis temporal but not spatial processes are more significant in the cell population. This is what separates the clonal from the looping-out excision theory of erythropoiesis.

Recent data, however, contradict Kabat's hypothesis. DNA complementary to γ-globin mRNA was synthesized with the aid of an avian virus reverse transcriptase. Hybridization of this DNA with DNA of differentiating erythroblast cells of humans showed that the DNA of these cells has γ-globin genes (Mitchell and Williamson, 1977). The presence of genes for β^A-, β^C- and γ-globins was found in sheep erythroid DNA which synthesizes only I^C-globin (Benz et al., 1977). Thus, unexpressed globin genes remain until the end of erythroidal cell differentiation.

In mice, erythropoiesis starts in the yolk sac on the 8th to 11th day of embryo development. At this time, Hb-F is synthesized. On the 9th day of development, erythrocytes begin to circulate in the blood. Between the 12th and 16th day, erythropoiesis occurs in the liver and, according to some (Fantoni et al., 1967, 1969), only adult Hb is synthesized. After the 16th day, the erythropoietic organ is the bone marrow. There are data, however, to suggest that embryonic cells can transform into adult cells in the liver (Niewisch et al., 1967, 1970; Vogel et al., 1969). In fact, microelectrophoresis of separate erythrocytes from the liver of 12–14 day embryos of C57BL mice has shown that both embryonic and adult Hb cells are present. This observation may be due to the changes in gene activity in those genes which regulate Hb synthesis. In mammals, then, the replacement of "embryonic" erythrocyte populations by "adult" populations is more complicated than it is in amphibians (Maximovsky and Korochkin, 1973).

Regulation at the translational level can also affect Hb synthesis. It is known that globin chain synthesis is reduced when heme and its precursors (ferrum) are absent. The addition of ferrum to reticulocytes stimulates Hb synthesis. Moreover, the α-chains are not released from the polysomes until the β-chains are synthesized (Baglioni, 1966; Colombo and Baglioni, 1966). Also, the composition of methionine and leucine tRNA populations differs in embryonic and adult reticulocytes of chickens. It is possible that the transition from γ to β polypeptide synthesis occurs at the translation level and depends upon the composition of the tRNA population. If the cell has both γ and β polypeptide mRNA's, then only the RNA with the necessary tRNA anticodon as an initiator codon would be translated. The discrimination between the γ and β polypeptide mRNA's could also occur during the transport of the RNA's from the nucleus to the cytoplasm (Medvedev, 1972a, b). Apparently, α- and β-globin synthesis is coordinated in a certain way in the differentiating erythroidal cells. For instance, during the differentiation of cultured mouse erythroidal cells, there is an accumulation of two m-RNA species. The ratio of their concentrations equals one during many stages of development. If differentiation is disorganized (as with thalassemia, hemolytic stress, or viral transformation of erythroid cells), only one of the two mRNA's accumulates (Maniatis et al., 1978). The mRNA preparations were used to synthesize equimolar amounts of α- and β-globins in wheat embryo systems (Stewart et al., 1977). In the reconstructed cell-free system which contained components from rabbit reticulocytes, there was competition between α- and β-globins in wheat embryo systems (Stewart et al., 1977). This factor demonstrated different degrees of competency to these RNA's (Kabat and Chappell, 1977).

What determines the direction of development of stem cells which are predecessors of three types of erythropoietical cells, i.e., erythroidal, myeloidal, and megakaryoblastoidal? It was suggested that developmental character is determined not only by internal but also by external factors. For example, erythroid colonies of anemic mice with the ff genotype are small. Differentiation begins later than in corresponding colonies of healthy mice. The organ stroma that the differentiated blood cells go to can also play an important role. Bone marrow from the same donor differentiates in an irradiated animal's spleen in the direction of erythroid cells. In the bone marrow these cells differentiate in the direction of granulopoesis. In different tissue environments, blood-forming cells become sensitive to different inductors (Curry and Trentin, 1967). Relative concentrations of inductors can be important factors in determining final cell specializations. Among the inductors are erythropoietin, leucopoietin, and thrombopoietin (Fridenstein and Tchertkov, 1969).

Humphries et al. (1976) used a sensitive method to hybridize globin mRNA to DNA. Globin RNA sequences were found not only in erythroidal cells but also in the cells of other tissues, i.e., fibroblasts, liver, and brain. In these cells, however, globin mRNA is not transported from the nucleus into the cytoplasm. Thus, there is no Hb synthesis. Therefore, the development of cells with erythroidal characteristics is determined mostly by gene modifiers which act on the post-transcriptional and possibly the translational levels. Knöchel and Kohnert-Stavenhagen (1977) found that preparations of nuclear RNA from chicken embryo brain are contaminated with erythrocyte RNA. This RNA composed 0.41% of the total RNA population. Globin mRNA was not found in the cytoplasm of brain cells.

There is only a small probability that globin mRNA is transcribed in all tissues. It is, however, possible that it is present in various tissue elements – especially mesenchymal derivatives. Large gene blocks (clusters) may be successively activated during development and determine the synthesis of specific organ proteins. After this, gene modifiers perfect the biochemical parameters of the cell by acting inside each differentiating cell of the embryonic layer. This adjustment may occur at all levels of the regulation of gene expression.

In the light of this assumption, it is interesting to note that some antigens which are characteristic of differentiated tissues in mouse embryos have been found (Stern et al., 1975; Wiley and Calarco, 1975). Solter and Schachner (1976) found a cell surface antigen – nervous system antigen: 4 (NS-4) – in dividing mouse embryos. NS-4 was then seen in the trophoblast and the inner cell mass of the mouse blastocyst.

Therefore, differential gene expression during the ontogenesis of erythroidal cells (as well as other specialized cells) involves multiple levels of regulation – from molecular to the tissue level.

In general, the process of cell differentiation is a result of (1) the differential activation of genes which spezialized in the syntheses of proteins and enzymes which have structural and functional significance in the specialized tissue and (2) the differential action of gene modifiers controlling the rates of syntheses and degradation of various inhibitors and the formation of multienzyme complexes.

2.3 Genetic Regulation of Ontogenesis on the Organismal Level

2.3.1 Hormones and Development

As mentioned by Smalgausen (1938) "...the endocrine system is the basic integrating factor of the definite stages of morphogenesis." The nervous system acts, in part, in the same direction. Actually, the conditions of the organism, especially the hormonal system, affect the realization of the hereditary information from the very beginning of development. The establishment of the structural-chemical organization of the ova is an example of such events. For instance, hormones stimulate protein synthesis in insect oocytes (Chapman, 1969). Hormones also induce the synthesis of yolk protein in the bird liver. This protein is later accumulated in oocytes. The corpus allatum and neurosecretory system of mosquitos produce hormones which are necessary for egg maturation (Lea, 1972). Amphibian oocytes cannot induce DNA synthesis in nuclei transplanted into them before ovulation. Such an ability is possible only after stimulation with hypophyseal hormones and the destruction of the germinal vesicle. According to Gurdon (1967) this ability, which plays a crucial role in the initiation of development, "is the result of the action of hypophyseal hormones on the mature oocytes". It is obvious that hormonal action is a necessary condition for the occurrence of important events in the nucleus and cytoplasm of the maturating amphibian egg as well.

According to modern knowledge, a hormone similar to progesterone is synthesized in the follicular cells under the influence of gonadotrophins. In turn, this hormone stimulates a number of events which affect oocyte maturation, the flowing out of the contents of the germinal vesicle, and membrane disintegration and cortex maturation. All these processes can be inhibited by puromycin (Fig. 88; Detlaff and Skoblina, 1969). No oocytes matured when the follicular membrane was removed – even if hypophyseal suspensions were purified. Oocyte maturation was obtained only when follicular cells were present (Masui, 1972).

The transition from one larval stage to another is controlled by the ratio of concentrations of juvenile hormone and ecdysone (Novak, 1960). Apparently, there is a specific system which regulates this ratio. This system includes specific fractions of carboxyl esterases or molting esterases (Whitmore et al., 1972; Berger and Canter, 1973). The activity of these esterases is noticeably increased in *Hyalophora* pupae following juvenile hormone injection. These esterases are products of the fat body (Whitmore et al., 1974). The hormones stimulate RNA and protein syntheses during a 6 h period. Labeled carboxyl esterases were found in experiments with isotopes. In the light of these observations, the suggestion was made that inactive enzyme is stored in the fat body. Introduction of juvenile hormone causes its release. It is known that the hemolymph proteins are accumulated as specific

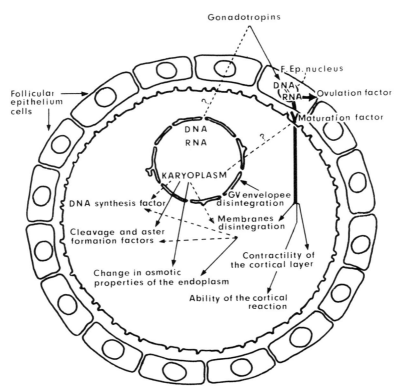

Fig. 88. Changes in oocytes during maturation revised tentative scheme. *thick line* puromycin sensitive processes; *dotted line*, the supposed part of the process. (After Detlaff and Skoblina, 1969)

granules in the fat body during the larval stage. This process is also hormone-dependent. Granule formation is correlated with a reduced titer of juvenile hormone. It is possible that the mRNA and protein synthesis, which are induced by hormones in the fat body, are associated with activation and modification of the preexisting enzyme molecules (Whitmore et al., 1974). The hormonal role in the regulation of metamorphosis in amphibians is well known and has been discussed in a number of reviews (Weber, 1967; Tata, 1971).

The implantation of mammalian blastocysts depends upon complex neurohumor interactions and upon various hormone concentrations. The hypothalamus affects the hypophysis which produces gonadotrophins. Gonadotrophin, in turn, controls the function of the ovary. The follicular phase and estrogen secretion follow the luteal phase (Jelesnjak, 1970). Apparently, a certain coordination is necessary during the changing of the hormonal balance and the development of the fertilized egg for its implantation.

The level of hormones in the developing organism significantly affects the establishment of various organ systems. The differentiation of the genital system is a good example. On the basis of this example, Zavadovsky (1922) formulated the

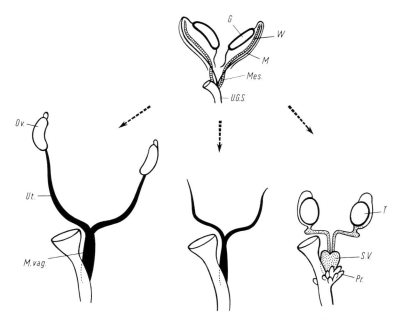

Fig. 89. Sexual differentiation of the sex duct in the rabbit fetus. From the undifferentiated condition (*upper part*), as shown in fetuses 19–20 days old, may arise either the female structure (*lower left*) or the male structure (*lower right*) or the gonadless structure in castrated fetuses of either sex (*lower middle*). *G*, gonad; *M*, Müllerian duct; *Mes*, mesonephros; *M. vag.*, Müllerian vagina; *Ov.*, ovary; *Pr.*, prostate; *S. V.*, seminal vesicle; *T*, testis; *U.G.S.*, urogenital sinus; *Ut*, uterine horn; *W*, Wolffian duct (*stippled*). (After Jost, 1961)

principal proposition about hormonal regulation of the development of sexual characteristics. According to this, "The whole complex of sex characteristics morphogenetically depends upon the sex hormones: feminizine (F) in females and masculinizine (M) in males...".

The differentiation of the sex organs of males or females from indifferent primordia is demonstrated in Fig. 89 (Jost, 1961). The differentiation of the genital tract occurs after the differentiation of gonads. Hormonal concentrations determine the character of development. At the present time it is known that the development of female sex structures in mammalian embryos does not require sex hormones or ovaries. In contrast, the male genitals develop under the influence of male sex hormones. These determine the differentiation of the Wolffian duct and the reduction of Muller's ducts (Witschi, 1967; Levina, 1974).

There are crucial periods when the differentiating genitals are sensitive to hormonal action. Genital development is relatively independent of hormonal influences before and after this crucial period (Jost, 1961) and is characterized by certain morphogenetic interactions with the surrounding tissues (similar to embryonal induction). This period occurs on the 19th–20th day of development and continues for 2–3 days. There are temporal differences in the response of different layers to hormonal action: the Muller's ducts start their regression on the 20th day of embryogenesis; the anterior prostate in the Wolffian duct begins differentiation on the 20th–21st day; external sex organs begin developing on the 23rd day and excretory ducts on the 24th day. Obviously, these physiological-embryological peculiarities are the bases for the explanation of some of the abnormalities in the development of the sex system in humans (see Mitskevich, 1966; Levina, 1974). The effect of various hormones on other morphogenetic processes is also known (Verne and Hebert, 1949; Ross and Goldsmith, 1955; Smith, 1956; Moog, 1959; Eayrs, 1960, 1961; Altman and Das, 1965; Moscovkin, 1975).

Among the many hormonal factors which regulate development, the formation of the hypothalamo-hypophyseal regulating system is significant in ontogenesis. The establishment of this system is accomplished by the prenatal period (or at least in the early postnatal period). Defects in its maturation induce nonreversable and severe disorders in development. For example, noise activates the hypothalamo-hypophyseal system. Mouse embryos demonstrate abnormalities in osteogenesis, have turning of hind limbs, omphalocoele, bilateral eye absence, and other damages, if the females during pregnancy were constantly subjected to noise (74–79 dB, 2.5 h daily). The degree of abnormal development depends upon the age of the pregnant female. Young females (120–130 days old) are more resistant to noise stress and only 37% of their progeny demonstrated abnormalities. On the other hand, the progeny of old mice (280–320 days old) demonstrated higher levels of damage – 74.8%. Apparently, alterations of calcium and phosphorous metabolism in mothers is the cause of these abnormalities. In turn, the changes of metabolism in the pregnant mice are caused by disorders in the hormonal balance which result from the noise stress. It is known, also, that the parathyroid glands, which take part in the regulation of ossification, have a function in the embryo. Their development depends upon the calcium and phosphorous metabolism in the mother (Geber, 1973).

It should be mentioned that sex differences in gonadotrophin activity (continuous in males and cyclical in females) are not associated with the hypophysis (which has an equal potential in both sexes), but are associated with the hypothalamus. It was found that the male pattern of gonadotrophin secretion is formed several days after delivery as a result of an effect of testicular hormones on nerve structures in the brain above the hypothalamus (Yazaki, 1960). Castration or neutralization of the sex hormones by reserpine injection (which eliminates testicular hormones from newborn males) affects hypothalamus action (Levina, 1974).

In all likelihood, the embryonal mortality and reduction of fecundity caused by changes of photoperiods are both consequences of a disorder of hormone balance. Additional lighting increased fecundity in a mink. There was also a mortality reduction (Belyaev et al., 1963). The light effect in this case was explained through the increase of hormonal function of the yellow bodies which produce progesterone. This suggestion was confirmed by analysis of the correlation between the number of yellow bodies and the level of embryonal mortality in this animal (Belyaev and Jelezova, 1968). It was supposed that the level of embryonal mortality, especially during the first stages of embryogenesis, depends upon an abnormal pool of hormones in the maternal organism and in the developing embryo. The influence of light on estrus duration was demonstrated for pigs: additional light lengthened estrus and, as a result, the number of mating females is higher compared

Fig. 90. 3.5-year-old fox, homozygous for "georgian white" mutation. (After Belyaev et al., 1973)

with controls. The effect found in experiments with pigs is different from that observed in mink. The length of the light regime influenced the start and finish of the estrus cycle in mink: there is a reduction of cycle number and correspondingly of the total number of matings during breeding season (Belyaev et al., 1969).

Additional lighting during the pregnancy of fox which are homozygous for a recessive lethal mutation, "georgian white," eliminates the preimplantant phase of the lethal effect of this mutation and, as a result, homozygotes for this lethal mutation are born (Fig. 90). Dissection of the pregnant females demonstrated that some of the homozygous embryos can overcome the implantation barrier. In this case the critical period is birth and the first 1.5–2 months of life. The action of the light factor specifically increases the viability of the corresponding homozygotes. The actual fecundity and preimplantation embryonal mortality remained at the same level as with natural photoperiodism. It is possible that there are some competitive interactions of embryos at the preimplantation stage which are regulated by maternal status (Belyaev et al., 1973).

What is the role of the genetic system in the regulation of such a status? This role has two meanings, as demonstrated by numerous data: firstly, examples of genetically controlled polymorphism in endocrine functions is well known and, secondly, hormones can act as genetic inductors and affect the activity of the nuclear apparatus during ontogenesis.

2.3.2 Genetic Regulation and the Hormonal System

Investigations of genetically controlled polymorphisms of various organs of the endocrine system have recently begun. At the present time, numerous examples of genetic control of morphological and functional aspects of the endocrine organs are well known (Shire, 1974). One of the first papers describes the character of degeneration of the juxtamedullar zone of the adrenal cortex of various mouse strains (Spickett et al., 1967). It was shown that there are interstrain differences in the starting time of certain degenerative abnormalities. These differences are noticeable from the 11th week in postnatal life in females on the CBA strain, whereas in mice of the A line, they can be visualized as early as the 6th week. Hybrids between these strains demonstrate the A strain character and in the F_2 there is a 3:1 segregation. Apparently, then, these differences are controlled monogenically and the allele of the A strain is dominant. The adrenal hormone reaction on testosterone varies between strains. For instance, the CBA mice are completely resistant to its effects, whereas in "Swiss" white mice, the degeneration of the juxtamedullar zone under the same conditions is reduced (Spickett et al., 1967).

The monogenic control of interstrain differences in corticosteroid control was observed in a number of cases (Badr and Spickett, 1965). There are known defects of synthesis of adrenal cortex steroids. These are caused by recessive autosomal genes with clinical expression in homozygotes (Prader et al., 1962). It was demonstrated, also, that the individual differences in response to a stress irritant are genetically determined. This occurs because of differences in the levels of hormone production and use (Hamburg and Kessler, 1967).

Inherited illnesses which are caused by abnormalities in the adrenal cortex are known. Hyperplasia of the adrenal cortex and the changing of their function was found in mice with inherited obesity and hyperglycemia (ob/ob). Increased levels of corticosteroids in blood plasma were found in obese animals (compared to normal). Both strains respond similarly to stress and the intravenous injection of ACTH — the elevation of serum corticosteroids. However, the half-life of hormones in obese animals was shorter (Naeser, 1974).

Estrogen secretion is reduced in fat rats (fa/fa). This is accompanied by hypothalamo-hypophysial abnormalities and by the inhibition of gonadotropin secretion (Saiduddin et al., 1973). It was also demonstrated that there is a correlation between the cholesterol content of mouse adrenals and the ability to produce corticosterone. These are in turn correlated with the amount of lipid in the adrenal cortex. At the same time there is evidence that in some mouse lines, two or more interacting loci result in adrenals which are exhausted of fat and have abnormal function (Doering et al., 1973). For various mouse strains the ratio of 11-deoxycortisol/corticosterone is controlled monogenically (Badr, 1970). The degradation rate of corticosteroids is regulated, also, by a small number of genes (Shire, 1974). The genetically determined variations in the renin-angiotensin system in mice and rats affect the level of aldosterone (Rapp, 1965, 1969). The genetic determination of interstrain differences in the function of the adrenal medulla was described for mice. For instance, the activities of three enzymes which control adrenaline synthesis (tyrosine hydroxylase, dopamine β-hydroxylase and phenylethanolamine N-methyltransferase) in subline J mice of the BALB/c strain are twice as high as in the N subline of this strain. The differences in enzyme activity for each of these are controlled by one gene which demonstrates a codominant type of inheritance. It was suggested that there is a gene regulator which coordinates the activity of these three genes (Ciaranello et al., 1972).

The degeneration of Langerhans' pancreatic islets was seen in homozygous mice (db/db) with genetically inherited diabetes. These mice were of the C57BL/KSJ strain. The same db gene manifests its effect slightly differently in another genetic background. Its action is expressed as hypertrophy and hyperplasia of Langerhans' islets in C57BL/6J mice (Boquist, 1974). Moreover, in the sky strain these structures are characterized by a reduction of insulin production upon the administration of glucose, theophylline, and iodine acetamide. This reflects the early degeneration of β-cells (Gunnarson, 1974). The response of the Langerhans' islets to the same compounds is different in the 6J line of mice. Here, there is an increase of insulin secretion and moderate hyperglycemia (Boquist, 1974). The different expression of the db gene in various genomes has been demonstrated by other researchers also (Hummel et al., 1972).

Mutations which influence the development of some endocrine organs can pleiotropically affect the differentiation of other glands. These pleiotropic effects could not only be the direct consequence of gene action in various endocrine organs, but could also be the result of a secondary effect of an endocrine organ on the development and function of other organs. In the pygmy mice (dw/dw), for instance, the reduction of growth hormone and prolactin concentration is accompanied by a disproportionate ratio of cortex and medullary matter in the adrenals compared with normal mice (Shire, 1974). The nature of the genetic control of development and function of endocrine glands in humans is discussed in a monograph by Rimoin and Schimke (1971).

It is necessary to mention that the final effect of hormones on the morphogenetic processes is determined not only by the genetically controlled conditions of the endocrinal system, but also by the genetically controlled ability (competency) of the target-tissue which reacts to the hormone (Shire, 1974). For instance, liver tyrosine aminotransferase (TAT) is an enzyme which is induced by steroids both in vivo and in vitro in cultured cells. Sublines of hepatomas were described in which the level of activity of this enzyme was normal, but which were unable to react to steroid treatment by increasing TAT activity (Levisohn and Thompson, 1972). Interstrain differences of TAT activity were obtained in mice after starvation. In C57BL/b mice, the activity of this enzyme significantly increased, whereas in DBA/2J mice it was lower than control levels (Blake, 1970).

Interstrain differences in response to hydrocortisone treatment were also demonstrated. The injection of 4 mg of hormone on the 11th–14th day of pregnancy induced splitting of the palate in A/Sa embryos. CBA/T6T6 mice were not affected. In additional experiments it was demonstrated that low concentrations of hydrocortisone inhibited the development of the palate tissue from A/Sa embryos in vitro but did not affect the same morphogenetic stages in CBA embryos (Saxen, 1973).

There is a mouse strain that is resistant to insulin, which is characterized by the hypophyseal hypersecretion of growth hormone (Roos et al., 1974).

The abnormal photoperiod reaction of mink which is accompanied by differences in hormone secretion is also genetically determined (Klochkov, pers. comm.).

Recently, interest was created with demonstrations that various stresses (as simple handling) affected the pregnant females that, in turn, influence the progeny behavior (Treiman et al., 1970; Daly, 1973). It was demonstrated that the effect of prenatal stress on the offspring behavior depends upon the maternal genotype because intensity and length of the mother's reaction to stress is genetically determined.

Finally, the direction of the change in behavior in such experiments depends upon the progeny genotypes (Joffe, 1969). The hypothesis of Thompson and Sontag (1956) is one of many which explains the effect of irritants on pregnant females and her progeny. They suggested that the contraction of the uterine vessels causes embryo anoxia. The reaction of the embryo depends upon the time and duration of stress and upon the genetically determined resistance of the embryo to oxygen deficiency. Treiman with co-workers analyzed the corticosteroid concentration in the blood plasma of C57BL/10y and DBA/2 mice as a response to electrical shock or handling. They also tested hybrids from the reciprocal crosses. The DBA/2 genotype increased the corticosteroid concentration. However, keeping progeny with DBA/2 mothers acted in the opposite direction, i.e., decreased this hormone concentration. It was suggested that such an interaction represented a specific buffer system, which determined the optimal hormonal concentration in animals and which modified the behavioral response to various stress situations (Treiman et al., 1970).

The molecular-genetic mechanisms of action of some hormones were discovered with a series of experiments by Ohno and collaborators. It was demonstrated that ADH activity in mouse (and human) liver increases earlier than in kidney. Also this increase occurs first in females where it is four times greater than in the kidney. The activity is twice as high in male livers as in kidney. The ADH activity in the kidney becomes similar to that of liver after the injection of 10 mg of testosterone to castrated males and females. This experiment suggests that ADH activity is under the control of a secondarily repressed genetic mechanism. Taking into consideration that testosterone also induced an increase of β-glucuronidase activity and of a number of other enzymes, the workers suggested the presence in mammalians of a genetic regulatory system similar to the operon of prokaryotes. According to Ohno, the process of enzymatic induction under the control of testosterone (or 5-L-androstandione) is realized in two steps: (1) the induction of the enzymatic activity as a result of association of testosterone with corresponding translational repressors and, (2) the concomitant elimination of the translational block, and the hypertrophy of the reacting system.

Ohno (1971) suggested that this system is the basis for sex determination in mammals (Fig. 91) and includes two gene regulators. One of these determines the development of indifferent gonads in the testis or ovary direction and another controls the formation of secondary sex characteristics. The expression of the male or female phenotype depends upon the presence or absence of testosterone. Müllerian ducts are differentiated into fallopian tubes and the uterus, and Wolffian ducts are reduced if XY embryos are castrated before the critical period in development. Testosterone injection into XY and XX embryos stimulates development of Wolffian ducts into seminal ducts, epididymis and seminal vesicles and determines the differentiation of the urogenital sinus into the prostate and penis. However, in this experiment the total reduction of Müllerian tubes was not obtained and the castrated embryos demonstrated the hermaphroditic phenotype. It is possible that Müllerian tube degeneration is caused by the presence of another male hormone

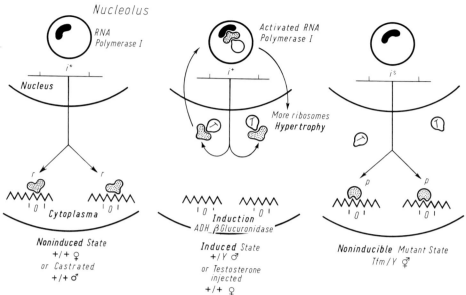

Fig. 91. Single gene control of mammalian sexual phenotypes. i^+ represents the wild-type allele of the X-linked *Tfm* locus. Its product (r) serves as a translational repressor of alcohol dehydrogenase (ADH) and β-glucuronidase in the cytoplasm until the arrival of testosterone (T). When it becomes bound to testosterone or its metabolites (such as DHT), it moves into the nucleus and activates RNA polymerase I in the nucleolar region, thus causing increased production of ribosomes and hypertrophy. i^s represents a mutant *Tfm* allele; its product (P) has lost the binding affinity to testosterone. The o (operator) region in each RNA (*wavy line*) signifies a homologous base sequence shared by ADH and β-glucuronidase mRNAs which is recognized by the product of *i* gene. (After Ohno, 1971)

("X" factor) which is produced by the embryonal testis (Jost, 1961). Ohno interpreted the results of these experiments as evidence that differentiation in the male or female direction reflects inducible or noninducible conditions of the same regulating system. The product of the gene regulator, according to the author, has to have competence to testosterone and/or its metabolites (Ohno, 1971).

Tfm sex-linked male mutants have lost their inducibility, although they have testis and are phenotypically similar to females (testicular feminization). The mechanism of expression of this mutation is associated with a product which is controlled by the Tfm locus. As a result of lost competence to testosterone, Tfm/Y males are characterized by noninducibility (Ohno, 1971; Ohno and Lyon, 1970; Ohno et al., 1970). Later it was demonstrated that the corresponding operator (O^s) of a noninducible mutant of β-glucuronidase is associated with the slow isozyme of this enzyme. The result of this experiment confirmed that the translation repressor determined by the Tfm locus recognize the translated segment of mRNA (Dofuku et al., 1971).

On the basis of numerous molecular experiments, the conclusion could be reached that the presence of a protein hormone receptor in the target tissue is necessary for the reaction of these tissues to steroid hormones. Such receptors were

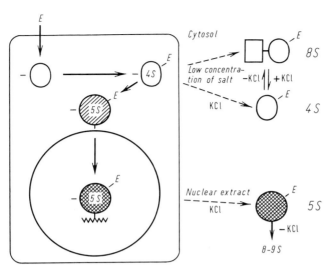

Fig. 92. Estradiol action in the target-tissue cell. (After Jensen and DeSombre, 1972)

found in various target tissues and are associated with both the cytoplasm and nucleus. They are thermolabile proteins with a sedimentation coefficient of about 8–10 S in low ionic strength and 3–5 S when the ionic strength is higher than 0.3 M KCl. The interaction of hormones with the receptor compounds can be subdivided into two steps: (1) the binding of hormone with the cytoplasmic receptor, and (2) the transport of this complex into the nucleus and the association of the hormone with the nuclear chromatin. The transport of the hormone-receptor complex from the cytoplasm to the nucleus in the case of estrogens, is accompanied by the hormone-induced transformation of the receptor protein-hormone complex from 4 S to 5 S. Such a transformation was not obtained in the case of progesterone (Fig. 92; Tomkins and Martin, 1970; Jensen and DeSombre, 1972). The critical periods during the ontogenesis of target-organ responses to hormones involve the formation of receptors in these target organs. Those processes under hormonal control can be listed as follows:

1. Increased incorporation rate of labeled RNA precursors.
2. Increased transcriptional activity of chromatin isolated from tissue which was treated with hormone.
3. Appearance of new RNA fractions or increasing of RNA fractions which already existed in tissues that were treated with hormone. It should be mentioned that hormonal induction is inhibited by inhibitors of RNA synthesis (Wingaarden, 1970).

   The induction of tyrosine aminotransferase synthesis in the rat liver by glucocorticoids is accompanied by an increasing amount of corresponding cytoplasmic mRNA (Diesterhalf et al., 1977). The induction of tryptophan2,3-dioxygenase by glucocorticoids in rat liver is also associated with the initiation of transcription of the corresponding structural gene (Killewich and Feigelson, 1977). Estrogen and progesterone affect the transcription of ovalbumin and avidine (O'Malley et al., 1975).

The activation of vitellogenin synthesis in liver cells of the rooster after injection of 17β-estradiol is a consequence of the induction of mRNA synthesis. The amount of mRNA increases, in this case, from 0–5 molecules to 5000 molecules per cell. Therefore, in this case, vitellogenin synthesis included the activation of a dormant gene in a totally differentiated and metabolically active cell (Deeley et al., 1977).

All these facts allow us to consider many hormones as genetic inductors. The introduction of hormone inductors stimulates not only mRNA synthesis, but also the synthesis of rRNA (ribosomal) and proteins, tRNA, and the recycling of the acceptor terminal part of tRNA (CCA). It was demonstrated in a number of cases that hormones "switch on" not only the synthesis of specific mRNA's but also the synthesis of all parts of the protein-synthesis apparatus (Salganik, 1968). Some authors suggested that there are protein fractions which are inducible or noninducible by hormones. Such fractions were found in the tyrosine aminotransferase (TAT) enzyme of rat liver with regards to cortisone and for hexokinase with regard to insulin (Mertvezov et al., 1974; Chesnokov et al., 1974). However, there are data which suggest that TAT fractions of rat liver are not isozymes but are conformers of the same protein, the synthesis of which is controlled by one gene. In this case, the changing distribution of total TAT activity among its fractions occurs at the post-translational level under hormonal control (Johnson and Grossman, 1974).

Cells lost the ability to respond to hydrocortisone when rats were injected with heavy doses of this hormone (5 mg per 100 g of living weight every day for a month). At the same time, the reaction to another inductor, for instance insulin, remained. Salganik (1968) suggested that the pathological hyperproduction of hormones could cause the "exhausting" induction and "disbalance" of regulatory genetical mechanisms in cases of various endocrine diseases. Apparently, the overtreatment of patients with hormones can induce the disfunction of the regulatory system of the organism.

The phase of the cell cycle plays an important role in the hormonal effect on the proliferative system. Differences in reaction were obtained for target tissues and nonspecific tissue. This follows from the next scheme (Epifanova, 1965):

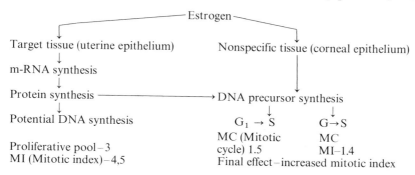

It is known, also, that tyrosine aminotransferase synthesis in the hepatoma cell culture is induced by steroid hormones only in certain stages of the cell cycle (late G_1 and S). It was suggested that gene expression in this case is controlled on the transcriptional and post-transcriptional level (Sellers and Granner, 1974). Post-transcriptional mechanisms of hormonal action were demonstrated for plants, e.g., gibberellic acid hormone which induces α-amylase and protease synthesis in the aleurone (Carlson, 1972).

The presence of two genes – structural and regulatory – which in the proliferating cells could be in alternative activity states, was postulated using the example of hormonal induction of tyrosine aminotransferase synthesis. In the inducible period of the cell cycle, both of these genes are actively transcribed regardless of inductor presence. Both genes are not active in the noninducible periods (G_2, M or early G_1) and cannot be activated by steroid inductors. The translation of mRNA that was synthesized by the structural gene is inhibited in the inducible period by labile post-transcriptional repressor (R) – the product of the regulatory gene. In synchronous cell populations it was demonstrated that the repressor is synthesized until the beginning of the inducible period of the cell cycle and disappeared at the end of this period. Steroids are the antagonists of the post-transcriptional receptor and therefore stabilized the active condition of the mRNA and promote its accumulation (Tomkins and Martin, 1970).

The cyclic AMP system is a nonspecific step in protein hormone action. This system is active in various target tissues (Table 41). The competence of the membrane receptor which transfers the signal from the hormone to cyclic AMP is, however, specific for every system. The biochemical basis for the action of this system can be illustrated by the example of the hormonal stimulation of the glycogen-glucose transformation. The reaction sequence in this case is as follows:

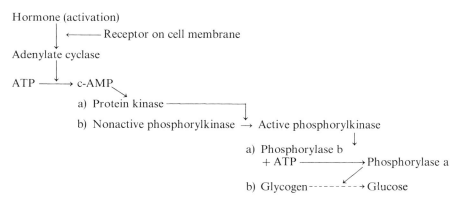

Disorders of cyclic AMP metabolism could be the cause of many abnormalities. For instance, cholera bacteria produce a toxin which stimulates the intestine cells to excrete large amounts of water and salts. This, in turn, leads to severe dehydration of the organism. It was demonstrated that this toxin causes the accumulation of the extra c-AMP. This distorts cell function. The reverse process – the expressed reduction of c-AMP concentration – was obtained during transformation of normal cells into tumor cells. It is interesting that when tumor cells are growing in the presence of c-AMP, they demonstrate some normal characteristics which were previously abnormal (Pastan, 1972).

Therefore, hormones have not only an integral significance for the realization of various processes on the level of the whole organism but, also, they are able to

Table 41. Examples of various hormones acting on the target-tissue through system of cyclic AMP. (After Sutherland, 1972)

Tissue, organ	Hormone	Process affected
Skin (frog)	Melanocyte-stimulated hormon	Darkness
Bone	Parathyroid hormone	Calcium resorption
Skeletal muscle	Epinephrin	Glycogenolis
Fat tissue	Epinephrin, adrenocorticotropic hormone, glucagon	Lipolysis
Brain	Norepinephrin	Lypolysis
Thyroid	Thyroid, hormone stimulating	Lipolysis
Cardiac muscle	Epynephrin	Thyroxine secretion
Liver	Epinephrin	Increasing of contractility
Kidney	Parathyroid hormone	Glycogenolysis
Adrenals	Adrenocorticotropic hormone	Phosphate excretion
Ovary	Corpus luteus hormone	Hydrocortisone secretion
		Progesteron secretion

affect the pathological development as well because of the universality of the hormonal influence on various tissues.

Organ reactions to hormones (similar to the response during differentiation) include events on the cellular and tissue levels. Naturally, the estrogen induction of phosphitine synthesis in the bird liver is determined by the specific proliferation of cell clones which are competent for this synthesis (Talwar et al., 1973). Hydrochloric acid synthesis in the rat stomach is induced by pentagastrin in a two-step process. First, hormone stimulates histamine synthesis in the histamine-containing cells; then these cells secrete histamine into the surrounding medium. This affects the system of epithelial parietal cells which produce HCl (Salganik, pers. comm.). In general, it is possible to suggest that the regulatory functions of hormones, (which occur at various levels of organization) are controlled by specific gene systems. However, the precise mechanisms of gene interactions which are active in these events are not yet clear.

2.3.3 Other Factors in the Regulation of Gene Activity on the Organismal Level

The nervous system is also an integrating apparatus for the organism. Certainly, in the early stages of development the nervous system does not play such a role, but rather takes part as a regular morphogenetic factor in the processes of primary and secondary embryonal induction. Moreover, the nervous system, apparently, does not serve as an integrating factor in the so-called prefunctional period of organism development, when the various systems are morphologically and physiologically maturing in directions which are necessary for the realization of specific functions. It is well known, for instance, that in the case of what is called extragastrulation in amphibians, when the developing embryo does not develop a nervous system, the differentiation of numerous organs and tissues is not disturbed. Mesoderm formed muscle, mesenchyme and cartilage and, in the trunk region, somites, pro- and mesonephros ducts, muscles of the intestine and a heart without blood which is able to contract rhythmically. Endoderm is, also, able to differentiate when the nervous system is absent. Here, the endodermal part of the mouth cavity, the pharynx with endodermal parts of the gill clefts, thyroid gland, esophagus, stomach, lungs, liver, pancreas, and intestines were all formed. Various parts ot the esophagus are characterized by normal histological composition. The process of synthesis and secretion-excretion by the intestinal cells is occurring at the same time periods as in normal embryos (Hexly and De Beer, 1936; Lopashov, 1937).

It is known, also, that mechanisms of regulation which are necessary for swimming are formed in amphibians under conditions of deep narcosis. Larvae which developed under this condition swim as well as control animals. Rabbits which were raised in complete darkness for 6 months did not demonstrate any defects in the development of the visual system – including the proectional zone of the cortex. Thus, the development of structure occurs independently of any anticipated functional necessity (Coghill, 1934; Sperry, 1960). The integrative significance of the nervous tissue constantly increases with the beginning of various system functions, especially in the postnatal period. Apparently, the nervous system is important for the maintenance of the differentiated morphophysiological condition. In the famous experiments of Cannon and co-workers, it was demonstrated that denervation of various tissues resulted in both an increased excitability and an alteration of many physiological characteristics in these tissues (Cannon and Rosenblut, 1951).

At the same time, structures with accelerated growth and differentiation could be found. These structures, during embryogenesis, determined the ability of newborn animals to survive under specific conditions. In this case, the coordinated nerve center played an important role in the formation of such a morphofunctional system. Anochin (1968) designated such a tendency in the development "systemogenesis". It would be interesting to discover the gene activation which determines such organization of differentiation and, also, the corresponding sequence of regulatory events on the cellular and tissue levels. On the other hand, the organospecific determination of the blastema of amphibian limbs is determined during regeneration by the remaining parts of the organ and its immediate surroundings. Further development of the limb and its final formation during regeneration is determined by the relationship between the level of regenerative development and the organism as a whole. This connection is a kind of control for regenerating growth and it is realized through nervous system control. Thus, limb regeneration depends upon a certain level of innervation. Only the absolute number of nerve fibers in the regenerated structure has any determining significance. The number of sensory nerves is usually higher than the number of motor nerves. However, a limb can regenerate without the sensory nerves if the motor innervation is increased (Vorontzova and Liosner, 1955). The nature of the chemical compounds in the nerve tissue which regulate regeneration is not clear. It was suggested that all neurons synthesized neurohumoral compounds which speed in all directions from the nerve fibers and promote the growth of the innervated parts (Parker, 1962). Growth is most intensive in those parts of the blastema which are most highly innervated. The separate axons are in tight contact with the blastema cells (Salpeter, 1965).

During embryogenesis all muscles contain a pool of identical myofibrils which have a prolonged contraction time that is characteristic of slow fibers. Only after innervation do physiological differences appear between fast and future slow-contractive muscles (Close, 1964; Gordon and Vrbova, 1975; Kugelberg, 1976). Rubinstein et al. (1977) studied the myosin types which are in the developing fast and slow muscles of chicken embryos. During development, *m. pectoralis*, a fast muscle in adults, contains the heavy chains and two-three light chains that are characteristic of adults. However, the *latissimus dorsi* muscle, also a slow muscle in adults, contains the fast myosin light and heavy chains. Only after innervation of this muscle there is synthesis of the slow myosin heavy and light chains. A number of investigators have suggested that gene expression in muscles may be under nervous system control (Guth, 1968; Guthman, 1976). The transformation of fast (according to physiological, biochemical, and histochemical parameters) muscles into slow muscles (and vice versa) in experiments with reciprocal innervation could prove this theory (Buller et al., 1960a, b). The genetic and molecular aspects of neural regulation are extremely interesting but not completely understood.

The system of mutual immunological interactions between the embryo and mother is another system which can play a role in the realization of the inherited information of the whole organism. It was suggested that there are special mechanisms which regulate these interactions. Particularly, there is a hypothesis that the placenta somehow neutralizes the mother's isoantibodies which can damage embryonal tissues (Vyasov, 1962). Nevertheless, cases are known where isoantibodies disrupt embryo development. Serum against connective tissue has an inhibitory effect on the growth of corresponding tissue. Similarly, antiorgan serum inhibits the growth in vitro of pieces of kidney, brain, and epithelial tissue (Vyasov, 1962). Serum against chicken tissues inhibited the embryonal development of stages of the primitive streak. The growth of embryonic heart is suppressed by serum against actomyosin (Johnson and Leone, 1955). Keeping in mind the close hormonal relationship between the embryo and mother it is possible to suggest that similar relations exist between animals – parabionts. For instance, partial hepatectomy in one of the parabiotic rats induces an increase in mitosis in the other rat. Clinical and experimental data suggest that some embryonic organs are developed and functional earlier than normal when certain maternal organs functioned inadequately (Vyasov, 1962). However, a more detailed investigation is needed.

2.3.4 The Genome and the Function of Tissue Systems

Endogenous daily rhythms apparently have an inherited basis. This was demonstrated in experiments where treatment of parents with unusual alternations of dark and light did not result in abnormal rhythms in the rat offspring (Hemmingsen and Krarup, 1937). Growing *Drosophila* under continuous weak light did not change the endogenous daily rhythm of the offspring (Bünning, 1935). Mice can survive for many generations without any changing of external factors (alternation of darkness and light). Normal endogenous daily periodicity was demonstrated

(Aschoff, 1955). Progeny between two strains with different durations of periods demonstrated an intermediate duration. Mendelian segregation in the next generation was not demonstrated (Bünning, 1961).

Numerous cases of individual variability of morphological and functional features of various organs of humans and other animals are listed in Williams (1960). These variables are apparently determined genetically. It is known that among the six inbred mice strains C57BL/CrgL, A/Cal, DBA/2, BAL/2, BAL/c, and C3H/HeCrgL, the C57BL/CrgL mice consume more alcohol than others. The genetic analysis demonstrated that this predisposition to alcohol is polygenic (McClearn and Rogers, 1959; Krushinsky et al., 1968). There is a large variability in alcohol preference among hybrids of these strains. The pathophysiological basis of this trait is inherited, and genes of high and low alcohol preference strains are dominant over genes which cause an intermediate degree of alcohol preference.

The genetic regulation of nervous system activity produced a great amount of interest. Differences in animal learning capacity were found in training experiments with various strains and races of animals. Bagg (1916) was the first to investigate the character of inheritance of the capacity of five mouse strains and their progeny to learn a maze. Significant interstrain differences were found despite the large variability observed for each strain. McDowell (1924) showed that rats from different races learn at different rates. Tolman (1924) observed polymorphisms in fast and slow maze learners in laboratory animal populations.

The genetic analysis of psychological peculiarities of the rat has been published in Russia. "One of the important results of the study of rat behavior was that within the same species various individuals demonstrate different capacities. From here the problems are defined: the genetic investigation of psychological capacities, the isolation of various inherited types of psychological capacities, the selection of these capacities over a number of generations, and the study of the Mendelian laws of their inheritance" (Sadovnikova-Koltzova, 1925).

Extensive investigations of the role of the genotype in determining the capacity to learn active avoidance using the shuttle box method (Warner, 1932) were performed by a group of Italian scientists (Bovet et al., 1968, 1969). They demonstrated that behavior could be variable among animals of the same colony of fish, bird, rat, and monkey. This was demonstrated with the conditioned reflex, maze training or operant behavior. The expressed individual differences in the Swiss Webster mice from one colony were demonstrated in special experiments with shuttle boxes (Fig. 93). The comparison of individual curves of learning demonstrated that each individual has its own level of avoidance.

In our laboratory (Shumskaya et al., 1975) polymorphism was found in the rate of establishment of foodprocuring in populations of laboratory rats. The criterion for making a conditioned reflex was the number of negative reactions which was not higher than 10% from the total number of signals. The mean number of conditioned reflexes for a population of 60 rats was 22.8; the variability for individuals was from 3 to 93. Thus, we obtained a significant variability in the animal's capability to learn.

The question now arises whether this variability depends upon the environment or is it genetically determined?

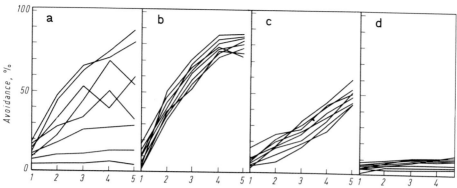

Fig. 93. Avoidance learning in **a** a heterogeneous population of Swiss mice; **b** DBA/2J mice; **c** BALB/c mice; **d** CBA mice. Each *curve* represents the individual performance of a mouse during five avoidance sessions of 100 trials each. (After Bovet et al., 1969)

The presence of polymorphism in various behavioral traits stimulated investigations to prove the genetic cause of such polymorphisms. One of the first investigations was by Tolman (1924) who selectively bred fast and slow maze learners. However, these experiments did not demonstrate positive results, because the expected segregation was not obtained in the F_2 progeny.

Many authors reported successive selection of animals which differed in the learning rate of active avoidance in mazes. After 2, 8, and 17 generations the desired strains were obtained which were significantly different in their ability to learn. A number of new facts proved the genetic determination of the food-procuring reflex (Sadovnikova-Koltzova, 1925). It was demonstrated that (1) there is a correlation between the capabilities of the parents and their progeny to find food in a maze. This correlation becomes stronger from generation to generation; (2) the distribution of this capability in a population has three peaks. This suggests that there are three genes affecting this capability; (3) the high learning capability in one maze did not transfer to other experiments using different conditions.

We tried selecting two rat strains that were different in the rate of the conditioned food-procuring reflex. The analysis demonstrated a normal distribution of this trait in the population. The heritability coefficient in those populations was 0.34 and it was calculated on the basis of double coefficients of correlation for the progeny and parents. Out of the progeny 49% had a low rate of learning, while 22% had a high rate of learning when the parents had a low rate of learning. It is possible to suggest a paternal type of inheritance because 42% of the progeny had a high rate of learning when the fathers had a high rate of learning (Shumskaya et al., 1975; Korochkin and Shumskaya, 1976). The data suggest, however, the inheritance of this type of behavior.

The method of diallelic crosses was used to analyze another type of learning – the conditioned reflex of avoidance. The influence of parental heredity on progeny learning was demonstrated. Maternal influence on this behavior was not found. Quantitative analyses of all crosses demonstrated that this trait has a dominant and overdominant mode of inheritance.

Comparison of various facts from the above experiments suggests that a complex polygenic system participates in the formation of learning capability. It should be mentioned that animals with a low capability to learn an action can demonstrate the specific capability to learn and to remember other actions. In this case, comparative genetic, biochemical, and morphological investigations would be interesting and beneficial.

Memory processes are accompanied by the activity of specific nerve cell groups. There is insufficient data to show that functional and biochemical processes have specific peculiarities which result in the synthesis of specific molecules. However, the activation of the genetic apparatus of neurons was found during their regular activity. This activity is associated with the necessity to supply proteins which are expended during normal nerve action. This was demonstrated in many laboratories and especially in Hyden's (Sweden), Brodsky's, Pevsner's and in our laboratories (U.S.S.R.).

Many facts suggest the activation of the genetic system associated with nervous system function:

1. The intensification of RNA and protein syntheses and RNA transport from the nucleus into the neuroplasm was demonstrated for active neurons. Neurons with higher levels of excitability are characterized by more intensive RNA synthesis. There is a reason to make a parallel between the neuron function and the enzyme induction which accompanies the activation of many genes (Meerson, 1963; Pevsner, 1963; Brodsky, 1966; Gratcheva, 1968; Hyden, 1968; Korochkin, 1972c). RNA which was synthesized in the neuron nucleus is transported by axoplasmic streaming into the synapses where it serves as a matrix for the synthesis of synaptosomes and, therefore, could promote the organization of new synaptic bonds (De Larco et al., 1975). However, it is not known whether new genes are activated or the activity of already active genes is increased. The latter seems to us more likely although the observed data show contradictions (Uphouse and Bonner, 1975).

2. The suppression of RNA and protein syntheses with specific inhibitors (actinomycin, puromycin, cyclohexamide) induced abnormalities of the brain in some cases (Karasik et al., 1971; Flexner et al., 1963; Agranoff et al., 1965). The results of these experiments should, however, be interpreted with care because the antibiotics used are toxic even in small doses (Karasik et al., 1971, 1976).

3. RNA synthesis can be accompanied by improvement of neural functions (Brodsky, 1966). Moreover, some mediators which take part in neural functions stimulate RNA synthesis.

It should be mentioned that the participation of the genetic system in normal neural cell function occurs mainly at the post-transcriptional and later levels. There is a small possibility that the different functional conditions of cells in some tissue systems were affected by the differential activity of their genes. Such differences determine the various paths of tissue system development and their differential conditions.

It was demonstrated in very accurate experiments (Edström and Grampp, 1965) that normal nerve cell function is not accompanied by significant changes in nuclear function. The investigators experimented with isolated giant neurons from the stretch receptors of the crayfish. They simultaneously registered electrophysiological and biochemical parameters and hence compared the functional cell characteristics with the activity of their nuclei. It was demonstrated that normal neuron function does not accompany changing RNA synthesis or constitution, even within 24 h. Moreover, almost total inhibition of RNA synthesis in the neuron by actinomycin does not affect cell activity.

The constant synthesis of RNA molecules (and naturally the maintenance of transcriptional activity of loci which control this synthesis) is required for the synthesis of acid mucopolysaccharides in the differentiated cells of the pancreas. At the same time, the activity of some enzymes remains the same, even when RNA synthesis was totally inhibited (Davidson et al., 1963).

Sometimes, RNA synthesis inhibition with actinomycin can cause the activation of cell functions, i.e., heart cell contraction in in vitro cultures. It is possible that the RNA fractions (the synthesis of which was blocked in these experiments) control the synthesis of some inhibitors that act on the translational and posttranslational levels (McCarl and Shaler, 1969).

The course of many physiological processes (spread of the nerve impulse and, sometimes, the lack of nerve connection, contraction of the muscular fiber, etc.) goes so fast that direct participation of genes in regulation in such processes is simply impossible. Apparently, genes regulate and determine only the degree of reaction and the limits of variability in the realization of functions which are different for various organisms. This regulation specifically directs the process of cell differentiation and the establishment of cellular interactions. Systems already differentiated function within the limits which are determined by the genotypes and realized during development as a result of genotype environment interaction. The regulation of the function itself occurs primarily at the level of translational and posttranslational events.

What are the features of the brain that could play important roles in learning?

1. The mass of brain (primarily, the number of neurons). Principally, large cell masses are able to form complicated intercellular connections which represent a physiological basis for learning (although a large number of neurons is not in itself enough for a successful memory). As a result of selection for brain weight (Wimer et al., 1969), strains were obtained which differed according to activity in open field and according to ability to learn. The differences in the brain weight (according to morphological parameters) were found in progeny of these strains. The weight of the brains of "smart" rats was higher than that of "stupid" rats. Nonsignificant increasing of brain weight in smart rats was obtained by others (Silverman et al., 1940) in analogous experiments. This tendency was analyzed in more detail in the progeny of both strains. The comparison of growth dynamics of the brain demonstrated that within 30–150 days the S_3 strain of rats (smart) have a greater growth rate than the S_1 strain rats (stupid). Further experiments dealing with establishment of correlations between brain weight and rat behavior demonstrated that S_1 rats raised under conditions of frequent outside communication and which received additional training had greater brain weights (mainly due to an increase of the cortex) than S_1 rats raised under isolated conditions. Similar results were obtained in experiments with S_3 rats, but the difference in weight was less significant. Therefore, the S_1 brain was more labile according to morphological, physiological, and biochemical parameters when compared with the S_3 brain (Rosenzweig et al., 1958).

Brain mass was increased in experiments with the African fish *Tilapia macroce-phala* (Bresler and Bitterman, 1969). This increase in brain mass correlated well with an increased ability to learn complex behaviors.

2. Neurons have the capacity to form new branches and new synapses. One of the first hypotheses to explain the mechanism of memory as a cell function was the theory of the conditioned reflex (Pavlov, 1937; Hebb, 1949).

It is known that every cell of the CNS is associated with many nerve fibers and on its surface and the surface of branches there are 1000 or more contact-synapses (Sholl, 1956). Experiments have demonstrated that using synapses over a long period of time can increase their size (Eccles, 1953).

Apparently, not all cell contacts are able to transfer an impulse at any given moment. These contacts function only at a certain time of animal learning. In other words, the establishment of contacts between neurons does not mean there is a functional association between them. Additional structural-physiological transformations in the synaptic apparatus are necessary for its function. These transformations initiate the beginning of the synapsis function.

The memory process and the formation of conditioned reflexes, then, should be associated with the initiation of the already existing but nonfunctional synapses rather than with the establishment of a new interneuronal association of neural fibers (the rate of neuron growth is 3–4 mm per day which is too slow and uneconomical).

The hypothesis of a dominant role for RNA in the memory processes is widely popular. This idea was formulated for the first time by Hyden (1959). He states that the replacement of certain nucleotides by others occurs in the RNA of neurons and glial cells. This replacement can occur as a result of the instability of the RNA molecules due to a change in the ionic equilibrium after a nerve impulse. Such RNA determines the synthesis of specific protein molecules which code the memory. A number of experiments were performed to prove this hypothesis (Hyden, 1968; Hyden and Lange, 1969). Later, Hyden (1974) suggested that participation of a neuron in the learning process stimulated the action of certain loci and, therefore, induced the synthesis of certain specific RNA fractions.

Rischkov (1965) proposed a mechanism of such stimulation based on experimental observations that various salt concentrations can stimulate the puffing of different loci in the giant chromosomes of the salivary glands of *Chironomus thummi* (Kröger, 1963). However, some investigators (Barondes, 1965; Stewart, 1973) are not satisfied with such an interpretation and have suggested that Hyden's experiments do not allow for the differentiation of specific and nonspecific nerve tissue. Finally, it is not known whether the nucleotide composition of RNA of neurons changes because of the memory process or because of an increase in neuron activity.

The hypothesis of a role for RNA in the establishment of memory has stimulated a number of experimental investigations of two types: (1) the influence on memory of various compounds which disrupt RNA synthesis (inhibitors, base analogs), and physiological processes in the CNS; (2) the influence of purified RNA on memory.

The results obtained using compounds that block RNA synthesis are contradictory. Actinomycin injection (intracerebral) had no effect on certain conditioned reflexes in mice. However, the investigators used high doses of this antibiotic. The animals died after 24–48 h (Barondes and Jarvik, 1964; Barondes, 1965). Similar results were obtained by Appel (1965). He reached the conclusion that changing the character of intercellular interactions depends upon the dynamic interaction of the postsynaptic cells with presynaptic cells. It does not depend upon constant macronuclear programs. Therefore, it cannot be conceived that a program is limited to the inside of one cell — it should be determined by the interaction of several neurons and can be expressed as the effect on electrical activity of changing biochemical conditions and vice versa. There is indirect evidence that the metabolic character of animals which differ in learning ability could be genetically determined. Electrophysiological and biochemical differences were obtained when "good" and "bad" learning rats were analyzed in shuttle boxes. Slow-learning rats were obtained after seven generations of inbreeding. Fast-learning rats were obtained from laboratory populations after preliminary tests. It was demonstrated that homo- and heterosynaptic potentials during training were not realized in rats of the first group, but were obtained for the second group. Rats with a low inherited learning ability have a lower rate of K^+ release from cells into intracellular space (Izquierdo et al., 1972). Increasing RNA contents — a process which depends upon K^+ concentration in the intracellular space — was obtained only in rats from the fast-learning group after 25 min of electrical stimulation of the cortex (Izquierdo, 1972).

To summarize, RNA undoubtedly influences neuron action. However, the mechanism of interaction is still not clear. Most likely RNA is associated with the heightened necessity of the neuron for increased protein synthesis. The active degradation of protein occurs constantly in neurons (Hyden, 1960, 1968; Brodsky, 1966). Therefore, the protein pool must be constantly renewed. Consequently, the nerve cell requires constant RNA synthesis. It is natural that any distortions of this synthesis induce abnormalities in neurocellular function.

Smith (1962) has drawn an analogy between memory and substrate induction in microorganisms. He pointed out a number of common features in these two events. Particularly, a nerve impulse induces an increase in the activity of some enzyme systems (e.g., esterases). This is caused by the stimulation of secretion and synthesis. The mediators of these increases can serve as specific inductors which initiate the increase of enzyme biosynthesis. Such induction and the associated success of training can depend upon the amount of substrate.

Immunological concepts have invaded neurology as well. Particularly, Hechter and Halkerston used these concepts to explain mechanisms of memory (Hechter and Halkerston, 1964). They considered the possible role of cyclic 3', 5'-adenosine monophosphate since this compound is the mediator of peptide hormone actions (glucagon, ACTH, vasopressin) and biogenic amines action (serotonin, noradrenaline, histamine, etc.). These cAMP's can also serve as messengers of information from biogenic amines to the reacting system. Apparently, biogenic amines associating with proteins of the neuron can play the role of specific antigens which induce antibody synthesis in nerve cells. This leads to an antigen – antibody reaction which is a molecular mechanism of recognition. According to Hechter and Halkerston, the sensor stimulus activates certain specific neurons that contain biogenic amines. These specific neurons are distributed in various parts of the brain — particularly in the limbic region. They are associated with specific nerve cells which are stimulated by acetylcholine and biogenic amines. Acetylcholine stimulates electrical activity and biogenic amines increase cAMP synthesis. Increasing RNA and protein synthesis in neurons is the consequence of increased cAMP production. The formation and maintenance of active neuron chains for 30 min or more (short-term memory) depends upon cAMP formation in nerve cells which are activated by specific neurons.

The connection between RNA and protein syntheses (especially the organospecific S-100 and 14-3-2 proteins) and the learning process was demonstrated by numerous biochemical methods (Hyden et al., 1973). This interdependency was seen at various times during the establishment of the conditioned reflex.

In our laboratory (Vyasovaya et al., 1975; Malup and Sviridov, 1978) a correlation was found in laboratory mouse strains between the capacity for training and the S-100 protein content of the brain. Mice which are most capable of learning (DBA/2, PT, and Re/Re strains) are characterized by a maximal S-100 protein content whereas the "stupid" mice (C57L, CC7BR and tftf strains) have a lesser content of this protein in the brain:

C57	27.8 ± 0.87	A/He	37.27 ± 0.6
tftf	30.63 ± 1.9	DD	38.47 ± 0.93
CC57BR	30.94 ± 1.9	Ttf/tc$^+$	38.57 ± 0.9
AKR	31.88 ± 1.55	Bal/Bc	38.57 ± 1.3
C3H	32.11 ± 2.75	Re/Re	40.21 ± 1.47
AU	32.18 ± 1.9	nu/nu	40.81 ± 2.04
C3H/He	34.17 ± 1.26	PT	42.36 ± 1.28
CBA	34.10 ± 1.5	DBA/1	46.54 ± 2.4
UT	36.56 ± 1.36	DBA/2	48.84 ± 1.6
C57BL	37.24 ± 1.98	(mg S-100 protein/1 g of total brain weight).	

The initial levels of transcriptional and translational processes and their role in the learning processes can be characterized when various strains of animals with genetically determined types of behavior are analyzed. Such work was done in Hyden's laboratory (Hyden et al., 1973). The analysis of specific protein (14-3-2) synthesis in the nervous system was performed on two Wistar strains of rat which were obtained after 16 generations of inbreeding. These strains differed in the rate of establishment of the conditioned reflex of active avoidance. The incorporation of H^3- and C^{14}-labeled precursors of valine into the 14-3-2 neurospecific protein was higher in the fast-learning rats compared with controls in four important parts of the brain: sensory and optic parts of the cerebral cortex, hippocampus, and enthorinal cortex. Animals from the slow-learning group did not demonstrate such differences. On the basis of these experiments, the researchers came to the conclusion that the metabolism of this protein reflects the process of learning and memory. Interstrain differences in behavior and learning capability are accompanied by differences in activity of a number of enzymes which participated in neurotransmitter metabolism (Oliverio, 1974; Will, 1977). Slow-learning C57 mice have a lower level of choline acetyltransferase (ChAc) and acetylcholinesterase (ACHE) in the temporal lobe when compared with SEC and DBA mice (Mandel et al., 1974). F_1 hybrids are similar with their dominant parents for ChAc but not for ACHE (Tunnicliff et al., 1973).

The various hypotheses on memory should be considered as mutually exclusive (Olenev et al., 1968). One must consider the hypothesis of the gradual development of memory.

As mentioned, the ontogenetic formation of memory system is a complex process which certainly should be polygenetically determined. Therefore, present morphophysiological data are in good agreement with genetic data which demonstrate the polygenic control of learning capabilities. The widely accepted opinion about "memory molecules" is difficult to fit with this concept because in this case the memory process would be more simple.

Some pathological events of nervous system function are also determined genetically, e.g., epilepsy (Krushinsky et al., 1968). Two strains of rat were raised. One of these was resistant to an acoustic irritant and another responded to such an irritant with an audiogenic epileptic seizure (Krushinsky et al., 1968). The pathological reaction of these rats was genetically determined. Some believe it is inherited as a monogenic recessive trait (Elkin, 1971). However, according to Krushinsky, the inheritance of this characteristic is more complicated. Interstrain differences in RNA metabolism in the cells of Deiters nucleus and part of the cortex were found in adult animals. The RNA content of neurons of Deiters nucleus is higher in rats with an inherited epilepsy predisposition (KM) than in resistant rats (487 ± 12 and 437 ± 1 mkg per neuron, respectively). The RNA amount per neuron decreases after regularly repeated convulsions (Maximovsky, 1970). RNA synthesis and its transport into the cytoplasm is also higher in the Deiters neurons of KM rats (Raushenbach and Korochkin, 1972). The reduction of RNA content (20% less) in neurons of Deiters nucleus after intracerebral injection of actinomycin D (1–1.5 mkg) was accompanied by a termination at the convulsion response to an auditory irritant. This reaction was reversed after the RNA concentration returned to normal concentrations (Karasik et al., 1971, 1976; Korochkin, 1972c, 1973). A comparison of distribution of Deiters neurons according to their RNA content demonstrated that the changes in the mean amount of RNA per cell is determined by increases of RNA content in a limited group of cells called duty cells, but not in all cells of a neuron population. This, perhaps, is reflective of the ratio of resting and active neurons in the center of excitation (Maximovsky, 1970). Cell function is accompanied by the reduction of protein contents. The activation of the cellular processes of transcription and translation is necessary for the restoration of the used cellular resources. This was demonstrated by a number of authors using nerve tissue, heart, kidney, etc. (Meerson, 1963; Brodsky, 1966; Hyden, 1968; Geito, 1969; etc.). As mentioned above, animal learning is accompanied by increased RNA synthesis and by the increased concentration of S-100 protein in certain parts of the brain (Hyden and McEwen, 1966). It is possible that the animal strains, which differ according to their capacity for learning, are characterized by genetically determined differences in intensity of RNA and specific protein synthesis. It was demonstrated that some types of neurons (i.e., neurosecretory) are characterized by a cyclic RNA and protein synthesis. During intensive RNA synthesis the neurosecretory product is secreted from the perikaryon, whereas increase of neurosecretory protein synthesis is accompanied by decreasing RNA synthesis in the corresponding cells (Belyaev et al., 1966).

In the second case, genetically controlled changes in the proportions of various cells were observed. Particularly, diet changes are accompanied by gradual changes of enzyme contents of the digestive juice. This occurs as a result of the shedding of secretory cells of the intestine and their replacement by new cell populations which synthesize the necessary enzymes (Ugolev, 1972).

The possibility is not excluded that such fundamental processes as learning and memory are associated with the natural death of certain nerve cells (perhaps, inhibitory cells). It is known that many nerve centers contain extra numbers of cells during early stages of development and differentiation. During specialization "extra" cells degenerate and the definitive organization of the corresponding neuroregulatory complex is formed (Hughes, 1968). It is possible also, that cell elimination is a natural event which occurs at the tissue level and is necessary for successful function.

Thus, the final level of development of nervous tissue (as well as other tissue systems) is determined by a number of genetically controlled processes:

1. The mass of tissue and the number of cells that composed it. In turn, the cell number depends upon the rates of proliferation and physiological degeneration of the cell during ontogenesis.
2. Regional characteristics of cell distribution (qualitative and quantitative) inside a given organ
3. Adaptive and regulatory capabilities of cells of a given organ.
4. The capability of cells to form branches and intercellular connections (in the case of nerve tissue).
5. The capability to establish new synapses for normal function.
6. The capability of certain cellular groups to degenerate rapidly in one morphophysiological complex when functionally necessary.

The study of the genetic mechanisms of the processes is an important aspect in phenogenetics. Answers will help us to understand the basic tendencies of ontogenesis and tissue and cell activities.

2.4 The Organization of Systems Which Control Differential Gene Expression

2.4.1 The Organization of Gene Regulatory Systems in Bacteria and Phage and the Function of These Systems in Ontogenesis

Analysis of the genetic mechanisms of control of development is connected with the problem of regulation of gene activity. Naturally, cell specialization is due to an organized sequence of function of various genetic systems. A knowledge of the "anatomy" and functional organization of the genome is required for an explanation of the regulation of genome action. Moreover, systems must be present in higher organisms which control the mosaicism produced by differential gene expression in cells and tissues that develop in different directions. Such a system should include gene interactions which affect differential transcription, and gene complexes which can potentially control certain superstructures at the post-translational level.

Our knowledge of bacterial and phage genome organization and function is considerable. Analysis of these systems is didactic because such an analysis allows us to extrapolate some of the principles of prokaryotic genome organization to eukaryotes.

Let us first discuss modern concepts about the genetic regulatory systems of bacteria and phage. The basic ideas were proposed by Jacob and Monod (Jacob and Monod, 1961; Jacob and Wollman, 1962; for review see Benzer, 1963; Kumar et al., 1970; Szybalski, 1972; Ratner, 1975; Flint, 1977; Smith, 1977; Nover et al., 1978). It was demonstrated that in bacteria there are blocks of contiguous structural genes which code for specific proteins and which are assembled into functional units or operons. These units participate for instance in the regulation of the synthesis of enzymes which in turn participate in lactose utilization (Fig. 94).

The operon has a promotor (P) located at its 5' end. RNA-polymerase may associate with this promotor at one of two sites: (1) the site for a protein of catabolic repression and cAMP, and (2) the site where RNA-polymerase attaches to the operon. Near this promotor is an operator (O) which is not transcribed but is able to associate with a specific repressor protein and thereby inhibit RNA synthesis.

If the operator is translocated to another part of the chromosome, RNA-polymerase goes through it, even if this operator is associated with a repressor. Thus, the repressor inhibits RNA-polymerase attachment to the promotor but does not inhibit its motion along the template. The next DNA region, which is located immediately before the block of structural genes, is called the starter because it is here that transcription is initiated. The RNA-polymerase of the lac operon moves from the promotor to the starter (without transcribing this region) in preparation for initiation.

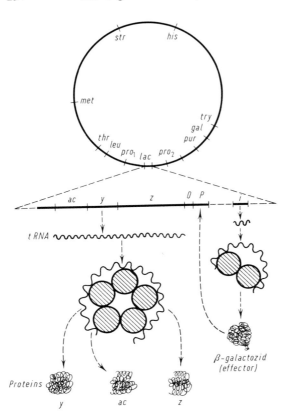

Fig. 94. Jacob and Monod's scheme of lactose utilization. *z*, α-galactozidase gene; *y*, permease gene; *ac*, transacetylase gene; *O*, operator; *P*, promotor. (After Bresler, 1973)

The operator of the lac operon in *E. coli* (Fig. 94) is composed of four sites — each of them five nucleotides in length — which are separated by one–two nucleotide spacers. The synthesis of repressor proteins is controlled by a special gene known as the regulator (i). The protein repressor for each operator is constantly being synthesized, but at a very low rate. No more than 10–20 molecules are present in the cell at any time. The attachment of the activated operon protein (CAP), which is an acceptor of cAMP, and the removal of repressor protein from the DNA of the operator under the control of substrate (lactose) concentration, are both necessary for the initiation of transcription. Cyclic AMP does not associate with the activator (CAP) if its concentration is low in the cell. Hence, the operon remains in the repressed state (Blattner and Dahlberg, 1972; Blattner et al., 1972; Khesin, 1972; Bresler, 1973). The termination of transcription is also determined by a special site on the DNA.

There are two mechanisms for operon regulation — negative and positive — and there are two types of control — induction and repression. Therefore, there are four possible variants in the regulation of genetic systems: negative induction, negative repression, positive induction, and positive repression (Ratner, 1975).

The lac operon in *E. coli* is an example of negative induction (Fig. 94). This operon contains three structural genes — *z*, β, and *ac* — which code for β-galactosidase, permease, and transacetylase, respectively. Permease participates in the regulation of lactose penetration into the cell. Once in the cell, the lactose is split into glucose and galactose by β-galactosidase. The function of transacetylase is unknown. If lactose is added to the bacteria cultivation medium, this substrate binds the protein repressor which is blocking the operator of the operon. As a result, transcription of the structural genes of the operon is induced. In this operon, then, there is negative feedback between the substrate concentration in the medium and the level of enzyme production.

Negative repression was investigated in the tryptophan (trp) synthesis operons of *E. coli* and *Salmonella typhimurium*, and the histidine synthesis operon of *Salmonella*. The trp operon is totally derepressed in the presence of nonsurplus amounts of tryptophan. Here, transcription and translation go at maximal rates. Enzymes of the tryptophan system synthesized it in amounts which are totally utilized during translation. This condition is known as turn on. The activity of the first enzyme in the sequence of tryptophan synthesis — anthranilate synthetase — is inhibited within several seconds after the addition of surplus tryptophan. The aporepressor becomes the repressor and inhibits transcription of the operon (negative repression) after it associates with the effector, tryptophan. This association occurs if the surplus of tryptophan exists for a sufficient amount of time. After several minutes, most of the resulting mRNA is degraded and the rate of synthesis is noticeably reduced (known as the turn of condition).

Positive induction has been described for the arabinose operon (ara) in *E. coli*. The regulator produces a repressor protein tetramer which inhibits the ara operon by affecting the operator. The product of the regulator cistron has not only lost its properties of repression, but becomes a positive inductor, i.e., an activator of transcription after it associates with the effector, arabinose (Jacob and Monod, 1964; Bresler, 1973).

The regulator gene synthesizes activating protein in the case of positive repression, which is partially or totally inactivated when it interacts with the final product of the operon. Such a system, for instance, controls riboflavin synthesis in *Bacillus subtilis* (for details, see Ratner, 1975). Also, in the λ phage it has been demonstrated that antirepressors which both inhibit repressor synthesis and inactivate it, and antiterminators which eliminate termination may play a role in the turn-on of new operons (Szybalsky et al., 1972; Khesin, 1972).

What are the modes of function of these regulatory systems in the ontogenesis of prokaryotes? For an analysis of this problem, the reviews of Ratner (1970, 1974a) and Szybalsky et al. (1972) will be used. The λ phage genome contains about 48,000 base pairs — i.e., about 50 cistrons for 50 types of polypeptides (Fig. 95). There are two periods of phage ontogenesis: (1) the synthesis of specific products as a result of gene action, and (2) the assembly of structures from these products without the influence of these structures on the genome.

After it invades the bacterial cell, the phage genome exists as either a prophage — i.e., it is incorporated into the bacterial chromosome — or it goes in the direction of lysis (vegetative multiplication).

There are four paths of ontogenesis possible for the λ phage: (1) lytic growth after invasion, (2) lysogeny after invasion, (3) lytic growth after prophage induction, and (4) lysogeny after induction. The sequence of steps of ontogenesis have

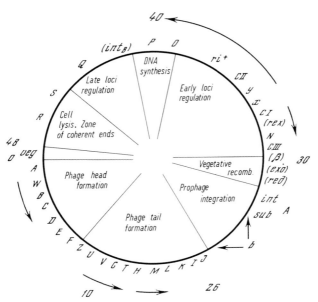

Fig. 95. Genetic map of lambda phage projected on the circle molecule DNA. *Numbers* thousands of nucleotide pairs. *Arrows* point to the polar effect of some mutations, which could, apparently, be explained by the presence of common transcriptional units. (After Ratner, 1970)

been proposed as: (a) those which occur very early and involve bacterial RNA polymerase recognition without participation of any product of phage cistrons; (b) those that occur moderately early and involve the antiterminator product of the phage N cistron: (c) those which occur late and involve the phage Q cistron product — a σ' factor which is part of the RNA polymerase.

In general, then, phage ontogenesis is characterized by the function of various stages and by the presence of specific genes which regulate by turning on or turning off the structural cistrons.

The prophage genome is almost totally repressed in that it does not function and is passively reduplicated together with the bacterial DNA. This repression is determined by the synthesis of the protein repressor C_1 which either directly or indirectly inhibits the rest of the gene systems (Fig. 96). The λ phage particles inject their linear DNA molecules into the bacterial cell. This DNA is rapidly converted into circular molecules. The transition of prophage of lytic growth terminates the inhibitory action of the C_1 repressor and turns off the system of its synthesis (CI regulator). As a result, transcription of certain phage cistrons occurs. Figure 96 shows the scheme of regulation for this transition.

Early period gene transcription starts before the conversion of linear DNA molecules to circular molecules and is due to bacterial RNA polymerase. Within several minutes, enough molecules of antiterminator (product of the N cistron) are synthesized. This stimulates transcription at many loci and determines the transition to the moderately early period.

Fig. 96. Interaction and sub-ordination of the lambda phage loci during ontogenesis. A hypothetical system controlling together with CI regulator the selection and the regulation of subprogram of lytic growth or prophage conditions. (After Ratner, 1970

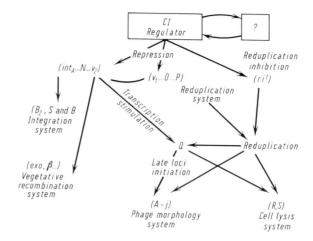

Three loci of the recombination system, two loci of integration, one locus of bacterial metabolic suppression, two loci of duplication, and the C_{II} and C_{III} loci which control the selection of developmental pathways (lytic growth vs lysogeny) begin to function within the first 5–7 min. Transcription of the late loci is possible only after circularization of the phage DNA. This conversion of DNA structure, as mentioned above, occurs very early and requires the aid of bacterial ligase.

If conditions in the bacterial cell are favorable, phage will select the lytic path-way and the Q locus is activated. This locus produces the σ' factor which replaces the σ factor of the host RNA polymerase. This replacement initiates transcription of all cistrons necessary for phage morphogenesis. By 15–20 min after infection, when the late loci begin to act, there is a high concentration of antirepressor (from the locus tof or cro). Now the N cistron is turned on and controls the transcription of certain morphogenesis genes. Transcription of the C_I DNA segment is termi-nated, so prophage maintenance does not occur. Once complete, the lytic enzyme (from the S cistron) breaks the bacterial membrane and the phage progeny are re-leased.

The choice between the lytic and lysogenic paths of ontogenesis is determined by the ratio between C_{III} and tof proteins. Together these proteins make a nonac-tive complex. A surplus of tof antirepressor dictates the transition to lysis while a surplus of C_{III} protein leads to lysogeny.

With lysogeny, rapid transcription is initiated in the C_I region. Increasing con-centrations of C_I protein repressor turns off transcription in all operons involved with lysis. At the same time, the loci which determine lysogeny and are involved in incorporation of the phage DNA into the bacterial chromosome are transcribed. Ultimately, however, only the small region of DNA containing the C_I locus re-mains active.

It is of interest to note that no specialized genes were found which control the temporal sequence of events in the genetic program of ontogenesis. The sequence of morpholological periods is determind by the sequence of specific associations of protein complexes during the phage assembly process. This sequence of morphogenetic periods is not determined directly by the activation and inactivation of specific genes according to a defined timetable.

The functional organization of the bacterial and phage genomes allows the temporal control of development of the inherited information. Thus, immediately after infection, the σ subunit of the *Bacillus subtilis* host RNA polymerase directs transcription of the early phage genes. Then, the protein product of gene 28, a regulatory gene (and one of the early genes), turns on the moderate genes by inhibiting the host transcriptase. The precise in vitro transcription of the moderate genes by RNA polymerase containing the gene 28 does not require the host σ factor. Rather, the transcription is dependent upon a novel *B. subtilis* protein, δ, or upon the assay conditions of high ionic strength.

The protein products of genes 33 and 34, which are also regulatory genes (they belong to the group of moderate genes), associate with the host RNA polymerase. This leads to the activation of late genes. Therefore, there is a specific, regulated sequence of gene activations which control the stages of phage development (Gage and Geiduschek, 1971; Pero et al., 1975; Tjian et al., 1976).

Ratner (1970, 1974a) suggested that morphogenetic genes act simultaneously and their products form successive chains of interacting structures. An estimation suggested that the number of different structural phage proteins is half the number of morphogenetic genes (Khesin, 1972). In phage, however, there are specialized gene systems to control the formation of structures and, in particular, there are genes that control the production of enzymes.

It was demonstrated for *E. coli* that the genes controlling glucose catabolism are grouped into four regions. This grouping is statistically highly significant. On the circular *E. coli* genetic map, each of these four groups is 90° or 180° with respect to each other. This nonrandom gene position suggests a relationship between the position and expression of certain groups of genes that control related biochemical reactions (Riley et al., 1978).

The coordinated gene action in prokaryotes is apparently not determined by the spatial organization of these genes. Ribosomal RNA genes are located in different chromosomal regions. However, they are closely coordinated in function (Morgan and Kaplan, 1976; Yamamoto et al., 1976). The regulation of rRNA synthesis is determined by several genes. Mutations in these genes lead to an altered control of rRNA synthesis. The participation of guanosine tetraphosphate in the regulation of this synthesis has been demonstrated with the rel A mutation. This compound interacts with RNA polymerase and transforms it into a form which has a low efficiency of formation of open complexes at the promotor site for rRNA genes (Cashel, 1969; Edlin and Broda, 1968). It should be mentioned that some of the organization of the gene regulatory systems in prokaryotes is seen in eukaryotes.

Thus operon for histidine utilization has a unique property compared to the lac and ara operons of *E. coli*. In the lac and ara operons, specific proteins regulate transcription. These proteins are specialized for this regulatory role and do not perform any other function. In the case of the histidine utilization operon, however, the complex enzyme glutamine synthetase interacts with the regulatory elements of the operon. Hence, this enzyme serves as the regulatory protein (Tyler et al., 1974).

In rare cases, regulation of repressor synthesis in phage was found to act at the level of translation (Steitz and Jakes, 1975).

Ratner (1970) suggested that in phage there is a special dynamic specificity which, together with a structural specificity, guides the interaction of the genetic system (in a manner analogous to the relationship between the repressor and operator or between RNA polymerase and the promotor, between DNA polymerase and the replicator, or involving the recognition of protein subunits by each other, etc.). A two-operon model serves as a possible example of dynamic specificity. When one operon is active, it interacts with the other operon and initiates repression of this second operon.

A question which needs to be answered is how similar is the prokaryote genome organization and function to that in eukaryotes?

2.4.2 Hypotheses on the Organization of the Gene Regulatory Systems in Eukaryotes

It could be argued that the genetic apparatus of eukaryotes has its own specificity and has a different functional organization from prokaryotes. This follows from the fact that in eukaryotes the inside environment is maintained in a constant condition (homeostasis) regardless of changes in the outside environment. In prokaryotes, however, the opposite occurs. In these cells, there is a maximal change intracellularly in response to changes in the environment. Eukaryotic genomes have the following characteristics: a high percent of chromosomal protein, a chief role for various regulatory proteins in morphogenesis, an important role for post-transcriptional and post-translational levels in the realization of the genetic information, and a significant amount of DNA that is concentrated into one gene. At the same time, however, scriptons, the unit of transcription, cistrons, the unit of translation, and replicons, the unit of replication, are characteristic for eukaryotes as well as prokaryotes. It is not clear, though, whether eukaryotes have the same regulation of these units that has been demonstrated for prokaryotes. In connection with this the classical scheme for the regulation of gene activity in bacteria has been modified in the hypotheses for genome organization in eukaryotes. All of these hypotheses, however, originated from the principles which were formulated by Jacob and Monod.

A key question with respect to the regulation of gene activity in eukaryotes is whether systems exist in eukaryotes which are analogous to the prokaryotic operons. Also, are there systems of substrate induction in eukaryotes such as those which are common in bacteria? Evidence for either type of system in eukaryotes is lacking. Similarly, specific gene regulators analogous to the gene regulators of bacteria and phage have not been found in eukaryotes. Although Ohno and others (see above) have noted the possible presence of such genes, the nature and mechanisms of the regulatory processes were not clarified.

Nevertheless, some interesting aspects of gene interactions in eukaryotes can be discussed in connection with the Jacob and Monod scheme. The example of the white locus has been discussed earlier. The most illustrative examples, however, are the experiments by Lewis on the bithorax locus in *Drosophila*, the corn regulatory system Ac-Ds which has been described by McClintock, and the work of Stern on the scute locus in *Drosophila*.

2.4.2.1 Bithorax Model

This model deals with a series of bithorax pseudoallelic mutations in *D. melanogaster*. The normal configuration of fly segments is presented in Fig. 97. The bithorax mutation, being a homeotic mutation, induces the transformation of one body part into another. The effects of various mutations on fly development is demonstrated in Fig. 98. The a (bithorax) mutation determines the transformation of the anterior metathorax into the anterior mesothorax (AMT → AMS). Such flies have two normal and two defective wings which form as a result of the transformation of the anterior part of the halter disc into the anterior part of the wing.

The e (postbithorax) mutation occupies the other end of the bithorax mutation series. With this mutation, the posterior metathorax is transformed into the posterior mesothorax (PMT → PMS). Progeny with this mutation and the a mutation demonstrate a complete transformation of the metathorax into the mesothorax. There is a double mesothorax with four wings.

The d (bithoraxoid) mutation induces the transformation of the first abdominal segment into the posterior mesothorax (PMT → PMS; AB → AMT). Such flies have eight instead of six legs. The c (ultrabithorax) mutation is generally lethal in the homozygous condition. However, the flies which survive show all the above described abnormalities with variable degrees of expression.

The b (contrabithorax) mutation is characterized by the transformation of the posterior mesothorax into the posterior metathorax (PMS → PMT).

All the described genes are closely linked and are located in the region of 58.8 on the third chromosome. It has been suggested that these genes control the synthesis of specific substances which influence morphogenesis. The e^+ gene controls the synthesis of Se substances which inhibits the ability of the metathorax to devel-

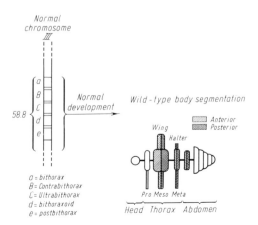

Fig. 97. Representation of *Drosophila* segmentation. (After Lewis, 1963)

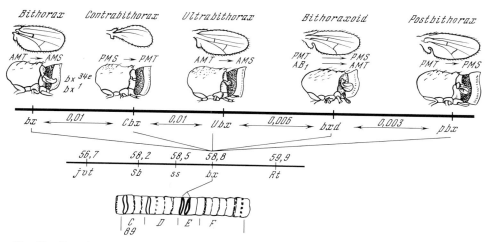

Fig. 98. Genetic map, bithorax locus structure and phenotypes of *D. melanogaster. Arrows* indicate the transformation of loci as a result of mutations; *AMT*, anterior metathorax; *PMT*, posterior metathorax (gray); *AMS*, anterior mesothorax; *PMS*, posterior mesothorax (white); *AB*$_1$ the first abdomenal segment (black). (After Swanson et al., 1969)

op into the mesothorax. The d$^+$ gene controls the synthesis of Sd substances which inhibits the transformation of the abdominal segment into the metathorax. Similarly, the a$^+$ gene inhibits the transformation of the metathorax into the mesothorax. The mutant alleles of these genes cannot synthesize these compounds. If there is at least one wild-type allele, the mutant alleles will not affect the phenotype.

There is a well defined polarity effect in the bithorax system: the transheterozygote d+/+e demonstrates the transformation characteristic for e but not for d. The a+/+e heterozygote shows the e transformation but not a, and the c+/+d heterozygote shows the type d transformation. Thus, each locus influences the loci to the right of it.

The existence of a specific inducible substances, X, has been supposed for the bithorax "operon". This substances is evenly distributed in the egg at a very early stage of embryogenesis. A concentration gradient for substances X occurs, however, due to the differences in mitotic rates in the anterior and posterior parts of the embryo. Specifically, the relative concentration of substances X will be highest in the posterior cells and lowest in the anterior cells of the embryo. The concentration of X is low in the presumptive cells of the mesothorax in the developing embryo and larvae. The synthesis of Sa, Sd, and Se substances does not occur in these cells. The concentration of inductor in the metathorax, however, is high enough to maintain the synthesis of the Sa and Se substances. This inhibits the ability of the metathorax to develop into the mesothorax. Cells of the first abdominal segment contain enough inductor to activate Sd synthesis which, in turn, inhibits the potential for thorax development in this region. According to this hypothesis, the b mutation is similar to the O^c constitutive operator mutation in the lac operon of *E. coli* (Lewis, 1963,1964). It should be remembered that all of this is speculation. The confirmational experimental evidence is, for the most part, lacking.

2.4.2.2 Transposable Genetic Elements and the Ac-Ds System in Corn

Endosperm coloration in maize is governed by an unusual system which has been described by B. McClintock. There are three types of endosperm coloration in corn: colorful, colorless, and spotty (color spots on a colorless background). The character of coloration depends upon the interaction between Ac (activator) and Ds (dissociator) genes. These genes control pigment synthesis. Both these controlling elements are located in heterochromatic regions. They could, however, be on different chromosomes.

The dissociator gene (Ds) is located near the coloration gene and inhibits its expression. The Ds gene, however, can induce chromosomal breakage at the border of the coloration gene. This gene has a capacity to be displaced to other chromosomal regions within the same chromosome or in other chromosomes. When this occurs, the corns demonstrate various degrees of coloration, depending upon the moment of dissociation during development. If Ds dissociates before corn formation, both the corn and the plant that develops from it will be completely colored Displacement of Ds during the beginning of seed development results in the appearance of large color spots. These spots are small if Ds is displaced during the later stages of seed ontogenesis when only a few cell divisions have occurred.

The activator gene (Ac) acts as a gene regulator. When this gene is absent, color spots are not formed and the seeds are white. In the presence of Ac genes, spots are formed. The more Ac genes present, the smaller the spot sizes. Apparently, the Ac genes delay Ds gene action and therefore not only reduce the ability of Ds to induce a break, but also regulate the moment of breakage (McClintock, 1950, 1956, 1961).

Dooner and Nelson (1977) investigated flavinoid glucosyltransferase variability and induction by the controlling element UDP. The synthesis of this enzyme in the endosperm is determined by the structural gene bronze (bz). A number of bz mutants were analyzed by transposing the Ds controlling element to the bz gene. Three of these mutations had no enzyme at any stage of endosperm development. Two had an altered enzyme. Two other bz mutations showed different patterns of ontogenesis.

In addition to flavinoid glucosyltransferase, another case of Ds-induced enzyme variation has been described for ADP-glucose pyrophosphorylase in the sh2m mutation (in sh2m, there is an association of the dissociation locus with the sh2 locus) (Hannah and Nelson, 1976). There are two possible mechanisms of regulation of the structural gene. First, the controlling element could inhibit structural gene activity on the transcriptional levels (the regulatory phenomenon): protein production is either reduced or terminated totally. Second, Ds association or integration could change the information encoded within the gene. Thus, the protein could change its catalytic activity (the mutation phenomenon). Peterson (1970) compared the controlling elements in maize with the regulatory systems of bacterial operons and with accessory genetic material (i.e., bacterial episomes) at the same time. Fine structure mapping of the controlling elements inside the wx locus demonstrated that these elements do not lie on one end of the wx locus. Rather, they may be transposed into the gene and induce the repression at the gene level (Nelson, 1968).

There may be a system similar to this in *D. melanogaster* (Sandler and Hiraizu-mi, 1961). Ordinarily, a 1:1 ratio of wild-type to white-eyed flies is obtained when homozygous white-eyed females (cn and bw, chromosome II) are crossed with dou-ble heterozygous males, $++/$cn bw. However, if the unmarked second chromo-some from the natural population is used, then 93%–99% of the progeny are wild-type. It has been suggested that this change in segregation is due to functional dif-ferences in the number of male $++$ gametes as compared to cn bw gametes which participate in fertilization. This was determined by a disturber of the segregation of SD which is located in the heterochromatin region of the right arm of the second chromosome, not far from the centromere. SD^+ was on the cn bw chromosome. SD induces changes which either causes the loss of the cn bw homolog or renders the sperm with this genotype incapable of fertilization. The SD chromosomes in this case predominate in the F_1. The synapsis of SD with SD^+ is necessary to block SD^+ action. Inversions, which prevent such synapsis, reduce the SD effect. The presence of two other second chromosome factors has been supposed. The first of these is the SD stabilizer St(SD), which determines the stability of the distorted seg-regation. This factor works both cis and trans with the SD gene. A second factor is necessary for SD action and is active only when cis with the SD locus. This factor is designated Ac(SD).

The mechanism of Ac-Ds action became clearer after the discovery of unstable genes and transposable genetic elements (Nevers and Saedler, 1977). The presence of such transposable elements or insertion sequences (IS) were first described for prokaryotes (Saedler and Starlinger, 1967; Adhya and Shapiro, 1969). Five non-homologous classes of such sequences are distinguished (IS1-IS5). The molecular sizes of these elements (800–1400 nucleotide pairs) are equivalent to one or two genes which are too small for phage DNA. IS sequences demonstrate a polar effect because gene activity is inhibited at the transcriptional level (Starlinger et al., 1973). These elements may be excised and, as a result, a reversion to wild-type occurs. The frequency of such reversions is one in ten million (Saedler and Starlinger, 1967).

Some investigators have suggested that the action of the IS segment is similar to that of the rho protein in bacteria and phage. The rho protein recognizes specific terminator sequences of DNA and, in doing so, terminates transcription. It is pos-sible that the IS elements contain such terminator sequences, and this leads to rho-dependent termination of transcription (Nevers and Saedler, 1977). The Ds-con-trolling elements of maize may be IS segments.

Cases of gene instability in *D. melanogaster* may be explained by IS segments (Berg, 1974; Golubovsky et al., 1977). A series of singed bristle (sn) mutations has been described. On the basis of the unusually high reverse mutation rate, ten of these mutations have been designated as putative insertion mutations. The rever-sion to wild-type in unstable strains is 10^{-4}–10^{-3}. This was explained by the succes-sive transcription and excision of the IS segment in a specific region of the X chromosome. Changing the time of ontogenic expression of the genes some of the IS segments may play an important role in cell differentiation. Such a shift in time can result in asynchronization of maturation of the inducible and competent tis-sues. Therefore, it could affect the normal events of the morphogenetic processes.

2.4.2.3 The Scute Locus in Drosophila

Another well-studied genetic system is that which controls bristle formation in *Drosophila*. Bristle development occurs according to the following (Less and Waddington, 1942):
1. Determination of the bristle cells (which is controlled by the ac, sc, eb [41c] and
↓ h genes).
2. Proliferation (spl and D genes).
↓
3. Orientation according to the surface (spl and D genes).
↓
4. Regulation of the placement of trichogenic (bristle-forming) and formogenic
↓ (auxiliary) cells (spl, H, Sb, sv genes).
5. Growth (ss and mr genes).
6. Secretion of cuticular compounds (sn, f, Bl, sv genes).
7. Darkening and hardening of cuticular compounds (sn, f, H, sv, Sb, y, e, stw, b
   and other coloration genes).

   The phenogenetic mechanisms of localized differentiation of bristles was investigated using scute-achaete mutations. Two explanations are possible to describe this process (Stern, 1954, 1968): (1) Genes which control bristle manifestation established regional differences which precede structural differentiation. (2) The structure (chaete pattern) which forms is the result of various genetically determined cell responses to factors which are nonrandomly distributed in the undifferentiated epithelium. Stern designated these factors as prestructural (Stern, 1954). The question to be answered is whether mutant genes change the distribution of the (prestructural) controlling factors or whether these genes determine the sensitivity of cells (which carry different alleles of the scute locus) to these factors.

   To distinguish between these two hypotheses, experiments with mosaics were performed. Stern (1954) analyzed mosaic flies with achaete (ac) tissues. The formation of three of eleven chaetes is reduced in flies homozygous for the sex-linked ac gene. The X chromosome carrying the ac gene was also marked by the gene which effects bristle coloration and by the sn gene which determines bristle shape. The second X chromosome was a ring chromosome and had the wild-type, dominant alleles of these genes. The recessive genes y, sn, and ac showed up phenotypically only when the ring chromosome was lost. Bristles formed by cells with the ac genotype are yellow and twisted, whereas bristles from wild-type cells are dark and straight. If the chaete was not formed, then the color and shape of the surrounding microchaete could be used to determine the genotype of a certain region of tissue. In Stern's experiments, yellow spots of various sizes formed on the thorax. The size depended on the stage of development when the ring X chromosome was eliminated. It was demonstrated that the ac⁺ tissue always forms chaetes regardless of the size of the spot and the place of bristle formation in the mosaic flies where the ac genotype predominates and a small region of ac⁺ tissue is present. In other words, chaete formation is autonomous and is determined by the genetic composition of the cells in the spot. Apparently there is no special prestructure which is characteristic for the ac epidermis; otherwise the small spots with the ac⁺ genotype should not form chaetes. Mosaics with ac spots on an ac⁺ background lost chaetes as often as flies homozygous for ac. In this case also, then, the ac$^+$ genotype does not establish its specific prestructure (otherwise, the bristles would not form). According to Stern, these data can be explained by supposing that the imaginal discs of both wild-type and mutant phenotypes have similar prestructures. The effect of the mutant ac gene is to change the ability of the cell to respond to this prestructure. Cells with the normal, wild-type ac allele are competent, on the other hand, and can respond to the prestructure by forming chaetes.

Thus, chaete formation is the result of the interaction of two factors: (1) the genetically determined prestructure, and (2) the genetically determined competence of the tissue to react to the prestructure. It is possible, however, that a third factor – the diffusion of gene-dependent substance which modify the final morphogenetic effect – also affects bristle development. In Stern's experiments where small ac spots were on an ac$^+$ background in the region of the posterior dorsal chaete, this chaete developed normally in spite of temperature conditions which regularly inhibit its formation. It may be that the diffusion of such an hypothetical substance induces the capability of the ac cells to respond to an invariant prestructure.

Among the many genes controlling bristle formation, the scute-achaete genes which induce chaete reduction have earned special interest due to their well-known mechanism of phenogenetic action. In addition, using the multiple alleles of the scute-achaete locus as an example, Serebrovsky and co-workers formulated the hypothesis of step allelomorphism (Serebrovsky and Dubinin, 1929; Dubinin, 1932a,b; 1933). They demonstrated that the scute locus controls chaete reduction in various parts of the fly body. Different mutations affect the reduction of various bristles (Fig. 99), and they are recessive with respect to wild-type. Two alleles act in conjunction to effect the reduction of different chaetes. The phenotypic traits, in the case of the scute phenotype, can be "subdivided" into many "subcharacteristics".

The genetic control of such "subdivisions" can be explained by two different models. The first model is based on Serebrovsky's hypothesis (Fig. 99) and was developed long before the modern conception of fine gene structure. This model states that the scute locus is an operon composed of an operator and a group of structural cistrons. Each of these cistrons is responsible for a specific chaete.

The second model was proposed by Ratner (Fig. 100). According to his hypothesis, the scute locus is an operon consisting of one structural cistron and a complex operator with both overlapping and nonoverlapping recognition sites. For each chaete there is a regulator protein which is specific for one site in the operator. Activation of specific sites results in the production of specific products and metabolites. Point mutation in the operator will inactive specific sites (Ratner et al., 1969). Neither this model nor the model based on Serebrovsky's hypothesis, however, have been proven experimentally.

It is possible that various scute mutations occur at different sites of the same structural cistron. The realization of the mutant phenotype (i.e., the reduction of a chaete in a specific body region) is determined by the nature of the product formed (i.e., a protein necessary for chaete development) and by external macromolecular conditions. Spatial differentiation would occur as a result of different conditions in different regions. Depending on the region of the protein in which the amino acid substitution occurs, the protein product would have differing functional properties in different parts of the developing fly.

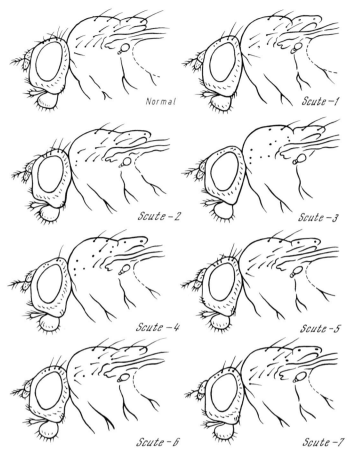

Fig. 99. Influence of various scute mutations on bristle development in *Drosophila*. (After Serebrovsky, 1938)

In addition to such spatial differentiation, the temporal specificity of the scute gene in various cells could influence chaete development. This possibility is explained by the temporal hypothesis proposed by Goldschmidt (1961). The time of scute gene activation could differ in various body parts. Specifically, the time of initiation of transcription for mRNA's that code for specific proteins can be an important determinant of the morphogenesis of bristles. Similarly, the intracellular conditions necessary for translation may occur in specific cells at specific times. A combination of spatial and temporal specificity in the intracellular environment may be the basis of the phenotypic variability of the scute mutations. In connection with this, it is interesting to note the results obtained by Ratner's group (Furman et al., 1977). The development of 12 scute mutant strains was studied at 11 °, 22 °, and 30 °C. From each strain 4 to 500 females and males were analyzed in addition to 200 heterozygous females from all direct and reciprocal matings. Using the data

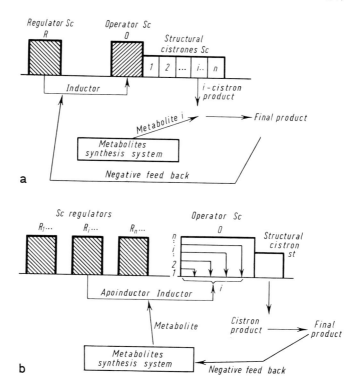

Fig. 100 a, b. Hypothetical schemes of the scute locus regulation. **a** Product of "i"-cistron corresponds to the "i"-bristle. "i"-product interacts with the "i"-metabolite, which specifically synthesized in cytoplasm of corresponding cells. Mutants with the defective "i"-cistron lost ability to respond on the action of this metabolite. **b** Regulator Ri corresponds to the "i"-bristle. This regulator is active only in the corresponding cells. The activated Ri product interacted with operator. Activation occurred under control of nonspecific (for bristles) metabolite. Mutants with the defect segment of operator specificity do not respond to the metabolite action. (After Ratner et al., 1969)

obtained, a mathematical model was built to describe the structure and function of the scute locus. The following results were obtained:

1. While the number of transcribed cistrons is unknown, this number is apparently small. Also, there is an unknown amount of aggregation between the products of the cistrons involved.

2. The specific response of the scute locus to prestructure occurs at the post-translational level. The protein subunits aggregate to form specific multimeric proteins. These proteins, in turn, associate with specific factors in the prestructure. Such an association initiates chaete formation in specific regions of the fly body (Fig. 101).

In this model, then, the scute locus acts by turning on the system which is responsible for the formation of a specific trait. The locus is responsive to various metabolic signals (Furman et al., 1975).

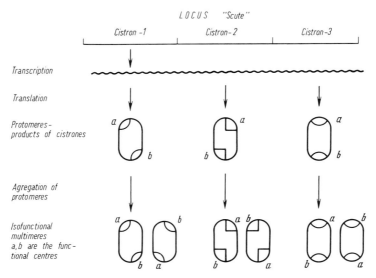

Fig. 101. Specific response of the scute locus on the prestructure. (After Furman et al., 1977)

None of the models cited above contradicts a monocistronic model. Data exist which suggest that certain large loci in *D. melanogaster*, which are considered complex loci, are monocistronic. This interpretation has been proposed for the rudimentary (Green, 1963; Carlson, 1971), Notch (Welschons, 1965) and maroon like (Chovnik et al, 1969) loci.

The Abruptex (a member of the Notch pseudoallelic series) locus and recessive alleles of the Notch locus have been studied in detail by Portin (1975, 1977). Using the following argument, he suggested that the Notch locus is monocistronic: visible recessive mutations in this region fall into two complementary groups of mutations: eye and wing mutations. This confirms that this locus has at least two functions. There is a large group of amorphic lethal mutations, however, which are dispersed through the gene and complement neither each other nor the eye and wing mutations.

The pattern of complementation between visible mutations and lethal Abruptex mutations of the Notch locus depends upon temperature. At low temperatures, the Abruptex mutations are complementary to all recessive mutations and their expression is high. At high temperatures the Abruptex mutations complement virtually none of the recessive mutations, and the Abruptex phenotype is very weak. These data suggest that the Abruptex mutation may produce a gene product which is not functional at high temperatures. These data also support the idea that only one mRNA is transcribed from Notch. Hence, this locus should be thought of in terms of one cistron. Similar complementation data to those cited above have also been obtained by Shellenbarger and Mohler (1975) for three temperature-sensitive Notch alleles: $1(1)N^{ts1}$, $1(1)N^{ts2}$, and $1(1)Ax^{ts1}$.

In another series of experiments with Abruptex mutations (Portin, 1975), it was shown that these mutations can be divided into two groups on the basis of viability:

one group contains the Notch suppressors (alleles 28 and 9B2) and the other group has alleles with a neutral effect in Notch (71d) or with an ability to increase the wing-nicking phenotype of Notch (E2 and 16172). The suppressors were designated as Ax^{SoN} (the Notch suppressor) and Ax^{EoN} (enhancement of Notch). Both Ac^{EoN}/Ax^{EoN} and Ax^{SoN}/Ax^{SoN} females are viable and demonstrate the Abruptex phenotype. Heterozygous Ax^{SoN}/Ax^{EoN} females have a reduced viability and, as a rule, are lethal. Alleles complementary to one group do not complement the other group. Rather, these alleles show a strong negative complementation. There could be several explanations for this:

1. The Abruptex gene is active twice during Drosophila development. Ax^{SoN} is active earlier than Ax^{EoN}. The heterozygotes accumulate abnormalities and consequently, negative complementation occurs.

2. The Abruptex locus is a sequence of nucleotides repeated in tandem which codes only one polypeptide. The final product of such a gene consists of two subunits which may or may not be identical. Ax^{SoN} codes for one subunit and Ax^{EoN} for the other subunit. In heterozygotes, a hybrid dimer forms which has an activity that is below some critical level.

3. The Abruptex (Notch) locus has regulatory functions and may be similar to the integrator gene of Britten and Davidson's model for gene regulation. The Abruptex gene has the characteristics of such a gene since the Abruptex gene has a large amount of pleiotropism and has multiple functions during ontogenesis. It should be mentioned that the Notch locus is very large — 0.13 map units — whereas the rosy cistron (the locus for xanthine dehydrogenase), for instance, occupies only 0.005 map units (Gelbart et al., 1976). Thus, although the Notch locus is 26 times longer than the rosy locus, both are considered as single cistrons. This represents a paradox which can be explained in terms of our knowledge about the molecular organization of the genome of higher organisms.

2.4.3 Molecular Organization of the Genome of Higher Eukaryotic Organisms

As mentioned above, the analogy between complex loci and the bacterial operon has not been proven convincingly (see Mglinez, 1973). It should also be noted that there is no proof of the existence of genes in higher eukaryotes which are analogous to the regulatory genes in bacteria. Nevertheless, various hypotheses on gene regulation in eukaryotes have been proposed by many investigators. Most of these hypotheses are complex variations of the Jacob and Monod scheme. Paul (1972) has contributed one such hypothesis. Since it is impossible to discuss all the current hypotheses, we will confine our discussion to those which are the most well-known and well-founded. Among these are the theories of both Britten and Davidson and of Georgiev. Actually, these two theories are very similar to each other. Both of these hypotheses attempt to explain the high molecular weight RNA (pre-mRNA) which is partially degraded inside the nuclei. These hypotheses also attempt to explain the role of repeated DNA nucleotide sequences.

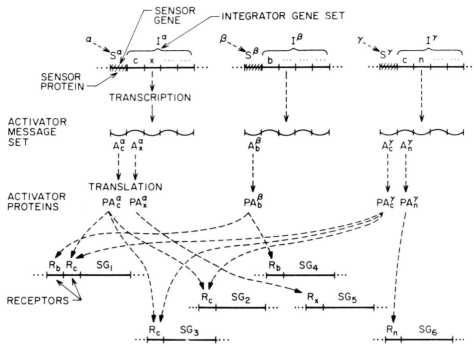

Fig. 102. Possible interrelations in the Britten-Davidson regulatory system using protein activator molecules. The diagram is read from top to bottom. Beginning at the *top*, the model elements which appear are: (1) Effectors, denoted by Greek letters α, β, γ, which are external signals, each producing a pleiotropic response of structural gene activations. (2) Sensors, the target regions for effectors. These are denoted S′ where i represents the α, β, or γ effectors. The sensors consist of sensor gene DNA and sequence-specific sensor proteins, each of which specifically binds its appropriate effector. (3) Integrator gene sets, the sequences coding for sequence-specific activator proteins. Transcription of the integrator gene sets follows as a result of effector-sensor interaction. Each integrator gene is denoted I_j where i again denotes the particular sensor to which the integrator gene responds and j denotes the particular activator protein for which the integrator gene codes. Only some of the integrator genes in each set are named. (4) Activator message sets, transcribed from the integrator gene sequences. (5) Activator proteins, translated from the activator mRNA's. These are denoted PA_j where the meaning of i and j are as above. (6) Receptor sequences, which are recognized in a sequence-specific manner by the activator proteins, and which are denoted R_j. (7) Structural genes, denoted SG. Since these are visualized as single-copy elements, they are indexed simply by numerical subscripts. Transcription of the structural genes may occur when their receptor sequences are bound by activator proteins. Each segment of the genome shown may be part of a larger related region; e.g., a given structural gene may or may not be part of a polycistronic complex, and this indicated by *dots* at either and of each segment. (After Davidson and Britten, 1972)

In addition to structural genes, Britten and Davidson distinguished between sensory genes, integrator genes, and receptor genes (Fig. 102). The gene sensors are recognized by repressors and derepressors. The integrator genes synthesize an activator RNA which either interacts directly with the receptor genes or is used to translate a product which in turn interacts with the receptor genes. In this scheme,

the RNA which is turned over rapidly (i.e., has high rates of synthesis and degradation) belongs to the class of activator RNA (Davidson and Britten, 1973). The activation of certain receptor genes induces the activation of specific corresponding
structural genes (Britten and Davidson, 1969).

Recently, Davidson et al. (1977) have suggested that heterogeneous nuclear
RNA has regulatory functions. They believe that the structural gene transcript is
not included in this class of RNA, but is rather formed independently. This idea
is supported by the observation that in the sea urchin genome, no more than
10%–15% of the total single-copy sequences are included in structural genes (Galau
et al., 1976). About 30% of the single copy DNA sequences are represented in the
heterogeneous RNA of the 600 cell gastrula (Hough et al., 1975) and only 2.7%
was found in polysomal complexes (Galau et al., 1974). On the basis of these data,
it is possible to conclude that the majority of unique sequence RNA in the heterogeneous RNA fraction is not the product of structural genes and is not present in
cytoplasmic polysomal mRNA. This same tendency probably exists in other animals.

The repeated sequences of nucleotides in the high molecular weight heterogeneous nuclear RNA are separated by spaces which are composed of unique sequences. What is the function of such unique spacers? Davidson et al. (1977) have
suggested that they may play an important role in the interaction of regulatory proteins with genome DNA. The spacer DNA sequences which surround a binding
site for a regulatory protein may influence the fraction of time that the site is bound
by protein. It is possible that these sites are promoter-like in that they are responsible for binding RNA polymerase. These sites might also be necessary for the normal processing of heterogeneous RNA, i.e., these sequences could be the site of en-
donuclease action.

The existence of activator RNA has not been demonstrated. The existence of
such an RNA is not necessary in the Georgiev model (1973). This model is simpler
and describes more economically the regulation of gene activity in eukaryotes (Fig.
103). There are three basic postulates to this model: (1) Each operon is composed
of two zones: the major, noninformation-containing acceptor zone, and the minor,
structural zone. (2) The acceptor regions are proximal to the promoter while the

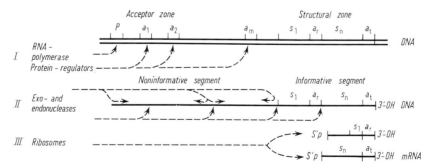

Fig. 103. Model of the transcriptional unit in eukaryotes. *I*, transcription; *II*, transformation; *III*, translation; *P*, promotor; $a...a_m$, acceptor loci; a_r,a_t other functional segments;
$S_1...S_n$ structural genes (Georgiev, 1973)

structural region is distal to the promoter. (3) Most of the acceptor site corresponds to operators, i.e., that part of the genome which interacts with regulatory proteins.

According to Georgiev's hypothesis, the acceptor region can function both as a positive and a negative regulator. In the former case, the regulatory proteins associated with the DNA are derepressors. In the latter case, these proteins are repressors. The high molecular weight, untranslated RNA which is degraded in the nucleus is transcribed exactly from the regions represented by repeated nucleotide sequences. It is difficult to explain, however, why these sequences are transcribed since it is known that bacterial RNA-polymerase can move through the operator without transcribing this region (Blattner and Dahlberg, 1972; Blattner et al., 1972). One must ask if it is possible and probable that eukaryotes differ from prokaryotes in this respect? If this is not feasible, then what role could the high molecular weight RNA, which is degraded in the nucleus, play? In addition, if the above data are correct, then the regulatory zone is larger than the structural zone. In turn, if this size difference exists, then is should be possible to obtain a large number of regulatory gene mutations in eukaryotes. This, however, has never been seen.

On the basis of our knowledge about chromosome structure — especially polytene chromosomes — Crick (1971) proposed a scheme for the organization of DNA in higher eukaryotes. He formulated three postulates:

1. The DNA sequences which code for proteins are located mainly in the interband regions of the polytene chromosome. According to Crick's calculations, the mean amount of DNA in the interband region in the *Drosophila* genome is enough to code for a protein with a molecular weight of 30,000–40,000.

2. The "recognition centers", where regulation in eukaryotes occurs, are regions of the DNA double helix which are represented by unconjugated threads (the unconjugation postulate). Thus, most of the DNA does not code for proteins, but rather serves for regulation.

3. The energy necessary for the maintenance of the unconjugated recognition centers may result from DNA-chromosomal protein (possible histones) associations. Thus, two DNA configurations were suggested: fibrillar (for the coding DNA) and globular (for the regulatory DNA). The globular DNA in turn consists of two types: a twisted, double thread and an untwisted loop which can serve for the recognition of certain regulatory elements. If Crick's three-dimensional model is reduced to two dimensions then it would be similar to Georgiev's model. According to Paul's hypothesis, the chromomeres of dipteran polytene chromosomes contain structural genes while the promoter and regions recognized by regulatory proteins are located in the interband regions (Paul, 1972).

A more complex version of this model was proposed by Zuckerkandl (1976). This model suggest the existence of a negative relationship between the DNA length and the rate of transcription of the coding sequences.

Fine structure cytogenetic gene localization (Sorsa et al., 1973) was used to associate the white gene of *Drosophila melanogaster* with the 3C2-3 bands and, perhaps, with the neighboring interband regions. Later, Sorsa (1978) tried to localize genes with the aid of electron microscopy. An attempt was made to localize the yellow and white loci on the X chromosome. Genetic doubling and tripling of these loci is accompanied by an increase in the number of bands. Selection of these loci, on the other hand, results in the loss of bands. Analysis by electron microscopy of the active loci which code for histones (39D) and 5S RNA (56F) demonstrated decondensation and puffing of these bands. These observations support the hypothesis that genes are located in the band DNA of polytene chromosomes. The absence of RNA synthesis in the interbands and the active synthesis and puffs from certain discs has been convincingly demonstrated (Ananiev and Barsky, 1978).

Keppy and Welshons (1977) have suggested that fasw6 in *D. melanogaster* is located in the interband between 3C5, 6, and 7. On the basis of radioautographic data, Zhimulev and Belayeva (1975b) concluded that most if not all interbands in *D. melanogaster* are characterized by active RNA synthesis. They suggested that genes are present in these regions of polytene chromosomes and that these genes code for enzymes which carry out the cell's general metabolic functions. Interband transcription was also suggested by Gersh (1975) and Skaer (1977).

Contradictory results were obtained in experiments using immunohistochemical localization of RNA-polymerase B in the polytene chromosomes of *D. melanogaster* (Plagens et al., 1976; Jamrich et al., 1977). Data from our laboratory are in agreement with those obtained by Ananiev and Barsky in experiments using hybridization of polytene chromosomes with poly-(A)-containing RNA extracted from *Drosophila virilis* ejaculatory bulbs during the active synthesis of esterases (Karasik et al., 1978). It was shown that four structural genes, which code for various esterases, are located in four neighboring bands but not in the interbands. Moreover, the interband regions were shown to contain no more than 3000 nucleotide pairs and therefore cannot code at least for some enzymes such as xanthine dehydrogenase which requires about 4500 nucleotide pairs.

If structural genes are located in the bands, then what is the functional significance of the rest of the DNA in the genome? As already mentioned, Davidson et al. (1977) have suggested that only about 10% of the unique sequence DNA of the genome codes for protein. The remaining unique sequence DNA is used to make heterogeneous nuclear RNA which has a regulatory role.

There are approximately 1.8×10^8 nucleotide pairs in the *D. melanogaster* genome. About 80% of these nucleotides are confined to unique sequences. Thus, the number of coding sequences equals $1.8 \times 10^8 \times 0.1 \approx 10^7$ nucleotide pairs. If we consider the average gene to be 1000 nucleotide pairs, then the total number of structural genes in *D. melanogaster* equals $10^7/10^3 = 10^4$ or 10,000. This number is in good agreement with the maximal estimate of 7000 loci in the *D. melanogaster* genome (see review, Golubovsky, 1977). If we consider that some bands — particularly those which contain ribosomal RNA genes — contain more than one gene (about 100 in the case of rRNA genes), and if we consider genes which average 1000 nucleotide pairs, then the total number of structural genes in *D. melanogaster* approximately equals the number of bands. Therefore, the "one gene–one band" hypothesis (Hochman, 1971; Judd and Young, 1974) might more accurately be stated as "one structural gene one band". Lefevre (1974) has noted that no data have demonstrated two or more functions to be associated with one band. However, 16 lethal and 4 viable mutations are located in the 1B region of *D. melanogaster*. This region contains 14 bands. In the 3C7-3D4 region, 8 bands are observable, but no mutations have been recovered from this region.

Other aspects of eukaryotic genome organization are being studied. It has been demonstrated that at least some eukaryotic genes are not continuous nucleotide sequences but instead are represented by coding regions that are separated by spacers

(Gilbert, 1978; Marx, 1978). Such spacers have been found in immunoglobulin genes (Brack and Tonegawa, 1977), ovalbumin genes (Breathnach et al., 1977; Lai et al., 1978), and globin genes (Glover and Hogness, 1977; Pellegrini et al., 1977; White and Hogness, 1977; Marx, 1978). These spacers have not been found in any bacterial genes, but have been observed in the late mRNA of adenovirus (Berget et al., 1977; Chow et al., 1977). Therefore, eukaryotic genes consist of two parts: the "intron" zone which is not expressed and the "exon" zone which is expressed. Both zones are transcribed together as one sequence, but following transcription "splicing" occurs. The intron regions are cut out by nucleases and degraded. This can be considered an alternative explanation for the existence of giant hetero-geneous RNA molecules (Davidson et al., 1977). The exon regions are "linked" to-gether by ligases. The final mRNA molecule is used in translation.

Each gene can contain one to several intron insertions. Each of these is 10–10,000 nucleotide pairs. The moderately repetitive DNA sequences of introns could create hot spots for recombination to rearrange exonic sequences. For rec-ombination to create such mosaic genes, the intron must contain five–ten times more DNA than the exon (Gilbert, 1978). The bands of polytene chromosomes contain sufficient amounts of DNA for this to occur. Thus, this observation is in good agreement with the one structural gene — one band hypothesis.

An important question is whether 5000 structural genes are sufficient to code for all the necessary proteins for growth and development in *Drosophila*. It seems to us that this is enough, because in cultures of differentiated muscle cells of chickens 2500 types of various mRNA's were present. Only 6 of these mRNA's are represented in the pool in high concentration (about 15,000 copies per nucleus). Apparently, this number of mRNA's is enough to maintain metabolic function and allow differentiation of the muscle cells (Paterson and Bishop, 1977). It was also shown that there is not a large number of immunoglobin genes as was suggested by early models of the organization of these genes (Rabbits and Millstein, 1977). No more than 20 immunoglobin genes are present in the mouse genome. These are actually the genes which code for the variable portions of the H- and L-chains (Mu-to, 1977).

Mutations have been located at less than 1000 gene sites on the *E. coli* genetic map. The lower limit of the number of enzymes necessary to carry out the main cell functions is 106 (Pevsner, 1974). These enzymes function in the following met-abolic processes:

Glycolysis	15 genes
Citric acid cycle	12 genes
β-oxidation of fatty acids	5 genes
Hexosemonophosphate shunt	13 genes
Oxidative phosphorylation	4 genes
Fatty acid synthesis	4 genes
Pyrimidine synthesis	10 genes
Purine synthesis	15 genes
Amino acid activation	20 genes
Methylation and metabolism of single carbon compounds	8 genes

Total = 106 genes

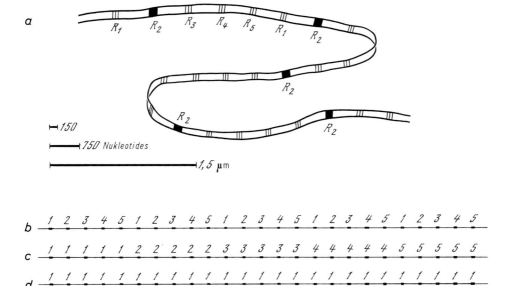

Fig. 104a–e. Some models of sequence organization in DNA of a *D. melanogaster* chromomere. Scales of 150 and 750 nucleotides, and 1.5 μm, refer to part **a**. Each long line in **b** through **c** represents about 7 μm DNA. *Heavy line segments* in **b** through **d** indicate short repeats (150 nucleotides) of sequence indicated by numbers 1–5. Thin regions indicate nonrepeat sequences.

Model **a** is an expanded form of **b**, indicating the 1.5 μm hypothetical separation of the same repeat (R_2, for example), and the 750 nucleotide unique sequences that separate different repeats R_1, R_2 etc. Model **d** shows all of the repeats in a chromomere as being similar. This does not seem to fit the data for polytene chromosome DNA. Model **e** based on the paucity of single-stranded regions (*gaps*) near ring closure sites. However, with this model it is difficult to arrange repeat sequences (*heavy lines*) such that ring frequencies decrease with small fragments and renaturation kinetic data are not contradicted.

These models are attempts to rationalize kinetic, electron microscope, and ring formation data. Critical tests in one species are necessary to demonstrate the validity of any or all of these possibilities. A unique and specific prediction of model **a**, for example, is that ring frequencies with salivary gland DNA will have peaks at multiples of the 1.5 μm repeat separation length, i.e., 3.0, 4.5 μm, etc. **a**. As we have no information on the function of these short repeat sequences, there is no logical a priori reason to discount any of the models. Indeed, different chromomeres may have quite different sequence organization. (After Laird, 1973)

Special attention has been given to the distribution of unique and repeated sequences along the chromosome and to the construction of polytene chromosomes. Hypotheses on chromomere construction in *Drosophila* have been based on data of DNA renaturation kinetics (Laird and McCarthy, 1969), formation of rings from DNA fragments (Lee and Thomas, 1973; Thomas et al., 1973), and electron microscopic analysis of repeated sequences (Wu et al., 1972). The models of chromomere construction are presented in Fig. 104 (Laird, 1973).

The "a" hypothesis is based on the presence of regularly alternating, short (750 nucleotides) repeated sequences. Recent data suggest that in *Drosophila* the repeated sequences are 5000–6000 nucleotides in length and are separated by unique sequences of more than 13,000 nucleotides (for review, see Edström and Lambert, 1975). Four different repetitive sequences and five different unique sequences are separated by the same repetition (R_2 on the scheme). It is postulated that two similar or identical repetitions are located 5000–6000 nucleotides from each other. The unique sequences (750 nucleotides) are separated by unsimilar repetitions. Therefore, unsimilar repetitions are located every 900–1000 nucleotides and similar repetitions are located every 5000–6000 nucleotides (Laird, 1973)

Model "b" is equivalent to model "a" except that the scale has been reduced. Models "c" and "d" are similar except that in "c" five similar repetitions are located together in a continuous sequence, while in "d" the unique sequences are separated by similar repetitions. This model was suggested by Bonner and Wu (1973). According to this model every chromomere contains 30–35 unique sequences of DNA where each unique sequence is 750 nucleotides long. Between each of these is a repeated sequence 125 nucleotide pairs in length. They are identical in each chromomere but differ between chromomeres. Bick et al. (1973) (see model "e") supposed that all repetitions are concentrated in one half of the chromomere DNA.

Further investigation, however, showed that the sizes of unique and repeated alternating sequences in *Drosophila* are different. Of the *Drosophila* genome 70% consists of repetitions 5000 nucleotide pairs in length interspersed with unique sequences of 13,000 nucleotide pairs in length (Manning et al., 1975; Davidson et al., 1975; Edström and Lambert, 1975; Crain et al., 1976a). The mosquito genome (Wells et al., 1976) and honey bee genome (Crain et al., 1976) are organized in the same manner. Another type of genome organization is characteristic of the domestic fly and butterfly genomes (Crain et al., 1976b; Efstratiadis et al., 1976). These are organized in the same manner as the *Xenopus* genome (Davidson et al., 1973; Chamberlin et al., 1975), the sea urchin genome (Graham et al., 1974), and a number of other sea invertebrate genomes (Angerer et al., 1975; Goldberg et al., 1975). In this case, the 300–400 nucleotide pair repetitions alternate with 800 nucleotide pair unique regions in *Xenopus*, 1000 nucleotide pairs in sea urchins and 2000 nucleotide pairs in molluscs. In most cases, more than 70% of the unique sequences are located near short repetitions.

Palindromes, or inverted repetitions are characteristic components of the genome. They are formed by two complementary sequences which run in opposite directions. These two sequences are on the same strand of DNA but can be either very close or at great distances from each other. Palindromes range from 50 to several thousand nucleotides in length. They can result in the formation of a loop or a double helical region (Hardman and Jack, 1977).

Several percent of the eukaryotic genome is in palindromic sequences. The estimated number of palindromes in humans is 2×10^6, in the loach, 5×10^5, in *Xenopus*, 10^5, in mice, 4×10^4, in *Drosophila*, 3.4×10^3 per haploid genome. The longest palindromes have been found in *Drosophila* (Schmid et al., 1975). About 80% of the palindromes in *Drosophila* are represented by hairpin loops which are 1300–1500 nucleotide pairs. In *Drosophila* and some fungi, palindromes can be grouped or spread along the chromosome (Cech and Hearst, 1975; Hardman and Jack, 1977).

It has been suggested that palindromes serve a regulatory function (see Lewin, 1975) or participate in recombination (Wagner and Radman, 1975). These sequences might also be involved in intragenome translocations in somatic cells (Perlman et al., 1976).

Some genes are organized as palindromes. For instance, the extrachromosomal amplificated copies of rDNA in *Tetrahymena* and the fungus *Physarum* exist as giant palindromes. The structural genes for the small subunit of rRNA are located inside a palindromic region. The large subunit rRNA genes are located alongside this palindrome. Transcription and replication begins in the center of this region and spreads to both sides (Engberg and Klenow, 1977).

Palindromes are located on both sides of the hypervariable regions of the variable genes (V-genes) of immunoglobulins in humans. This is not seen for the genes of the constant part of the immunoglobulins (C-genes) (Whilmart et al., 1977).

It is tempting to suggest that palindromes are located in the interbands because the number of palindromes in *Drosophila* is approximately equal to the number of interband regions in the polytene chromosomes. It is possible to illustrate polytene chromosome organization in *Drosophila* as follows:

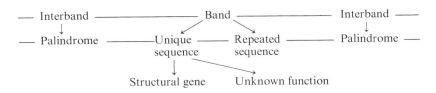

The part of the band which contains the structural gene is located near the interband. Part of the band contains repeated sequences which separate the unique sequences from each other. The giant palindrome which corresponds to the 5 S RNA genes in *Drosophila melanogaster* was demonstrated to be associated with a band on the polytene chromosomes (Cohen, 1976).

In addition to palindromes, many genes in *Drosophila* are present in multiple copies. This was demonstrated by cloning various *Drosophila* DNA fragments in plasmids. Two families of multigenic families were found: dispersed copies and tandem repeats. Genes that are transcribed to form mRNA belong to the first family while ribosomal and histone and genes induced by temperature shock are associated with the second family. There are 30 genes which are inducible by temperature shock. These are located on one chromosomal segment as a number of short segments repeated in tandem, which are separated by heterogeneous spacers (Lis and Hogness, 1977). A peculiar class of repeated genes which are unstable was localized in the heterochromatin. The region of the polytene chromosomes containing these genes differed in various strains of *D. melanogaster* and in various individuals of the same strain. The DNA of three λ phage vectors carrying various *Drosophila* genes bound to about 1% of the cellular mRNA of *Drosophila*. With the aid of in situ hybridization, about 60 sites of hybridization were found (Georgiev et al., 1977).

The hypothesis of a cascading regulatory mechanism is currently being revised. The originators of this hypothesis are attempting to take into account all the com-

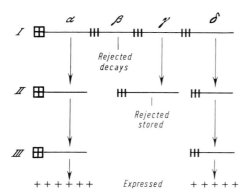

Fig. 105. Model of polycistronic unit of transcription. (After Scherrer and Marcaud, 1968)

plexities and peculiarities of the known genetic regulatory mechanisms in eukaryotes (Scherrer and Marcaud, 1968; Scherrer, 1973). They suggest that the units of transcription are polycistrons and can contain genes which are not associated functionally. Some nontranslated sequences are included in their composition as well. Thus, the newly synthesized RNA includes a lot of information which the cell does not seem to need (Fig. 105). This pre-mRNA (30 S–105 S) contains several units which are limited by oligoadenine nucleotide sequences. The half-life of this molecule is 20–30 min. The gradual cleavage of the nonfunctional segments from the coding segments occurs under the influence of endonucleases. Pre-mRNA of an intermediate size (25 S–45 S, mw $= 1.5 \times 10^6$d) is more stable and has a half-life of 3–6 h. The stabilization of the mRNA occurs with the association of specific proteins with this mRNA. Particles similar to informopheres are formed (Georgiev, 1973).

Thus, one type of post-transcriptional, pretranslational regulation involves the gradual reduction and maturation of the newly synthesized RNA (Table 42). The

Table 42. Regulation of informational transfer in animal cells. (After Scherrer and Marcaud, 1968)

Type of regulation		Primary regulation		Secondary regulation
		Post-transcriptional regulation		
Level of regulation	Transcriptional regulation	Intermediary regulation	Translational regulation	
Phase of information transfer	Transcription	Metabolism and transport	Translation	Phenotypic expression
Carrier of information	Nascent RNA	Transition-state RNA	Functional RNA	Polypeptide, protein
Association of carrier	Nascent RNP	Transfer RNP (informopher) (informosome)	Polysome (mRNP)	
Cellular component involved	Chromatin	Nucleoplasm and cytoplasm	Cytoplasm (reticulum)	Total cell

Flow of information from the gene to phenotypic expression →

multilevel character of regulation in eukaryotic cells has been noted. According to new data the constructive role of the high molecular weight RNA is probably a result of mosaic (exon-intron) structure of gene.

2.4.4 The Multilevel Character of Gene Regulation of Ontogenesis in Eukaryotic Organisms

The scheme of Scherrer and Marcaud is useful even though it is not complete and has shortcomings when used to explain the individual regulation of development. A general model for development of multicellular organisms has been designed by building on the basic concepts of the Jacob and Monod scheme (see Bonner, 1967). This model implies that development starts at some zero condition and consists in successive gene system repression and derepression. This in turn controls the synthesis of the specific products needed at precise moments by the differentiating cell. As such, the divergence of tissue primordia and cell specialization is the highest level of organization. The order of events is according to epigenetic principles.

When development is considered to start at a zero condition, the Jacob and Monod scheme is acceptable for describing the regulation of development. Here, the processes occurring in development are analogous to the metabolic events which are under genetic control in bacteria. The sequence of metabolite → regulator → new operon in bacteria may be analogous to the sequence of inductor→regulator→new operon in differentiating cells. As shown in Part I, however, development does not start from a zero condition and therefore does not fit the framework of a purely epigenetic process where there is a successive formation of new, specific compounds.

The egg ready for fertilization represents a system which in many ways is predetermined. It contains RNA's which are active at comparatively late stages of embryogenesis. It also contains protein inductors which are not important for cleavage or gastrula cell migration, but they are important for neural or mesodermal cell differentiation. What causes the early synthesis of these compounds? Neither the ova nor the embryo needs these compounds during early developmental stages. Such synthesis does not conform to the Jacob-Monod scheme since in this scheme, genes, which are controlled by inductors, are activated immediately before the initial interaction of the embryonic primordia which respond to the genes.

The synthesis of stable proteins and long-living mRNA's, which are not necessary functionally until a later period, is a widespread event (see Part I). Cases of information transmission into the egg from the maternal organism through a system of trophic cells has been described. This is a maternal effect. There is a transport of the compounds which determine normal embryonic development. Thus, the previously formed chemicals of the cell influence significantly the future activity of the genetic apparatus and create conditions for effective selection of RNA molecules to undergo translation. During differentiation and successive cell functioning, the previously synthesized proteins undergo post-translational modifications (for example, the transition of trypsinogen to trypsin and prothrombin to thrombin, etc.). Differences in proliferative activity can alter the phenotypes of groups of cells. As demonstrated in Part I, there is a specific system of regulatory genes which controls the realization of a trait at various levels, ranging from the transcriptional to post-translational levels. It is important to know the interrelationship between structural genes and genes which control the expression of these structural genes in eukaryotes. When isozyme synthesis was studied in eukaryotic systems, evidence of gene regulators was not found. In one specific example, the molecular basis of lethal and semilethal mutations in the G6PD locus of *D. melanogaster* was studied. Eleven mutations were found in the structural gene for G6PD: three are semilethal mutations which result in altered molecules that have decreased catalytic activity, and eight mutations were lethal and consisted of null alleles. The polypeptides produced by these lethals reacted with the antiserum against highly purified G6PD. Thus, these mutations can not be considered as mutations in gene regulators (Gvozdev et al., 1977).

Gelbart et al. (1974) demonstrated that all null mutations at the rosy locus lie within the structural gene for xanthine dehydrogenase. Similarly, Schwartz and Sofer (1976) noted that 11 of 16 ADH-negative mutations (induced by ethyl methane sulfonate) produce an inactive ADH enzyme. It is possible that the remaining five mutations are also located in the ADH locus. The authors have supposed the following possible explanations for the data observed:

1. There is no gene which regulates the ADH gene. When considering the tissue specificity of enzyme distribution and the dynamics of changes in isozyme patterns, however, this conclusion seems doubtful.
2. ADH in *Drosophila* may possess autogenic regulation (Goldberger, 1974). That is, ADH might influence the expression of its own structural gene in a manner which is independent of the catalytic function of this enzyme. In this case, regulatory mutations would lie within the structural gene.
3. The regulatory gene is in the genome but not associated with the ADH structural gene.
4. The regulatory DNA is protected against mutagenesis.
5. Mutations in the regulatory material are mostly silent mutations.
6. The exact base sequence of the regulatory DNA is not critical. If ethyl methane sulfonate produces primarily point mutations, then a single base change in a regulatory sequence would not have a profound enough effect to be detected in mutant isolation procedures.
7. The regulatory sequences are redundant.

There are indications that Chovnick's group has found a specific gene regulator in *D. melanogaster*. One isoallelic electrophoretic variant of xanthine dehydrogenase, which is coded for by the rosy locus, demonstrates a band which is four times more intense than that of wild-type. Titration experiments with anti-XDH serum demonstrated that more antiserum is needed to inactivate the extract in this band in mutants than in wild-type. Thus, the higher activity of the mutant variant is a result of the increased production of enzyme molecules. At the same time, the K_m of the darker band is similar to that of wild-type. These data might be explained by a mutation to the gene regulator, although the observed data do not exclude that there is a structural variant of the enzyme of which the rosy gene is composed of a number of sites which are separated by spacers. The site ry^h, which is responsible for the synthesis of the components in the heavy band, can be separated from a site determining an electrophoretic variant by recombination. It is dominant in the cis position and results in an increase in the production of the electrophoretic variant coded for by a structural gene on the same chromosome (Newlon et al., 1975; Newmark, 1976; Chovnik et al., 1977).

A cis-dominant control element for aldehyde oxidase gene activity has also been described in *D. melanogaster*. This regulatory gene is closely linked with the structural gene and causes interstrain differences in the specific activity of the enzyme prior to pupation. Although the changes of enzyme activity at other stages is the same, the specific activity prior to pupation is increased in some strains but not in others (Dickinson, 1975).

The mechanism of control of gene action by such regulatory elements is unknown. Thus, the discovery of the presence of gene regulators for the XDH and AO genes does not prove that eukaryote genomes are organized in the same way as prokaryotic genomes.

There are data to suggest that some gene regulators are not close to the structural genes they regulate. Some of the regulators may even be on a different chromosome from the structural genes being regulated. Glassman et al. (1968) have suggested that the ma-1, cin (X chromosome) and 1xd (chromosome III) genes are regulatory genes for the rosy locus (chromosome III). They noted, however, that these regulatory genes function differently from the gene regulators in bacteria. Moreover, they have suggested that eukaryotic organisms do not have bacterial-type operon systems to control phenotypic expression. Our data on the genetic regulation or organospecific esterases in *Drosophila*, and the data of Paigen's group on the genetic regulation of expression of the glucuronidase gene in mice also confirm the existence of numerous controlling elements which act on many levels and are not linked to the structural genes being controlled. This multilevel character of regulation occurs in a specific fashion both temporally and spatially. Previously synthesized products influence current gene activities which are essential for cell proliferation rate and development. Also, the multiple steps of regulation occur at different points in the realization of genetic information, i.e., transcription, RNA processing and transport, translation and post-translation.

In general, then, it is possible to suppose that the genic control of ontogenesis characteristic of multicellular organisms is different from that in bacteria and phage. Gene regulators are present in multicellular organisms, but there is no proof that these function in the same way as in bacteria. There are numerous facts concerning the regulation of enzyme synthesis. For instance, membrane proteins may be able to associate with specific enzymes and stabilize them. In this case, then, a structural gene product could be considered a gene regulator which functions at the post-translational level.

It has also been suggested that some genes which code for protein inductors (or histones) may be able to regulate the activity of another gene system by acting at the transcriptional level (i.e., dosage compensation). Therefore, some elements of the Jacob and Monod scheme for bacteria can be included in the scheme of control in eukaryotes. These similar elements, however, represent only one component of the complex system of eukaryotic regulation.

2.4.5 A Hypothesis for the Temporal Organization of the Gene Systems Controlling Development

Gene activation in various differentiation systems at different, specific periods of development could occur autonomously due to internal factors which are characteristic of the system. There can also be "dependent" gene activation due to the influence of external factors from surrounding tissues (i.e., transmission of activated compounds, changes in hormonal status, etc.). The basic processes which determine either morphogenetic effects (i.e., nerve tube formation), competence, induction or determination (i.e., neurulation in the process of primary embryonic induction or something analogous to this process in other organogenesis systems) have an independent genetic determination. In other words, the development of every process is under the control of an independent gene system. It is important to know at what functional-genetic level this determination occurs.

According to Neyfach (1963), there is a gradual accumulation of external information by cells during development. When there is a small amount of accumulated information, the cell can choose one of two–five developmental paths. Gradually, however, information accumulates and, as a result, determines which specific gene complex is activated. This scheme is analogous to dialing a phone number: with the dialing of each digit, differentiation is one step closer. Competence is represented by a cytoplasmic protein repressor (activator) which can be "switched on" by an effector-inductor (i.e., hormone, etc.). Ratner (1975) has postulated the existence of a special system of ontogenesis which controls the switching-on of a small number of structural cistrons that are responsible for the formation of one group of characteristics (i.e., enzyme synthesis, chaete formation, etc.) in various body parts at specific periods of time. This occurs as a response to various protein and metabolic signals. Through these operators, the same enzymatic system can be connected with various subprograms of ontogenesis in different cells. For example, two programs of ontogenesis occur in *Drosophila* — the pupae program and the differentiation of imaginal discs, which lead to the formation of structures in the adult. The second program does not depend upon the first program and is determined by approximately one/third of the *Drosophila* genome, or about 1000 genes. There are two alternative ways to express these programs: (1) the genes which control imaginal disc development are unique; (2) these genes are duplicated and the copies regulate the same processes; but operate in different larval programs (Shearn and Garen, 1974).

In which sequence are these genes activated during development? Hadorn has formulated a hypothesis of the stepwise action of genes in ontogenesis (Fig. 106). He has distinguished "vitally important loci" (the black rings on the scheme) — the absence of which leads to irreversible abnormalities of development and the eventual death of the embryo—and the second type of chromosomal loci which are less significant for development. Defects in these loci are not lethal. The switching on of vital loci is crucial for normal ontogenesis (black rectangle on the scheme). Development is terminated when there is a deletion in any of these loci (Hadorn, 1961).

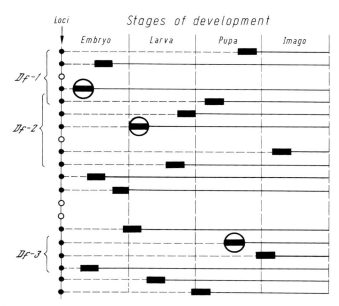

Fig. 106. Steplike onset of action of the genes in the development of an insect. Developmental pattern and lethal crisis for 3 deficiencies (*Df-1, Df-2, Df-3*). (After Hadorn, 1961)

What molecular mechanisms determine the stepwise switching-on of genes in autonomously developing systems? Manner (1971) has discussed successive gene activation as a result of the actions of specific compounds. These compounds act in specific periods of development to activate certain genes. It was suggested that all somatic cells of a developing embryo contain $G_1, G_2, G_3, ..., G$ genes. The G_1 gene begins to function before the G_2 gene, the G_2 gene before the G_3 gene, and so on. If, at time T_1 a specific activator S_1 is formed, then the G_1 gene is derepressed and the cell differentiates into the C_1 cell type. If at time T_1, on the other hand, the substance S_1 is absent, the G_1 remains in the repressed condition and the cell shifts into a new temporal phase, T_2. If the S_2 substance is made while the cell is in T_2, then the G_2 gene is activated and the cell differentiates into the C_2 cell type, etc. (Manner 1971 cited after Konyukhov, 1973).

What is the spatial organization of the genes which, when successively activated, influence cell determination on the molecular level? As mentioned, it is not necessary for the structural genes and controlling elements to be closely linked with each other. Genes which code for the products which act successively during development could be randomly distributed along the chromosome as opposed to being present in clusters. Simultaneous gene activation, in this case, could occur if the regulatory zones which recognize specific inductors are similar. The activation sequence of various gene groups in this case would be determined by the successive changes of protein regulators. While this has been observed in some cases, the principle is not seen universally.

Some research has shown that genes which influence the same morphological trait or successive events of ontogenesis tend to be clustered together. A statistical analysis of the organization of genes which control the development of morpho-logical characteristics was done on the genome of *D. melanogaster*. It was demon-strated that genes which are closely related functionally are usually on the same chromosome and clustered in one region of this chromosome (Elston and Glass-man, 1967). A more detailed analysis was done by Golubovsky (1972, 1977). He found that most wing mutations are located on the second chromosome (40%). About 26% of the wing mutations are located on the third chromosome. On the other hand, eye mutations (for shape, structure, and color) are distributed as fol-lows: 40% on the third chromosome and 29% on the second chromosome. Golu-bovsky (1977) also noted the presence of an intrachromosomal gene order in *D. melanogaster*. There are groups of genes which code for similar morphological ex-pressions. These conclusions have recently been confirmed by others (McCartney et al., 1977).

An analysis of known mutations in *D. virilis* (Alexander, 1976) has also dem-onstrated microgroups of genes on the X and fifth chromosomes. On the X chromosome there are two groups of genes which control wing development. The first group consists of 72.0 (we), 74.9 (ro), 78.0 (mt), and 78.1 (dy) while the second group consists of 94.5 (Bx), 96.0 (br), 100.4 (Ds), 102.9 (N), 103.3 (Ax), and 104.5 (T). Two small groups of genes are on the fifth chromosome. One of these deter-mines chaete formation: 44.2 (r), 44.5 (tt), and 44.5 (sb). The other group affects eye shape and color: 150.0 (Gl), 151.0 (po), and 151.5 (Ps) (Korochkin, unpub-lished).

Goldschmidt (1961) has proposed the special term "field of action" for the small chromosomal segments which have clusters of loci with similar functions or loci which are used in the same developmental pathway. The group of homeotic mutations which causes the replacement of one differentiated structure by another can serve as an example. In this group are the following: 47.0, extra sex comb; 47.7, proboscipedia; 47.8, antennapedia; 48.0, polycomb; 48.0, multiple sex comb; 48.0, nasobemia; 48.5, tetraltera; 58.5, spineless aristopedia; 58.8, bithorax. According to Shearn and Garen (1974), the genes which control imaginal disc development also form clusters on the third chromosomes in *D. melanogaster*. Thus, there seem to be two types of gene organization in *Drosophila*. In the first case, there is an as-sociation of genes which are simultaneously activated during ontogenesis and may possess an operon type of organization. This type of gene organization is known to occur in prokaryotes. In *Drosophila*, and probably most eukaryotes, such clusters occur, but are infrequent. For example, the genes for glucose-6-phosphate and 6-phosphogluconate dehydrogenases (two enzymes which control successive oxidative reactions of the pentose cycle) are located on opposite sides of the same X chromosome (MacIntyre and O'Brien, 1976). The genes for various *D. melano-gaster* esterase are located on the third chromosome: Est-6 at 36.0, ali-Est at 48.3, Est-C at 51.7, ACE at 62.0, and ACE-3 at 90.0 (MacIntyre and O'Brien, 1976). In *D. virilis*, a space of 10 map units separates the α and β esterase genes (Korochkin, 1975).

Cases are known, however, where *Drosophila* genes coding for one family of isozymes are linked. The classical example is the rudimentary locus (I-54.5). This locus contains genes which control the synthesis of the functionally related en-zymes aspartate carbamyl transferase (ACT), carbamyl phosphate synthetase

(CPS), and dihydrooratase (DHO). These enzymes are all involved in pyrimidine synthesis (Norby, 1973; Jarry and Falk, 1974; Falk, 1975, 1976). It was suggested that these genes are located in the following order: centromere-ACT-CPS-DHO (MacIntyre and O'Brien, 1976). It is possible that these enzymes are assembled into a multienzyme complex. Another example of this type of organization is the Amy-Hex gene assocation which codes for amylase and hexokinase synthesis (Elston and Glassman, 1967).

It was demonstrated, however, that the clustered genes function in an uncoordinated fashion. In other words, these genes are not organized into operons. For instance, the closely linked α_1-, α_2- and S-esterase genes in *D. virilis* are activated at different times of ontogenesis (Korochkin, 1975, 1978). Est-α_1 and Est-α_2 are active at the end of embryogenesis, and Est-S is active no earlier than pupation (Korochkin and Matveeva, 1974; Korochkin et al., 1976a). The activity of the linked Amy[3] and Amy[6] genes is also not coordinated (Doane, 1969). Neither is the activity of the lap-A and lap-D genes in *D. melanogaster* (Beckman and Johnson, 1964). Dickinson and Weisbrod (1976) investigated the changes in aldehyde oxidase and pyridoxal oxidase activities in *D. melanogaster*. The structural genes for these enzymes are 0.8 map units apart. Aldehyde oxidase activity increases during pupation and decreases during eclosion while the pyridoxal oxidase activity continues to increase in the imago, even after the aldehyde oxidase has decreased.

In the second type of gene organization in *Drosophila*, it is possible to suppose there is a cluster of associated genes which code for products appearing successively during development. There are facts which point to the possibility of such a temporal gene organization. The analysis of mutations which influence *Drosophila* embryogenesis show that there are blocks of genes located close to each other which control successive embryonic stages. This means that genes could be organized according to a temporal scheme. Another example is represented by five closely linked genes which control successive developmental stages during meiosis and the first 5 h of embryogenesis. These genes are located on the X chromosome of *D. melanogaster* near the scute locus. Three genes are located in the region of two map units on the second chromosome. These genes influence the 3rd to 16th h of development. Similarly, Dickinson (1975) described a specific gene which controls the initial time of aldehyde oxidase synthesis in *Drosophila* development and which is closely linked with the structural gene.

Kauffman et al. (1975) determined the time of the lethal effect of various lengths of X chromosome deficiencies. They showed that deficiencies on the left part of the chromosome caused an early lethal effect, while deficiencies on the right part became lethal at later stages of ontogenesis. Similar results were obtained by Hochman (1973). He investigated the effects of various fourth chromosome deletions on development in *D. melanogaster*. Deficiencies lethal during the embryonic period are on the left, deficiencies lethal in the larvae are in the middle, and deficiencies lethal in the pupae are on the right.

The products of the adjacent T and H-2 loci appear in the somatic cells of mice during early embryogenesis, late embryogenesis and in adult mice (Bennett, 1975). The transition from one type of globin chain to another could possibly be associated with a "transcription wave" which moves along the DNA molecule. It is difficult to explain the globin chain transition by assuming there is a loss of a specific

DNA region. This was initially suggested by Kabat (1972) in the "looping-out excision" hypothesis.

There is a similar correlation between the appearance of various immunoglobulins and the positions of their genes in the chromosomes (Lawton et al., 1975; Mage, 1975). In addition, the sequence of appearance of various esterase isozymes in *D. virilis* corresponds to the position of the esterase genes in the chromosome (Korochkin, 1975).

It is possible to suppose that each chromosome is subdivided into autonomously functioning segments. This would partially explain the above data. During development, a wave of transcription extends through each segment. As a result of this, morphological cell differentiation is accompanied (or determined) by the sequence of protein synthesis which, in turn, corresponds to the gene positions within a given functional segment (Korochkin et al., 1976b; Korochkin, 1978). The rate of transcription, as determined in certain *D. melanogaster* mutants, equals 30–60 nucleotides per second (Wright, 1970). This rate is in good agreement with the rate estimated for *E. coli*, which equals 43 nucleotides per second (Stent, 1974), and for *Culex* larvae which equals 25 nucleotides per second (Egyhazi, 1975).

Mechelke (1961) obtained a dispersion of maximal puffing activity along segments of the polytene chromosomes of *Acricotopus* (Fig. 107). Perhaps such moving of maximal puffing activity, which has also been demonstrated in *Drosophila melanogaster* (Zhimulev and Belyaeva, 1975a), reflects the successive transcription of different segments within the same gene that is in turn located within a given band (Fig. 108). The increase in length of the interband, which is sometimes seen, could be due to the activation of a gene near the neighboring band (Zhimulev, 1974; Zhimulev and Belyaeva, 1975a). This could be possible if the structural gene is located on the side of the band (see above). This, however, does not mean that there are several genes in one band, as some believe (Speicer, 1974; Zhimulev and Belyaeva, 1975a). It is interesting to note that the amount of DNA in the *D. virilis* nucleus is twice that in the *D. melanogaster* nucleus, and the life cycle of *D. virilis* is twice as long as that of *D. melanogaster*.

The odd arrangement of genes which control unrelated traits (i.e., eye coloration and wing shape in *Drosophila*) becomes understandable if the genome is organized according to a temporal principle. In this case it is not necessary to localize genes which control similar traits. For the determination of certain morphogenetic events, it is necessary only to activate simultaneously those regions of the autonomously functioning chromosome segments which control the synthesis of products. If this suggestion is true, then the successive appearance of proteins during development should reflect the order of genes inside functional segments (Fig. 109).

The temporal genome organization of eukaryotes could have evolved from the functional organization of the prokaryotic genome. The temporal sequence of protein synthesis in prokaryotes occurs from a stable, polycistronic template which is determined by the linear sequence of cistrons (Ohtaka and Spiegelman, 1963). For instance, the enzymes for histidine synthesis are in a sequence which corresponds to the linear gene sequence in the histidine operon of *Salmonella typhimurium*. There is a 20 min period between the appearance of the first and last (tenth) enzymes. The data can be explained on the basis of the successive synthesis of individual mRNA's or on the successive synthesis of a polycistronic message. This

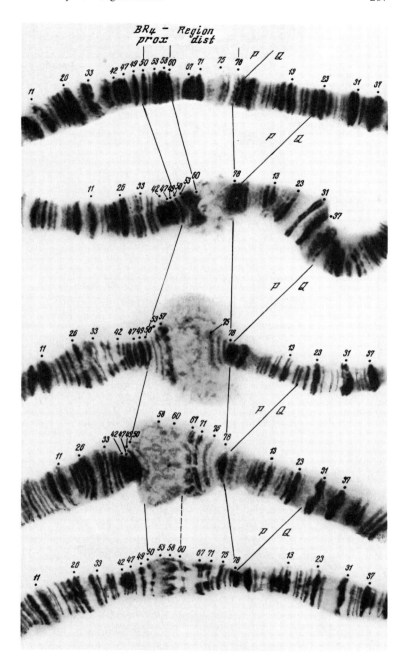

Fig. 107. Successive progress of puffing from distal to proximal end of BR$_4$ region in second chromosome *Acricotopus lucidus*. (After Mechelke, 1961)

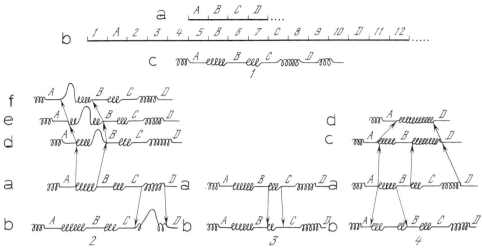

Fig. 108. Model of functional organization of chromosomes *1*, Hypothetical formation of the band structure: *a*, an eukaryot ancestor which has decondensated chromosomes; genes controlled the main metabolic processes in nondifferentiated cell (*A*, *B*, *C*, *D*); *b*, appearance of new genes in multicellular organisms (*1–12*); new genes controlled the tissue- and stage-specificity; *c*, inactivation of new genes and disc formation. *2–4*, an interpretation of some well-known phenomena from point of view of possible transcription in interband and of presence of more than one gene in single band: *a*, part of chromosome from the differentiated cell with certain band composition; *2b*, the puff formation by part of band (only some of genes in band are active); *2d,e,f*, "maximal puffing" transition inside one band; the successive activation of various genes in band; *3b*, the band lengthened as a result of the lateral gene activation; *4b*, two-band formation from a single band as a result of the medial gene activation in the middle of nonactive genes complex; *4c,d*, a "new bank" formation in the case when genes are nonactive in interbank. (After Belyaeva and Zhimulev, 1974)

would exclude the simultaneous synthesis of proteins from different cistrons (Goldberger and Berberich, 1965). The successive syntheses of sucrose, trehalase, ornithine transcarbamylase, aspartate transcarbamylase and threonine dehydratase have been described for *Bacillus subtilis*. This order corresponds to the gene order on the genetic map (Kennett and Sueoka, 1971).

A change in the type of synthesis was obtained in the *E. coli* and yeast life cycles. Here, the order of enzyme synthesis is determined by the gene order in the chromosome (Halvorson et al., 1971). An attempt to investigate the temporal genome hypothesis directly was made with *Neurospora crassa*. Jobbagy et al. (1975) proceeded from the assumption that the whole chromosome, rather than separate segments, is organized according to this principle. They used as examples genes separated by 20 morganids. The genes, naturally, could be located in different functional segments. This may explain the absence of a relationship between gene position and gene expression.

A better model for the investigation of this hypothesis is the gene system in *D. melanogaster*. Gene sequences which are located on the second and third chromosomes are especially convenient for analysis. Genes of alcohol dehydrogenase (Adh-50.1), phenol oxidase (Tyr-52.4), and kinurenin hydroxylase (cn-57.5) are lo-

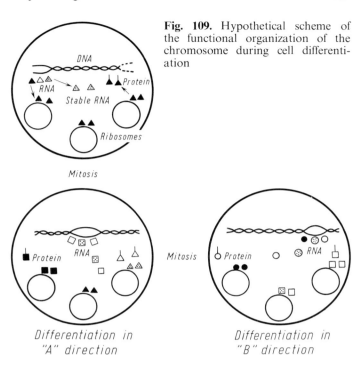

Fig. 109. Hypothetical scheme of the functional organization of the chromosome during cell differentiation

cated on the second chromosome, and genes for alkaline phosphatase (Aph-47.7), ali-esterase (ali-Est-48.3), esterase-C (Est-C-51.7), xantine dehydrogenase (ry-52.0), malic enzyme (Mdh-53.1), and aldehyde oxidase (Aldox-57.0) are located on the third chromosome.

The noticeable correspondence between the changes of the enzyme activities during *Drosophila* development and the order of the corresponding structural genes on the genetic map can be obtained (Dickinson and Sullivan, 1975; Korochkin, 1979). Similar data were obtained recently (Korochkin and Kaplin in preparation) using histochemical observations of the same enzyme in *D. melanogaster*. It was demonstrated also that the sequence of positive histochemical reactions for these enzymes during embryogenesis is constant for various organs.

Data which confirm the idea of the division of chromosomes into functionally autonomous zones have been obtained in experiments with in situ hybridization of polytene chromosomes with labeled poly-(A)-containing RNA extracted from the ejaculatory bulbs of *D. littoralis* at various stages of differentiation. The label was obtained in the group of bands which are packed together, e.g., only the adjacent bands were labeled. The part of chromosomes which separated these clusters was without label (Karasik et al., 1978).

It is important to notice that the time period between the appearance of various proteins does not correlate directly with the physical distance between corresponding genes. This lack of absolute correlation could be explained by the multilevel control of phenotypic expression of proteins and isozymes. As discussed above, the gene transcription does not mean that the phenotype controlled by this gene will appear immediately.

Specific mechanisms regulating time and rate of transcription in the hypothetical autonomous segments are required for the exact coordination of the activation time for various RNA templates in the cytoplasm of the differentiating cells. The repeated sequences of DNA and the intron spacers between the exon coding segments of mosaic genes could participate in such regulation. In this case, the highly repetitive, rapidly degrading nuclear RNA may serve as a peculiar biological clock, which regulates RNA synthesis in the corresponding structural genes. This type of nuclear RNA is not necessary after the terminating of transcription.

This mechanism probably evolved on the basis that the DNA molecules during prelife stages could effectively function only in very specific conditions:

1. For the appearance of "sense" sequences, the big enough number of "non-sense" nucleotide sequences, in the prebiological period, was necessary.
2. These molecules had to have mutational change points, e.g., nucleotide substitution, deletions or insertion, etc. As a result, sequences appeared which had sense for transcription.
3. The sense sequences should alternate with non-sense sequences which remain as a relict of the prehistoric DNA.

If this speculation is somehow true, it is possible to suggest that among the primitive life forms, organisms with very high DNA contents should be found. Actually, the DNA amount in some *Protozoa* is significantly greater than the DNA amount in some highly organized multicellular organisms. For instance, *Protozoa Astasilonga* has 1.5 picograms DNA per nucleus whereas in *Drosophila* there is only 0.2, in sea urchins 0.9, in fish 1.0, and in chickens 1.2 (Ris and Kubai, 1970).

The further fate of the non-sense sequences could be of two kinds:

1. Its elimination which occurred in the prokaryotic branch of the life tree.
2. Its conservation and successive utilization. This tendency is obtained in the eukaryotic branch.

What is the function of the extra DNA, which codes for neither proteins nor RNA of the ordinary types (ribosomes, tRNA)?

It could, first, determine the high concentration of total DNA in the nucleus. This increases the probability of the physical connection of the protein regulators with the corresponding DNA sites (Syr-Young Lin and Riggs, 1975). It could, second, become a regulator for the activation time of the structural gene transcription during development. This extra DNA could be either unique or repeated. The repeated sequences could, third, regulate nuclear functions during cell differentiation by affecting the spatial chromosomal organization in interphase. It was demonstrated that in *Drosophila* various zones of the polytene chromosomes participated in synapsis with a certain frequency (see review Hannach, 1951; Evgeniev, 1970). The regions of homologous chromosomes which are characterized by the high frequency of conjugation in the salivary glands demonstrated a high level of crossing-over in hybrid females. Later, regions of "strong synapsis" were described for the X chromosome, which are characterized by the high capability for homologous and nonhomologous conjugations. These regions are the latest reduplicated regions of the X chromosome and correspond to the position of intercalary heterochromatin (Jones and Robertson, 1970). It was suggested that the conjugation of the chromosomal segments is associated with the presence in those sequences of repeated homologous DNA sequences (Kulichkov and Belyaeva, 1975). The authors supposed that regions of ectopic (nonhomologous) conjugation formed the stable structures which determined the spatial configuration of chromatin in the nucleus and could have significance for the interphase nucleus functions.

Another regulator of transcriptional rate is, possibly, associated with the organization of the DNA replication processes. As mentioned in Part I, the transcriptional process is postponed in the late replicated regions.

The interrelation of DNA replication and the cell differentiation was demonstrated (Holtzer, 1970). It is possible that this relation is conditional in two ways. First, the replication and cell proliferation rates determine the rate of transcription and the peculiarities of "dispersion" of the "wave" of RNA synthesis along the chromosome. Second, the reduction of various inhibitor concentrations which act on the transcriptional and post-transcriptional levels way occur. Of similar significance is the cell polyploidization and polytenization. In this case, the number of the DNA segments, which the corresponding regulating proteins influenced, is significantly increased, whereas the number of the molecule regulators remained the same. It is obvious that the effect of such events will be the same as in the previous example, that of decreasing the inductor (inhibitor) concentration.

It is possible that each of the above listed mechanisms of the DNA activity during cell differentiation plays different roles (in comparison to others) in different organisms.

2.4.6 Phases of Gene Activity in Ontogenesis

The analysis of the development of the tissue systems during ontogenesis distinguished two periods of the genome transcriptional activity. The first period is when similar DNA sequences are transcribed in cells, which specialized in different directions, possibly as a result of more or less synchronous distribution of the transcriptional wave along chromosomes of various cells. Thus, the mutual metabolic basis is formed for successive specific differentiation. Already in this period, post-transcriptional, translational, etc. regulations could affect the quantitative differences in the enzyme and protein contents in the cells specializing for different functions. This is done through selection among templates and differential rates of their synthesis and degradation.

It is possible that the protein receptors which recognize the inducer substances and hormones and which determine the tissue competence are synthesized in this period, also. It is known, for instance, that in the differentiating embryonic and regenerating liver more RNA types are synthesized than are necessary for specialized mature cell. In the early period of the liver cell specialization there is a surplus of the RNA templates (Church and McCarthy, 1967). There are data that changing of the RNA amount is accompanied by the oscillation of the chromatin transcriptional activity (Thaler and Villee, 1967). Thus, the cell during development is as if it were searching for its own way. The second period is characterized by the transcription of the genome regions coding organospecific and cell-specific proteins. This is determined, perhaps, by hormonal or other organ and tissue interactions. The various cell clones and the complex system of their proliferation and interaction are formed in this period.

The transition from one phase of transcription to another could be induced by the changing of the multiple forms of RNA-polymerases (see Part I). Various RNA-polymerase fractions could be different in their adjustment to a site of the functional chromosomal segments which should play an important regulating role. It was demonstrated that the chromatin has a limited number of sites which can assemble with RNA-polymerase and there are specific restrictions of transcription in chromatin. However, there is no simple correlation between the peculiarities of such restriction and the chromosomal protein distribution on DNA (Axel et al., 1974). The multilevel, cascade system of the genetic control of the cell specialization needs special mechanisms for the initiation of the transcriptional processes and for maintenance of the differential peculiarities of the specialized cells. The first is regulated by the presence of specific RNA-polymerase contents or by the specific ratio of the activating and inhibiting proteins and other regulating compounds. The second is caused by the relatively stable histone (or acid proteins) association with DNA of the specific chromosomal regions. Such semistable ensembles of histone-DNA terminate the moving of the transcriptional wave at a certain point of DNA; such wave begins to be "stationary". Such a process could be facilitated by the repeated DNA sequences which form the peculiar heterochromatic "borders" between the active regions.

The part of the functioning genes diminished as long as specialization takes place. The data that a significantly large percentage of genomes functioned during early developmental stages, indirectly confirmed the possibility of such tendencies in cell development. It is possible that the hypothetical autonomously functioning chromosomal segments are activated at the start as a unity, and later the transcription is restricted within the limits of certain parts of these segments which are "selected" by the specific protein "restrictors".

In this connection, the facts about globin mRNA synthesis in *Xenopus* are of great interest. The globin mRNA of adult type started its function only after metamorphosis, which was demonstrated in the hybridization experiments of globin DNA in oocyte of *Xenopus* (Perlman et al., 1977). It was found that only 1000 out of 18,000 types of mRNA are represented by a large number of molecules, and the remainder (the globin mRNA among them) are represented only by a small number of copies. Thus, the useless (for cell) mRNA's copies could be synthesized in oocyte. Therefore, it could be supposed that either gene transcription is determined by its localization, but not by the action of gene regulators that are specific for this gene.

The specificity of the alternation of the heterochromatic and the actively transcribed chromosomal segments in various tissues and cellular types could be the physical basis for the maintenance of the differential conditions. The puffing in polytenic chromosomes could be considered to be a demonstration of such heterochromatin effect on the transport of RNA molecules from the nucleus into the cytoplasm by the regulation of the transportation time for specific templates.

The ratio of various inductors and inhibitors in cells played a very important role and their constant interaction with genome on various levels could determine the specific system controlling the specificity of a given cell type (Lopashov, 1965).

Such a system was given the terms cytogenotype (Korochkin, 1965) and epigenome (Ratner and Tchuraev, 1971; Ratner, 1974a, b). Authors designated two types of subsystem: the static, structural (all genome), and the dynamic, the simplest variant of which could be the trigger of the two operons system in the λ phage. A trigger condition could be transferred along the cell line which phenomenologically resembles inheritance. This phenomenon was first called "extra genome inheritance" and then more acceptably "epigenome".

It is possible, naturally, to ask what number of the active functioning genes compose such epigenome or "cytogenotype". The data obtained recently demonstrated that the polysomic mRNA extracted from various tissues are hybridized approximately with 1% of the unique DNA. Taking into account modern knowledge about mosaic gene structure, one may suggest that only 20% of the genome is active in the differentiated cell (Neyfach and Timofeeva, 1978).

An experiment was carried out with sea urchins in which mRNA of polysomes was hybridized with complementary DNA, which was synthesized with the aid of the reversal transcriptase. It was shown that approximately 37,000 genes out of a total of 40,000 of this species are active in the oocyte stage, about 30,000 in the blastula stage, 12,000–15,000 in the gastrula-pluteus stages, and only 3,000–5,000 in the adult animal tissues (Hough et al., 1975; Anderson et al., 1976). Similarly, 3,000–7,000 genes are active in the embryonic tissue culture of *Drosophila*, e.g., practically all the genes of fruit fly genome. Analogous data were observed with animals (Neyfach and Timofeeva, 1978). They apparently demonstrated that in various stages of animal development and in various differentiated cells, different groups of genes are active. However, altogether they covered all genomes of a given organism. In connection with these data it should be concluded that the suggestion of the presence of "silent" genes in genomes is doubtful. According to Zuckerkandl and Pauling (1962) the activation of such genes in the extremely stressed environmental conditions could have important evolutionary consequences and lead to an explosion of speciation.

Actually, limited groups of genes are active in each eukaryotic differentiated cell and in each stage of development. Every such group is different from any other and is specific for the type of cell and stage of development. It is impossible to say that this peculiarity is the privilege only of eukaryotes, since in the phage also different groups of genes are active in various stages of development (for instance, there are the periods of early and late proteins).

Because the process of individual development is varied and complex, the combination of genetics, biochemical and molecular biology, and embryology should be the basis for studying developmental genetics. These three parts of modern biology deal with and totally cover those tendencies which are united, concentrated, and realized in individual development events.

Conclusion

The individual development of the multicellular organisms is, as a final result, a matter of the growing heterogenization of the embryo, its division into complicating morphofunctional systems and specializing cell types, and also, the appearance of structures and factors affecting the integrational process.

Each step of the individual development is, at the same time, a result of the successive realization of the inherited information which is encoded in the DNA of chromosomes and, consequently, is controlled by genes, the specific activation/inactivation of which determines the cell specialization.

Apparently, it is possible to think that all genes of genomes are active in the multicellular organism. However, only a limited number of genes is active in every differentiated and developing cell. Therefore, the heterogenization of the developing living system is associated with the fact that various genes are active in various cell systems and tissues. In other words, there is differential gene activity. However, as was demonstrated recently, the differences between tissue and cells, which are specialized in different directions, cannot be considered to be differences in the activity of genes which code tissue and cell-specific proteins. Numerous gene modifiers regulating the gene expression on various levels — from transcriptional to post-translational events — affected the phenotypic expression of organospecific proteins. Therefore, the act of activation and transcription of either gene does not imply the appearance of coding by these gene products in the cellular phenotype. Numerous processes take place between gene and trait in the multicellular organism. Any gene regulator, in turn, could be the structural gene for another gene product. This is the principal difference between multicellular and prokaryotic organisms.

The embryonal development is thus determined by the number of gene interactions, which occur not only by way of direct effects of one gene on another (regulatory effects, position effect, etc.) but also via complicated interactions between the regulating products as well.

These gene interactions could take place not only on the level of separate differentiated cells, but also on the level of intercellular interaction, determining the character of the inductor-competent tissue relation, of proliferation and competition of various cell clones and other events, which determine the final morphogenetic effect. The role of temporary factors is important for cell differentiation and for intercellular and intertissue interactions. The mutations which change the activation/inactivation period of a given gene could affect the gene expression through the delay of the appearance of this gene product in the crucial moment of development.

Other mutations, also, could be the cause of the desynchronization of inductor synthesis and competent tissue maturation. The morfogenetic processes will be disordered as a final result, and various genetically inherited anomalies will appear. The importance of temporary factors for development has possibly a physical basis in the genome organization of eukaryotes. The nonrandom gene localization and the correspondence of the gene order in chromosomes and the time of their expression in a phenotype might be the evidence of such an idea.

The presence of such genome organization determines the successive activation of different gene clusters in various stages of ontogenesis. These events are associated with the transcription of numerous mRNA molecules, some of which could be useless for the given cell type in the given time.

Concerning cell specialization, the number of the transcriptionally active genes is diminished and the cell specialization in different directions becomes deeper and deeper according to the characteristics of the functionally active parts of genome. Finally, the absolutely different cytogenotypes are formed from the nondifferentiated primordial cell population. The patterns of genes of the differentiated cells determine the synthesis of general metabolic needs of the cell and the tissue- and stage-specific proteins. The number of such cells is not large, but only these genes are responsible for the synthesis of the organospecific proteins marking every specialized cellular type.

References

Abdel-Hameed F (1972) Science 178:864-865
Abelev GI (1971) Cancer Res 14:195-258
Adhya S, Shapiro J (1969) Genetics 62:231-247
Adler D, West J, Chapman V (1977) Nature (London) 267:838-839
Agostini A, Vergani C, Villa L (1966) Nature (London) 209:1025-1026
Agranoff BW, Davis RE, Brinte JJ (1965) Proc Natl Acad Sci USA 54:788-793
Akimova IM, Diban AP (1967) Arch Anat (Russ) 52:36-50
Albers J, Dray S (1969) J Immunol 103:163-169
Albers J, Dray S, Knight K (1969) Biochemistry 8:4416-4424
Alexander M (1976) In: Ashburner M, Novitski E (eds) The Genetics and Biology of Drosophila, vol 1c. Academic Press, London New York, pp. 1365-1419
Alexidze NG, Chaglid K (1970) Dokl Acad Sci USSR (Russ) 190:927-974
Alfageme C, Cohen L (1975/76) In: Inst Cancer Res 21st Sci Rep. Pa, Philadelphia, pp 125-126
Allfrey V, Faulkner R, Mirsky A (1964) Proc Natl Acad Sci USA 51:768-794
Allfrey V, Pogo B, Pogo A, Kleinsmith J, Mirsky A (1968) In: Histones and transport of the genetic information. Mir, Moscow, pp 59-80
Allfrey V, Teng CS, Teng CT (1972) In: Ribbons D, Woessner J, Schultz J (eds) Nucleic acid – protein interactions. North-Holland Publ Co, Amsterdam-London pp 146-171
Altenburg E, Browning L (1961) Genetics 46:203-212
Altman J, Das G (1965) J Comp Neurol 124:319-336
Amaldi P, Felicetti L, Campioni N (1977) Dev Biol 59:49-61
Ananiev E (1974) Investigation of the DNA replication and transcription in the polytenic chromosomes. Autoreferate of candidate dissertation M (Russ)
Ananiev E, Barsky V (1978) Chromosoma 65:359-371
Ananiev E, Gvozdev V (1974a) Genetica (Russ) 10:120-128
Ananiev E, Gvozdev V (1974b) Chromosoma 45:173-191
Ananiev E, Faizullin L, Gvozdev V (1974a) Genetica (Russ) 10:62-69
Ananiev E, Faizullin L, Gvozdev V (1974b) Chromosoma 45:193-201
Anders A, Anders F, Klinke K (1973a) In: Schroder J (ed) Genetics and mutagenesis of fish. Springer, Berlin Heidelberg New York, pp 33-52
Anders A, Anders F, Klinke K (1973b) In: Schroder J (ed) Genetics and mutagenesis of fish pp 53-63
Anderson D, Galau G, Britten R, Davidson E (1976) Dev Biol 51:138-145
Angerer R, Davidson E, Britten R (1975) Cell 6:29-39
Anochin PK (1968) Biology and neurophysiology of the preconditioned reflex, Medicina, Moscow, pp 1-547
Ansevin K (1965) J Morphol 117:171-184
Appel SH (1965) Nature (London) 207:1163-1166
Arfin S, Simpson D, Chiang C, Andrulis I, Hatfield G, Wesley A (1977) Proc Natl Acad Sci USA 74:2367-2369
Arms S (1968) J Embryol Exp Morphol 20:367-374
Arnott S, Fuller W, Hodgson A, Prutton I (1968) Nature (London) 220:561-564
Aronstam AA (1974) Investigation of the Est-6 expression in *Drosophila melanogaster*. Autoreferate of candidate dissertation. L (Russ)
Aronstam AA, Korochkin LI (1975) Genetica (Russ) 11:81-91
Aronstam AA, Kusin BA, Korochkin LI (1975) Genetica (Russ) 11:92-96

Arrigo P, Goldschmidt-Clermout M, Mirault M, Moran L (1977) Experientia 33:811
Aschoff H (1955) Naturwissenschaften 42:569-575
Ashburner M (1967) Chromosoma 21:398-428
Ashburner M (1969a) Chromosoma 27:47-63
Ashburner M (1969b) Chromosoma 27:64-85
Ashburner M (1969c) In: Darlington C, Lewis K (eds) Chromosomes today, vol 2. Oliver and Boyd, Edinburgh, pp 99-106
Ashburner M (1970a) Proc R Soc London, Ser B 176:319-327
Ashburner M (1970b) Cold Spring Harbor Symp Quant Biol 35:533-538
Astaurov BL (1927) J Exp Biol (Russ) Ser A 3:1-61
Astaurov BL (1968) J Gen Biol (Russ) 29:139-152
Astaurov BL (1969) Cytogenetics of the silkworm development and its experimental control. Nauka, Moscow
Astaurov BL (1974) Heredity and development, Nauka, Moscow
Atwood K (1969) Genetics Suppl 1, 61:319-327
Atwood S, Sullivan J (1943) J Hered 34:311-313
Auerbach R (1964) In: Selforganizing Systems vol I. Mir, Moscow pp 140-150
Auerbach S, Brinster RL (1967) Exp Cell Res 46:89-92
Austin C (1962) Acta Cytol 6:61-65
Austin C, Amoroso E (1957) Exp Cell Res 13:419-421
Axel R, Cedar H, Felsenfeld G (1974) Cold Spring Harbor Symp Quant Biol 38:773-783
Baba N (1970) Fed Proc 29:1406
Baba N, Sharma H (1971) J Cell Biol 51:621-635
Bachvarova R, Davidson F, Allfrey V, Mirsky A (1966) Proc Natl Acad Sci USA 55:358-365
Badr F (1970) A'in Shams Med J 21:259-270
Badr F, Spickett S (1965) Nature (London) 205:1088-1090
Bagg HJ (1916) Am Nat 50:222-236
Baglioni C (1960) Heredity 15:87-96
Baglioni C (1964) In book: Molecular genetics, vol I. Mir, Moscow, pp 443-513
Baglioni C (1966) J Cell Physiol Suppl 1, 67:169-184
Bahn E (1967) Hereditas 58:1-13
Bahn E (1971) Hereditas 67:79-84
Baillie D, Chovnick A (1971) Mol Gen Genet 112:341-353
Baker W (1963) Am Zool 3:57-69
Baker W (1967) Dev Biol 16:1-17
Baker W (1968) In: Caspari E (ed) Advances in genetics, vol 14. Academic Press, London New York, pp 133-169
Baker W (1971) Proc Natl Acad Sci USA 68:2472-2476
Balakier H, Czolowka R (1977) Exp Cell Res 110:466-469
Balinsky B (1965) An introduction to embryology. Saunders, Philadelphia London, pp 1-657
Baranov OK, Korochkin LI (1973) Genetica (Russ) 8:144-153
Barbett J, Maryanka D, Hamlyn P, Gould H (1974) Proc Natl Acad Sci USA 71:5057-5061
Barbiroli B, Monti M, Moruzzi M, Mezzetti G (1977) Biochim Biophys Acta 479:69-79
Barnes B, Birley A (1975) Heredity 35:115-119
Barnes B, Birley A (1978) Heredity 40:51-57
Barnes R, Tuffrey N, Dryry L, Catty D (1974) Differentiation 2:257-260
Barondes S (1965) Nature (London) 205:19-21
Barondes S, Jarvik M (1964) J Neurochem 113:186-195
Beçak W, Goissis G (1971) Experientia 27:345-346
Beçak W, Pueyo M (1970) Exp Cell Res 63:418-451
Beçak W, Beçak M, Ohno S (1966) Cytogenetics 5:313-320
Becker H (1959) Chromosoma 10:654-678
Becker H (1966) In: Moscona A, Monroy A (eds) Current topics in development biology, vol 1. Academic Press, London New York, pp 155-171
Beckman G, Beckman L, Tärvnik A (1973) Hereditas 73:31-40
Beckman L, Johnson F (1964) Hereditas 51:212-220

Beebe D, Piatigorsky J (1977) Dev Biol 59:174-182
Beebe D, Piatigorsky J (1978) J Cell Biol 79:356a
Beerman S (1959) Chromosoma 10:504-514
Beerman W (1961) Chromosoma 121-25
Beerman W (1964) J Exp Zool 157:49-61
Belicina NV, Gavrilova LP, Aitochojin ML, Neyfach AA, Spirin AC (1963) Dokl Acad Sci
    USSR (Russ) 153:464-467
Bello L (1969) Biochim Biophys Acta 179:204-215
Belyaev DK (1970) In: Genetics and selection of new breeds of domestic animals. Nauka,
    Alma-Ata, pp 30-44
Belyaev DK (1972) Second meeting of VOGIS name of Vavilov NI Nauka, Moscow, pp 8-
    12
Belyaev DK, Jelezova AI (1968) Genetica (Russ) 1:45-48
Belyaev DK, Klochkov DV, Jelezova AI (1963) Bull MOIP Biol Div (Russ) 68:108-125
Belyaev DK, Korochkin LI, Baev AA, Golubitza AN, Korochkina LS, Maximovsky LF,
    Davidovskaya AE (1966) Dokl Acad Sci USSR (Russ) 169:1227-1230
Belyaev DK, Klochkov DV, Klochkova AJa (1969) Agric Biol (Russ) 4:167-177
Belyaev DK, Trut N, Ruvinsky AO (1973) Genetica (Russ) 9:71-82
Balyaeva ES (1965) Specificity of the nucleoli formation as an actively functioning region
    of chromosome. Autoreferate candidate dissertation Novosibirsk (Russ)
Belyaeva ES, Zhimulev IF (1974) Genetica (Russ) 10:74-79
Belyaeva ES, Korochkina LS, Zhimulev IF, Nazarova NK, (1974) Cytologia (Russ) 16:440-
    445
Benbassat J (1974) J Cell Sci 16:145-156
Bennett D (1975) Cell 6:441-454
Bennett J, Owen R (1966) J Cell Physiol Suppl 1, 67:207-216
Bennett E, Diamond M, Krech D, Rosenzweig M (1964) Science 146:610-619
Benoff-Rend S, Bruce S, Skoultchi A (1977) Genetics 86 part 2:5-6
Benson S, Triplett E (1974a) Dev Biol 40:270-282
Benson S, Triplett E (1974b) Dev Biol 40:283-291
Benz E, Turner P, Barker J, Nienhuis A (1977) Science 196:1213-1214
Benzer S (1963) In: Molecular genetics. IL, Moscow, pp 11-32
Berdishev GD (1968) Adv Mod Biol (Russ) 66:226-246
Berdnikov VA, Gorel F, Argutinskaja S, Tsherepanova V, Kileva E (1976) Mol Biol (Russ)
    10:887-896
Berendes H (1965) Dev Biol 11:371-384
Berendes H (1966a) Chromosoma 20:32-43
Berendes H (1966b) J Exp Zool 162:205-217
Berendes H (1972) In: Beerman W (ed) Results and problems in cell differentiation, vol 4.
    Springer, Berlin Heidelberg New York, pp 181-207
Berendes H (1973) In: Bourne G, Danielli J (eds) Int Rev Cytol, vol 35, Academic Press,
    London New York, pp 61-116
Berg R (1974) Drosophila Inf Serv 51:100-102
Berg T, Blix A (1973) Nature New Biol 245:239-240
Berger E, Canter R (1973) Dev Biol 33:48-55
Berget S, Moore C, Sharp P (1977) Proc Natl Acad Sci USA 74:3171-3175
Berlowitz L (1965a) Proc Natl Acad Sci USA 54:476-480
Berlowitz L (1965b) Proc Natl Acad Sci USA 53:68-73
Bernhard H, Darlington G, Ruddle F (1973) Dev Biol 35:83-96
Berns A, Kranikamp M von, Blormendal H, Lane C (1972) Proc Natl Acad Sci USA
    69:1606-1609
Bernstein R, Mukherjee B (1972) Nature (London) 238:457-458
Bester A, Kennedy D, Heywood S (1975) Proc Natl Acad Sci USA 72:1523-1527
Beutler E (1964) Cold Spring Harbor Symp Quant Biol 29:261-270
Bhat N (1949) Heredity 3:243-248
Bianchi N, Molina J (1967) Chromosoma 22:391-400
Bick M, Huang A, Thomas C (1973) J Mol Biol 77:75-84

Bielka H (1977) Cell differentiation microorganisms, plants and animals, Jena, pp 156-157
Bier K (1963) J Cell Biol 16:436-448
Bier K (1965) Zool Jahrb Physiol 71:371-384
Bier K (1967) Naturwissenschaften 54:189 -195
Bier K, Kunz W, Ribbert D (1967) Chromosoma 23:214-254
Bishop J, Rosbach M (1973) Nature New Biol 241:204-207
Blake R (1970) Int J Biochem 1:361-370
Blanco A, Zinkham W (1963) Science 139:601-602
Blanco A, Zinkham W, Kupchyk L (1964) J Exp Zool 156:137-152
Blanco A, Zinkham W, Walker D (1975) In: Markert C (ed) Isozymes, vol 3. Academic Press, London New York, pp 297-312
Blattner F, Dahlberg J (1972) Nature New Biol 237:227-232
Blattner F, Dahlberg J, Boettiger J, Fiandt M, Szybalsky W (1972) Nature New Biol 237:232-236
Blomberg F, Raftell M (1974) Eur J Biochem 49:21-29
Bloom A, Choi K, Lamb B (1971) Science 172:382-384
Bogomolova VI, Korochkin LI (1973) Ontogenes (Russ) 4:420-423
Bonner J (1967) Molecular biology of development. Mir, Moscow
Bonner J, Pardue M (1976) Cell 8:43-50
Bonner J, Pardue M (1977) Cell 12:227-234
Bonner J, Wu J (1973) Proc Natl Acad Sci USA 70:535-537
Bonner W (1974) J Cell Biol 64:431-437
Böök J, Santesson B (1961) Lancet 11:318
Boquist L (1974) Acta Endocrinol 77:10-11
Bornstein P, Oudin J (1964) J Exp Med 120:655-667
Borovkov AJu, Karakin EI, Kuzin BA, Sviridov SM (1975) Dokl Acad Sci USSR (Russ) 2-3:751-753
Bovet D, Bovet-Nitti F, Oliverio A (1968) Brain Res 10:168-182
Bovet D, Bovet-Nitti F, Oliverio A (1969) Science 163:139-141
Bowman J, Simmons J (1973) Biochem Genet 10:319-331
Brachet J (1960) Biochemical cytology. IL, Moscow
Brachet J (1961) Biochemical cytology. IL, Moscow
Brachet J (1964) Nature (London) 204:1218-1219
Brack C, Tonegawa S (1977) Proc Natl Acad Sci USA 74:5652-5656
Braverman M, Katon A (1971) Nature (London) 230:392-393
Breathnach R, Mandel J, Chambon P (1977) Nature (London) 270:314-319
Breen G, Lusis A, Paigen K (1977) Genetics 85:73-84
Brent T, Butler J, Crathorn A (1965) Nature (London) 207:176-177
Bresler D, Bitterman M (1969) Science 163:590-592
Bresler E S (1973) Elementary processes of genetics, Nauka S L
Bretscher P, Cohn M (1968) Nature (London) 220:444-446
Bridges C (1922) Am Nat 56:51-63
Brien S, Gethmann R (1973) Genetics 75:155-167
Briggs R, Kassens G (1966) Proc Natl Acad Sci USA 55:1103-1109
Briggs R, King T (1959) In: Brachet J, Mirsky A (eds) The cell. Academic Press, London New York, pp 537-617
Brinster R (1965) Biochim Biophys Acta 110:439-441
Brinster R (1966) Biochem J 101:161-163
Brinster R (1967) Nature (London) 214:1246-1247
Brinster R (1968) Enzymologia 34:304-308
Brinster R (1971a) Roux Arch 166:300-302
Brinster R (1971b) FEBS Lett 17:41-44
Brinster R (1973) Biochem Genet 9:187-191
Britten R, Davidson E (1969) Science 165:349-358
Brodsky V Ja (1966) Trophics of cell. Nauka, Moscow
Brodsky V Ja, Agroskin A S, Lebedev E A, Marshak T L, Papajan G V, Segal O L, Sokolova G A, Jarigin I N (1974) J Gen Biol (Russ) 35:917-925

Brondz B, Roklin O (1978) Molecular and cellular bases of immunological recognition. Nauka, Moscow, pp 1-280

Brooker A, Tomaszewski A, Marcus P (1975/76) Inst Cancer Res 21st Sci Rep. Pa, Philadelphia, 177

Brown D (1964) J Exp Zool 157:101-114

Brown D (1966) Natl Cancer Inst Monogr 23: 297-317

Brown D (1972) J Mol Biol 63:75-83

Brown D, Blackler A (1972) J Mol Biol 63:75-83

Brown D, Dawid I (1968) Science 160:272-280

Brown D, Dawid I (1969) In: Annual review of genetics, vol 3. Academic Press, London New York, pp 128-154

Brown D, King R (1964) Growth 28:41-53

Brown D, Littna E (1966) J Mol Biol 20:81-94

Brown D, Wensink P, Jordan E (1972) J Mol Biol 63:57-73

Brown I, Church R (1972) Dev Biol 29:73-84

Brown S (1969) Genetics 61:191-198

Brumberg V A, Pevzner L Z (1975) Neurochemistry of isozymes. Nauka, L (Russ)

Bryant P (1970) Dev Biol 22:389-411

Bryant P, Schneiderman H (1969) Dev Biol 20:263-290

Buckingham M, Cohen A, Gros F (1976) J Mol Biol 103:611-628

Buckingham M, Whalen R, Gros F (1977) Biochem Soc Trans 5:474-477

Buller A, Eccles J, Eccles R (1960a) J Physiol 150: 399-416

Buller A, Eccles J, Eccles R (1960b) J Physiol 150:417-439

Bullough W (1969) Science J 5:71-76

Bullough W, Laurence E (1968) Nature (London) 220:134 135

Bullough W, Laurence E, Iversen O, Elgio K (1967) Nature (London) 214:578-580

Bünning E (1935) Jahrb Wiss Bot 81:411-418

Bünning E (1961) Rhythms of physiological processes. IL Moscow

Burnet F (1971) Cellular immunology, Mir, Moscow

Bush H (1965) Histones and other nuclear proteins. Academic Press, London New York, pp 1-286

Butler J, Johns E, Phillips D (1968) Prog Biophys Mol Biol 18:211-230

Cabada M, Darnbrough C, Ford P, Turner P (1977) Dev Biol 57:427-439

Cagianut B (1949) Z Zellforsch 34:471-491

Cahn R, Kaplan N, Levine L, Zwilling E (1962) Science 136:962-969

Callan H (1963) Int Rev Cytol 15:1-34

Callan H (1965) Genet Today 2:359-368

Callan H (1966) J Cell Sci 1:85-105

Campbell P, McIlreavy D, Tarin D (1973) Biochem J 134:345-347

Campo M, Bishop J (1974) J Mol Biol 90:649-663

Cannon W, Rosenblut A (1951) Increasing of sensitivity of denervated structures. IL, Moscow

Carbone C, Tjio J, Whang J, Block J, Kramer W, Frei E (1963) Ann Inst Med 59: 622-628

Carlson P (1971) Genet Res 17:53-81

Carlson P (1972) Nature New Biol 237:39-41

Carter T (1952) J Genet 51:1-6

Cashel M (1969) J Biol Chem 244:3133-3141

Castro-Sierra E, Ohno S (1968) Biochem Genet 1:323-335

Cattanach B (1974) Genet Res 23: 291-306

Ceccarini C, Maggio R (1969) Biochim Biophys Acta 190:556-559

Cech T, Hearst J (1975) Cell 5:429-446

Chalmers J, De Moss J (1970) Genetics 65:213-221

Chamberlin M, Britten R, Davidson E (1975) J Mol Biol 96:317-333

Champion M, Whitt G (1976) Biochem Genet 14:725-737

Chan L, Ingram V (1973) J Cell Biol 56:861-865

Chan L, Wiedmann M, Ingram V (1974) Dev Biol 40:174-185

Chandra H (1971) Genet Res 18:265-276

Chapman R (1969) The insect: structure and function. Ltd Press, London, pp 1-819
Chapman V, Wudl L (1975) In: Markert C (ed) Isozyme III. Academic Press, London New York, pp 57-65
Chapman V, Whitten W, Ruddle F (1971) Dev Biol 26:153-158
Chatterjee, S, Mukherjee A (1971a) Chromosoma 36:46-59
Chatterjee S, Mukherjee A (1971b) Basic Mech Radiat Biol Med Proc Symp. Bombay, 489-497
Chen P (1971) Biochemical aspects of insect development. S Karger, Basel, pp 1-208
Chen P, Farinella-Ferruzza N, Aelhafen-Gandolla M (1963) Exp Cell Res 31:538-548
Chesnokov VN, Mertvetsov NP, Salganik RI (1974) Biochimia (Russ) 39:294-297
Chetverikov SS (1926) J Exp Biol (Russ) 2:3-54
Chi J, Rubinstein H, Strahs K, Holtzer H (1975) J Cell Biol 67:523-537
Child Ch (1948) Role of organizators in development, IL, Moscow
Chlebodarova TM, Serov OL, Korochkin LI (1975) Dokl Acad Sci USSR (Russ) 224:1428-1430
Chlebodarova TM, Serov OL, Korochkin LI (1978) Genetica (Russ) 14:1213-1219
Chovnik A, Sihalet R, Kernaghan R, Krauss M (1964) Genetics 50:1245-1259
Chovnik A, Finnerty V, Schalet A, Duck P (1969) Genetics 62:145-160
Chovnik A, McCarron M, Gelbart W, Paudey J (1975) In: Markert C (ed) Isozymes, vol IV. Academic Press, London New York, pp 609-622
Chovnik A, Gelbart W, McCarron M, Osmond B, Candido E, Baillie D (1977) Genetics 84:233-255
Chow L, Gelinas R, Broker T, Roberts R (1977) Cell 12:1-8
Christy S, Jayaraman K (1975) Proc Symp Struct Funct Aspects Chromosomes Bombay, p 392
Chu E, Thulin H, Norby D (1964) Cytogenetics 3:1-18
Chung-chin-Huang (1971) Chromosoma 34:230-242
Church R (1970) In: Fraser F, McKusick V (eds) Congenital malformations. New York, pp 19-28
Church R, Brown I (1972) In: Ursprung H (ed) Results and problems in cell differentiation, vol 3. Springer, Berlin Heidelberg New York, pp 11-24
Church R, McCarthy B (1967) J Mol Biol 23:459-476, 477-486
Church R, Robertson F (1966) J Exp Zool 162:337-349
Church R, Schultz G (1974) In: Current topics in developmental biology, vol 8. Academic Press, London New York, pp 179-202
Ciaranello R, Barchas R, Kessler S, Barchas J (1972) Life Sci 11:565-572
Clayton R, Truman D, Hannan A (1974) Cell Diff 3:135-145
Clegg K, Piko L (1977) Dev Biol 58:76-95
Clever V (1966) Am Zool 6: 33-45
Clever V (1968) In: Roman H (ed) Annu Rev Gen vol 2 Annu Rev Inc, Palo Alto, pp 11-30
Clever V, Bultmann H, Darrow J (1969) In: Hanly S (ed) Problems in biology RNA in development. Univ Utah Press, Salt Lake City, pp 403-424
Close R (1964) J Physiol 173: 74-95
Coggins L, Gall J (1972) J Cell Biol 52:569-576
Coghill J (1934) Anatomy and problems of behavior. IL, Moscow
Cohen M (1976) Chromosoma 55:349-357
Cohn M (1967) Cold Spring Harbor Symp Quant Biol 32:211-221
Coleman D (1966) J Biol Chem 241:5511-5517
Coleman J, Coleman A (1968) J Cell Physiol Suppl 72:19-34
Collier J (1966) Curr Top Dev Biol 1: 39-59
Collier J (1977) Exp Cell Res 106: 390-394
Collins RL (1964) Science 143:1188-1190
Colombo B, Baglioni C (1966) J Mol Biol 16:51-66
Comings D (1967) J Cell Biol 35:699-701
Conclin J, Nebel E (1965) J Histochem Cytochem 13:510-514
Conway T, Dray S, Lichter E (1969) J Immunol 103:662-670
Cooper D, Van de Berg J, Sharman G, Poole W (1971) Nature New Biol 230:155-157

Cooper D, Johnston P, Muragh C, Sharman G, Van de Berg J, Poole W (1975) In: Markert
    C (ed) Isozymes, Academic Press, London New York, pp 559-573
Courtright J (1967) Genetics 57:25-39
Courtright J (1975) Mol Gen Genet 142:231-241
Courtright J (1976) In: Caspari E (ed) Adv Genet 18. Academic Press, London New York
    pp 249-314
Crain W, Eden I, Pearson W, Davidson E, Britten R (1976a) Chromosoma 56:309-326
Crain W, Davidson E, Britten R (1976b) Chromosoma 59:1-12
Crick F (1971) Nature (London) 234:25-27
Crick F, Lawrence P (1975) Science 189:340-347
Crippa M (1970) Nature (London) 227:1138-1140
Crippa M, Davidson E, Mirsky A (1967) Proc Natl Acad Sci USA 57:885-892
Crippa M, Tochini-Valentini G, Andronico F (1972) In: Biggers J, Schultz A (eds) Oogen-
    esis. Univ Park Press, Baltimore, pp 193-214
Crippa P, Lo Scavo A (1972) Biochem Biophys Res Commun 48:280-285
Croce C, Bakay B, Nyhan W, Koprowski H (1973) Proc Natl Acad Sci USA 70:2590-2594
Croce C, Litwack G, Koprowski H (1974) In: Davidson R, de la Cruz F (eds) Somatic cell
    hybridization. Raven Press, New York, pp 173-178
Croizat B, Berthelot F, Felsani A, Gros F (1977) Eur J Biochem 74:405-412
Crouse L, Kryl H (1968) Chromosoma 25:357-364
Curley L, Walters R, Barham S, Deaven L (1978) Exp Cell Res 111:373-383
Curry J, Trentin J (1967) Dev Biol 15:395-401
Dagg C, Coleman D, Fraser G (1964) Genetics 49:979-989
D'Alession J, Bagshaw J (1977) Differentiation 8:53-56
Daly M (1973) Br J Psychol 64:435-460
Daneholt B (1972) Nature New Biol 240:229-234
Daneholt B (1974) In: Bourne G, Danielli J (eds) Int Rev Cyt Suppl 4. Academic Press, Lon-
    don New York, pp 297-332
Daneholt B (1976) In: King R (ed) Handbook of genetics, vol 5. Plenum Publ, New York,
    pp 189-217
Daneholt B, Edström JE, Egyhazi E, Lambert B, Ringborg V (1969) Chromosoma 28:418-
    448
Daneholt B, Hosick H (1973) Proc Natl Acad Sci USA 70:442-446
Danes B, Bearn A (1966) J Exp Med 123:1-16
Danes B, Bearn A (1967) J Exp Med 126:509-521
Darlington G (1977) In Vitro 13: 149
Datta A, DeHaro C, Ochoa S (1978) Proc Natl Acad Sci USA 75:1148-1152
David C, Todd C (1969) Proc Natl Acad Sci USA 62:860-866
Davidson E (1972) Gene action in early development. Mir, Moscow
Davidson E, Britten R (1973) Q Rev Biol 48:565-613
Davidson E, Britten R (1979) Science 204:1052-1059
Davidson E, Mirsky A (1965) In: Genetic control of differentiation. Academic Press, Lon-
    don New York, pp 77-97
Davidson E, Allfrey V, Mirsky A (1963) Proc Natl Acad Sci USA 49:53-60
Davidson E, Allfrey V, Mirsky A (1964) Proc Nat Acad Sci USA 52:501-508
Davidson E, Haslett G, Finney R, Allrey V, Mirsky A (1965) Proc Natl Acad Sci USA
    54:696-704
Davidson E, Crippa M, Kramer F, Mirsky A (1966) Proc Natl Acad Sci USA 56:856-863
Davidson E, Hough B, Amenson C, Britten R (1973) J Mol Biol 77:1-23
Davidson E, Hough B, Klein W, Britten R (1975) Cell 4:217-238
Davidson E, Klein W, Britten R (1977) Dev Biol 55:69-84
Davidson R (1972) Proc Natl Acad Sci USA 69:951-955
Davidson R (1974a) In: Roman H (ed) Annu Rev Genet, vol 8. Annu Rev Inc, Palo Alto
    Calif, pp 195-218
Davidson R (1974b) In: Davidson R, de la Cruz F (eds) Somatic cell hybridization. Raven
    Press, New York, pp 131-150
Davis M, Calame K, Early P, Livant D, Joho P, Weisman I, Mood L (1980) Nature 283:733–742

Day T, Hillier P, Clarke B (1974) Biochem Genet 11:141-153

Deeley R, Gordon J, Burns A, Mullinix K, Bina-Stein M, Goldberger R (1977) J Biol Chem 252:8310-8319

DeLange R, Smith E (1971) In: Snell E (ed) Annu Rev Biochem, vol 40. Annu Rev Cuc, Palo Alto, pp 279-314

DeLarco J, Nakagawa S, Abramowitz A, Bromwell K, Guroff G (1975) J Neurochem 25:131-137

DeMoss J, Jackson R, Chalmers J (1967) Genetics 56:413-424

DeNayar Ph, Durieux B, Nissen-Cautier F, de Visscher M (1974) Life Sci 14:625-630

Denis H (1966) J Mol Biol 22:285-304

Denis H (1968) In: Krause G, Sander K (eds) Advances in morphogenesis, vol 7. Academic Press, London New York pp 115-150

Denis H (1970) In: Genetic information and development. Inst Cytol Genet, Novosibirsk, p 50

Deol M, Whitten W (1972) Nature New Biol 240:277-279

DeRobertis E, Gurdon J (1977) Proc Natl Acad Sci USA 74:2470-2474

Deryagin GV, Iordansky AV (1970) II Symposium on structure and function of chromosomes. Inst Cytol Genet (Russ), Novosibirsk, p 12

Detlaff T, Skoblina M (1969) Annu Embryol Morphogen Suppl 1:133-151

Detlaff T, Nikitina L, Stroeva O (1964) J Embryol Exp Morphol 12:851-873

Dewey M, Gervais A, Mintz B (1976) Dev Biol 50:68-81

Dewey M, Martin D, Martin G, Mintz B (1977) Proc Natl Acad Sci USA 74:5564-5568

Dewhurst S, McCaman R, Kaplan W (1970) Biochem Genet 4:499-508

De Witt W (1971) Biochem Biophys Res Commun 42:266-270

Diban AP (1952) Assays of pathological human embryology. Medgis L (Russ)

Dickinson W (1968) Genetics 60:173

Dickinson W (1971) Dev Biol 26:77-86

Dickinson W (1975) Dev Biol 42:131-140

Dickinson W, Sullivan D (1975) Gene-enzyme systems in Drosophila. Academic Press, London New York, pp 1-163

Dickinson W, Weisbrod E (1976) Biochem Genet 14:709-722

Diesterhaft M, Noguchi T, Hargrove J, Thornton Ch, Granner D (1977) Biochem Biophys Res Commun 79:1015-1022

Dingman W, Sporn M (1964) J Biol Chem 239:3483-3492

Doane N (1969) In: Problems in biology. RNA in development. Univ Utah Press pp 74-109

Doane W (1971) Isozyme Bull 4:46-48

Doane W (1975) In: Markert C (ed) Isozymes, Academic Press, London New York, pp 585-602

Dobrovolskaia-Zavadskaia N, Kobozieff N (1927) C R Soc Biol 97:116-118

Doering C, Shire J, Kessler S, Clayton R (1973) Biochem Genet 8:101-111

Dofuku R, Tettenborn V, Ohno S (1971) Nature New Biol 884:259-261

Donady J, Seekoff R, Fox M (1973) Genetics 73:429-434

Donahue R, Stern S (1970) J Reprod Fertil 22:575-577

Don Michele C, Masters C (1977) Int J Biochem 8:219-225

Dooner H, Nelson O (1977) Proc Natl Acad Sci USA 74:5623-5627

Doyle D (1971) J Biol Chem 246:4965-4972

Doyle D, Mitchell R (1975) In: Markert C (ed) Isozymes, III. Academic Press, London New York, pp 907-918

Doyle D, Schimke R (1969) J Biol Chem 244:5449-5459

Dray S (1962) Nature (London) 195:677-680

Dray S, Nisonoff A (1963) Proc Soc Exp Biol Med 113:20-25

Dubendorfer K, Nöthiger R, Kubli E (1974) Biochem Genet 12:203-211

Dubinin N (1932a) J Genet 25:163-181

Dubinin N (1932b) J Genet 26:37-58

Dubinin N (1933) J Genet 27:443-464

Dubinin NP, Sidorov BN (1934) Biol J (Russ) 3:307-331

Dubiski S (1967a) Nature (London) 214:1365-1368

Dubiski S (1967b) Cold Spring Harbor Symp Quant Biol 32:311-316
Dubroff L, Nemer M (1975/76) Inst Cancer Res 21st Sci Rep, Philadelphia PA, p 171
Duck-Chong C, Pollak J (1973) In: Pollak J, Wilson J (eds) The biochemistry of gene ex-
    pression in higher organisms. Reidel Publ Co, pp 305-319
Dunn G (1972) J Exp Zool 181:1-16
Dunn L, Gluecksohn-Waelsch S (1952) Genetics 37:577-578
Dunn L, Gluecksohn-Schonheimer S, Bryson V (1940) J Hered 31:343-348
DuPasquier L, Wabl M (1977) Differentiation 8:9-19
Easton T, Reich E (1972) J Biol Chem 247:6420-6431
Eayrs J (1960) Br Med Bull 16:122-126
Eayrs J (1961) J Endocrin 22:409-419
Eccles JC (1953) The neurophysiological basis of mind. Pergamon Press, Oxford, pp 1-314
Edelman G (1977) Cold Spring Harber Symp Quant Biol 41:891-902
Edlin G, Broda P (1968) Bacteriol Rev 32:206-226
Edström JE (1965) In: Locke M (ed) The role of chromosomes in development. Academic
    Press, London New York, pp 137-158
Edström JE, Grampp W (1965) J Neurochem 12:735-741
Edström JE, Lambert B (1975) Prog Biophys Mol Biol, vol 30. Pergamon Press, New York
    London, pp 57-82
Edström JE, Tanguay R (1974) J Mol Biol 84:569-583
Edström J, Lindgren S, Lonn U, Rydlander L (1978) Chromosoma 66:33-44
Edwards L, Hnilica L (1968) Experientia 24:228-230
Efron Y (1971) Mol Gen Genet 111:97-102
Efstratiadis A, Crain W, Britten R, Davidson E, Kafatos F (1976) Proc Natl Acad Sci USA
    73:2289-2293
Egyhazi E (1975) Proc Natl Acad Sci USA 72:947-950
Ekisashvili VK (1974) DNA-dependent RNA-polymerases of the loach embryos. Autoref-
    erate candidate dissertation Tbilisi (Russ)
Elgin S, Bonner J (1973) Biochem gene expression higher organisms. Proc Symp, Sydney,
    pp 142-163
Eliceiri G, Green H (1969) J Mol Biol 41:253-260
Elkin VI (1971) Genetics of epilepsy. Medicina, L
Ellem K, Gwatkin R (1968) Dev Biol 18:311-330
Elsdale T, Jones K (1963) R Soc Exp Biol Symp, no 17. Univ Press, Cambridge, pp 257-273
Elsevier S, Ruddle F (1976) Chromosoma 56:227-241
Elska A, Matsuka G, Matiash V (1971) Biochim Biophys Acta 247:430-440
Elston R, Glassman E (1967) Genet Res 9:141-147
Engberg J, Klenow H (1977) Trends Biochem Sci 2:183-185
Engel W (1972) Humangenetik 15:355-356
Engel W (1973) Humangenetik 20:133-140
Engel W, Kreutz R (1973) Humangenetik 19:253-260
Engel W, Wolf U (1971) Humangenetik 12:162-166
Engel W, Zenzes M, Schmid M (1977) Humangenetik 38:57-63
Ephrussi B (1935) J Exp Zool 70:197-204
Ephrussi B, Sutton E (1944) Proc Natl Acad Sci USA 30:183-197
Epifanova OI (1965) Hormones and development. Nauka, Moscow
Epstein C (1970) J Biol Chem 245:3289-3294
Epstein C (1975) Biol Reprod 12:82-105
Epstein C, Smith S (1974) Dev Biol 40:233-244
Epstein C, Wagienka E, Smith C (1969) Biochem Genet 3:271-281
Epstein C, Kwok L, Smith S (1971) FEBS Lett 13:45-48
Erickson R, Bethlach C, Epstein C (1974) Differentiation 2:203-209
Erickson R, Friend D, Tennenbaum D (1975) Exp Cell Res 91:1-5
Evans L, Jenkins B (1960) Can J Genet Cytol 2:205-214
Evans M, Lingrel J (1969) Biochemistry 8: 829-831
Evgeniev MB (1970) Genetica (Russ) 6:97-102
Fahey J, Finegold I (1967) Cold Spring Harbor Symp Quant Biol 32:283-289

Faizullin LZ, Gvozdev VA (1973) Genetica (Russ) 9:107-117
Falk D (1975) Genetics 80:29
Falk D (1976) Mol Gen Genet 148:1-8
Fambrough D, Fujimura F, Bonner J (1968) Biochemistry 7:575-585
Fantoni A, Bank A, Marx P (1967) Science 157:1327-1329
Fantoni A, De La Chapelle A, Marx P (1969) J Biol Chem 244:675-681
Fantoni A, Ullu E, Gambari R, Lunudei M, Farau M (1976) Ann Immunol C 127:881-886
Fa Ten Kao, Puck T (1972) Proc Natl Acad Sci.USA 69:3273-3277
Faust C, Diggelman H, Mach A (1974) Proc Natl Acad Sci USA 71:2491-2495
Felsenfeld G (1978) Nature (London) 271:115-122
Fieldman M, Yaffe D, Globerson A (1964) In: Cellular control mechanisms and cancer. Academic Press, London New York, pp 60-79
Filatov DP (1934) Adv Mod Biol (Russ) 3:440-456
Finger I, Onorato F, Heller C, Wilcox H (1966) J Mol Biol 17:86-100
Flavell R, McPherson F (1972) Biochem Genet 7:259-268
Flexner JB, Flexner LR, Stellar E (1963) Science 141:57-59
Flickinger R (1971) In: Cameron I, Padilla G, Zimmerman A (eds) Developmental aspects of the cell cycle. Academic Press, London New York, pp 161-189
Flickinger R, Coward S, Miyagi M, Moser C, Rollins E (1965) Proc Natl Acad Sci USA 53:783-790
Flickinger R, Greene R, Kohl D, Miyagi M (1966) Proc Natl Acad Sci USA 56:1712-1718
Flickinger R, Freedman M, Stambrook P (1967) Dev Biol 16:457-473
Flickinger R, Moser C, Rollins E (1967) Exp Cell Res 46:78-88
Flint J (1977) Cell 10:153-166
Ford P, Mathieson T, Rosbach M (1977) Dev Biol 57:417-426
Franke W, Scheer U, Spring H, Trendelenburg M, Krohne G (1976) Exp Cell Res 100:233-244
Fraser R (1964) Exp Cell Res 36:429-431
Frehel C, Dubray G, Wolf A (1974) Biochimie 56:583-598
Frenster J (1965) Nature (London) 206:1269-1270
Fridenstein AJa, Tchertkov IL (1969) Cellular basis of immunity. Medicina, Moscow
Friström D (1969) Mol Gen Genet 103:363-379
Fritz P, Vesell E, White E, Pruitt K (1969) Proc Natl Acad Sci USA 62:558-565
Fritz P, White E, Vesell E, Pruitt K (1971) Nature New Biol 230:119-122
Fritz P, White E, Pruitt K, Vesell E (1973) Biochemistry 12:4034-4039
Frota-Pessoa A, Gomes E, Galiccio T (1963) Science 139:348-349
Fudenberg H, Hirschorn K (1964) Science 145:611-612
Fujisawa H (1969) Embryologia 10:256-267
Fujita S (1974a) J Comp Neurol 155:195-202
Fujita S (1974b) Dev Growth Diff 16:225-235
Furman DP, Rodin SN, Ratner VA (1975) Research reports, 3 Novosibirsk Inst Cytol Genet, (Russ) pp 85-86
Furman DP, Rodin SN, Ratner VA (1977) Genetica (Russ) 13:1387-1397
Gage L, Geiduschek E (1971) J Mol Biol 57:279-291
Gaizchoki VS, Kiselev OI (1974) Adv Mod Biol (Russ) 78:385-403
Galau G, Britten R, Davidson E (1974) Cell 2:9-22
Galau G, Klein W, Davis M, Wold B, Britten R, Davidson E (1976) Cell 7:487-505
Galien K (1971) Ontogenes (Russ) 2:55-63
Gall J (1969) Genetics Suppl 61:121-132
Gall G, Callan H (1962) Proc Natl Acad Sci USA 48:562:566
Gall J, Pardue M (1969) Proc Natl Acad Sci USA 63:378-386
Gambrini R, Kacian D, O'Donnell J, Ramirez F, Marks P, Bank A (1974) Proc Natl Acad Sci USA 71:3966-3970
Ganschow R, Bunker B (1970) Biochem Genet 4:127-133
Ganschow R, Schimke R (1970) Biochem Genet 4:157-167
Garcia-Bellido A, Merriam J (1969) J Exp Zool 170:61-76
Garcia-Bellido A, Merriam J (1971) Dev Biol 26:264-276

Garcia-Bellido A, Ripoll P, Morata G (1973) Nature New Biol 245:251-253
Garcia-Bellido A, Ripoll P, Morata G (1976) Dev Biol 48:132-147
Gardner R, Lyon M (1971) Nature (London) 231:385-386
Garen A, Gehring W (1972) Proc Natl Acad Sci USA 69:2982-2985
Gargano S, Graziani F (1977) Atti Assoc Genet Ital 22:175-176
Gartler S, Liskey R, Campbell B, Sparkes R, Gant N (1972a) Cell Diff 1:215-218
Gartler S, Shi-Han Chen, Fialkow Ph, Giblett E, Singh S (1972b) Nature New Biol 236:149-150
Gartler S, Liskey R, Gant N (1973) Genetics Suppl 1, 74:589
Gazarjan (1972) Transcription and postranscription changing of RNA during growth and differentiation of animal cell. Autoreferate doctor dissertation, Moscow
Gearhart J, Mintz B (1972) Dev Biol 29:27-37
Geber W (1973) Fed Proc 32:2101-2104
Gehring W (1968) In: Ursprung H (ed) The stability in the differentiated state. Springer, Berlin Heidelberg New York, pp 136-154
Gehring W (1976a) In: Lawrence P (ed) Insect development. Blackwell Sci Publ, Oxford London Edinburgh Melbourne, pp 99-108
Gehring W (1976b) In: Roman M (ed) Ann Rev Genet, vol 10. Ann Rev Inc, Palo Alto Calif, pp 209-252
Gehring W, Nöthiger R (1973) In: Counce S, Waddington C (eds) Developmental systems in insects. Academic Press, London New York, pp 211-290
Geito J (1969) Molecular psycobiology. Mir, Moscow
Gelbart W, McCarron M, Panday J, Chovnick A (1974) Genetics 78:869-886
Gelbart W, McCarron M, Chovnick A (1976) Genetics 84:211-232
Gelderman A, Rake A, Britten R (1969) Carnegie Inst Washington Yearb 67:320-325
Gelderman A, Rake A, Britten R (1971) Proc Natl Acad Sci USA 68:172-176
Gell P, Sell S (1965) J Exp Med 122:813-821
Georgiev GP (1973) In: Advances of biological chemistry, vol X. Nauka, Moscow, pp 5-35
Georgiev G, Ilyin Y, Ryskov A, Tchurikov N, Yenikolopov G, Gvozdev V, Ananiev E (1977) Science 195:394-397
Gerace L, Schwartz M, Sofer W (1973) Isozyme Bull 6:30
German J (1964a) Science 144:298-301
German J (1964b) J Cell Biol 20:37-55
Gersh E (1975) J Theor Biol 50:413-428
Gibson G (1972) Experientia 28:975-976
Gibson C, Masters CJ (1970) FEBS Lett 7:277-279
Gilbert W (1978) Nature (London) 271:501
Gilmour R, Paul J (1973) Proc Natl Acad Sci USA 70:3440-3442
Gindilis VM (1967) Principles of human chromosomes identification. Autoreferate candidate dissertation Moscow (Russ)
Gineitis AA, Vinogradova IA, Nivankskas GG, Vorobijev VI (1970) Cytologia (Russ) 12:198-203
Gineitis AA, Vinogradova IA, Vorobijev VI (1971) Adv Mod Genet (Russ) 3:51-55
Ginsberg L, Hillman N (1975) J Embryol Exp Morphol 33:715-723
Ginter EK (1972) In: Problems of developmental genetics. Nauka, Moscow pp 41-57
Giudice G, Mutolo V, Moskona A (1967) Biochem Biophys Acta 138:607-610
Glass RD, Doyle D (1972) Science 176:180-181
Glassman E, Mitchell H (1959) Genetics 44:153-162
Glassman E, Shinoda T, Moon H, Karam J (1966) J Mol Biol 20:419-422
Glassman E, Shinoda T, Duke E, Collins J (1968) Ann N Y Acad Sci 151:263-273
Glenister T (1956) Nature (London) 177:1135-1136
Glover D, Hogness D (1977) Cell 10:167-176
Glueksohn-Schoenheimer S (1949) J Exp Zool 110:47-74
Glueksohn-Waelsch S (1955) In: Biochemistry of the developing nervous system. Academic Press, London New York, pp 357-396
Glueksohn-Waelsch S (1965) In: Genetics today, vol 2. Academic Press, London New York, pp 209-219

Glueksohn-Waelsch S, Cori C (1970) Biochem Genet 4:195-201
Goldberg, B, Grain W, Ruderman J, Moore G, Barnett G, Th, Higgins R, Golfand R, Galau G, Britten R, Davidson E (1975) Chromosoma 51:225-251
Goldberg E (1963) Science 139:602-603
Goldberg E (1965) Arch Biochem Biophys 109:134-144
Goldberg E, Cuerrier J, Ward J (1969) Biochem Genet 2:335-350
Goldberger R (1974) Science 183:810-816
Goldberger R, Berberich M (1965) Proc Natl Acad Sci USA 54:279-287
Goldschmidt R (1961) Theoretische Genetik. Akademie Verlag, Berlin, pp 1-546
Goldschmidt R, Katsuki K (1931) Biol Zentralbl 51:58-74
Goldwasser E (1966) In: Current topics in developmental biology. Academic Press, London New York pp 173-209
Golubitsa A, Korochkin L (1971) Folia Histochem Cytochem 9:19-21
Golubovsky MD (1972) Genetica (Russ) 8:143-156
Golubovsky MD (1977) In: Genetic problems in Drosophila research. Nauka, Novosibirsk pp 152-203
Golubovsky MD, Ivanov Yu, Green M (1977) Proc Natl Acad Sci USA 74:2973-2975
Goodman R, Goidl J, Richart R (1967) Proc Natl Acad Sci USA 58:553-560
Gordon T, Vrbova G (1975) Pflueger's Arch 360:199-218
Gorini L (1970) In: Roman H (ed) Annu Rev Genet, vol 4. Academic Press, London New York, pp 107-134
Gottesfeld J, Butler P (1977) Nucl Acids Res 4:3155-3173
Graham D, Neufeld B, Davidson E, Britten R (1974) Cell 1:127-137
Gräßman A (1970) Exp Cell Res 60:373-382
Gratcheva ND (1968) Autoradiography of the nucleic acids and proteins syntheses in nerve system, Nauka L
Green M (1955) Am Nat 89:65-71
Green M (1963) Genetics 34:241-253
Green M (1969a) Genetics 61:423-428
Green M (1969b) Genetics 61:429-441
Greenberg J (1975) J Cell Biol 64:269-288
Greene R, Flickinger R (1970) Biochim Biophys Acta 217:447-460
Grobstein C (1955) Ann N Y Acad Sci 60:1095-1107
Grobstein C (1964) Science 143:643-650
Grossbach V (1969) Chromosoma 28:136-187
Grossman A, Koreneva LS, Ulitskaja LE (1970) Genetica (Russ) 6:91-96
Grundbacker F (1972) Science 176:311-313
Grüneberg H (1958) J Embroyol Exp Morphol 6:424-443
Grüneberg H (1966) Genet Res 7:58-75
Gunnarson R (1974) Acta Endocrinol 77:18
Gurdon J (1964) In: Krause G, Sandler K (eds) Advances in morphogenesis, vol IV. Academic Press, London New York, pp 1-43
Gurdon J (1967) Proc Natl Acad Sci USA 58:545-555
Gurdon J (1969) Proc XII Int Congr Genet, vol III. Tokyo, pp 191— 203
Gurdon J (1970) In: Molecules and cells. Mir, Moscow, pp 19-37
Gurdon J (1971) Transplantation of nuclei and cell differentiation. Znanie, Moscow
Gurdon J (1973) In: Protein synthesis in reproductive tissue. Karolinska Symposia, Stockholm, Academic Press, London New York, pp 225-243
Gurdon J (1974a) Nature (London) 248:772-776
Gurdon J (1974b) The control of gene expression in animal development. Oxford Univ Press, pp 1-160
Gurdon J, Laskey R (1970) J Embryol Exp Morphol 24:249-255
Gurdon J, Speight V (1969) Exp Cell Res 55:253-256
Gurdon J, Woodland H (1970) In: Current topics in developmental biology, vol V. Academic Press, London New York, pp 39-70
Gurdon J, Woodland H, Lingrel J (1974) Dev Biol 39:128-133
Guth L (1968) Physiol Rev 48:645-687

Guthman E (1976) Ann Rev Physiol 38:177-216

Gvozdev VA (1968) Adv Mod Biol (Russ) 65:398-423

Gvozdev VA, Birstein VJ, Faizullin LZ (1969) In: Structure and genetic functions of bio-polymeres. M Inst Atom Energy (Russ), pp 138-165

Gvozdev VA, Ananiev EV, Gerasimova TI, Faizullin LZ (1972) 2nd meeting of VOGIS name of Vavilov N. I. Nauka, Moscow, pp 101-102

Gvozdev VA, Gerasimova TI, Birstein VJ (1973) Genetica (Russ) 9:64-72

Gvozdev VA, Gostimsky SA, Gerasimova TI, Dubrovskaja ES, Braslavskaja OJ (1975) Genetica (Russ) 11:73-79

Gvozdev VA, Gerasimova TI, Kogan G, Rosovsky J (1977) Mol Gen Genet 153:191-198

Hadorn E (1961) Developmental genetics and lethal factors. Academic Press, London New York, pp 85-104

Hadorn E (1965) In: Genetic control of differentiation, vol 18. Upton, New York, pp 148-161

Hadorn E (1967) In: Major problems in developmental biology. Academic Press, London New York, pp 85-104

Hagner R (1911) Biol Bull 20:237-259

Hahn ME, Haber S, Fuller J (1973) Physiol Behav 10:759-767

Hahn W, Laird C (1971) Science 173:158-161

Halvorson H, Carter B, Tauro P (1971) Methods Enzymol 21:462-470

Hamada T, Mishima Y (1972) Br J Dermatol 86:385-394

Hamashima Y, Harter J, Coons A (1964) J Cell Biol 20:271-282

Hamburg D, Kessler S (1967) In: Spickett S (ed) Endocrine genetics. Univ Press, Cambridge, pp 249— 270

Hamerton J (1964) In: Bourne G, Danielli J (eds) Int Rev Cytol, vol XII. Academic Press, London New York, pp 1-69

Hammerling J (1953) In: Bourne G, Danielli J (eds) Int Rev Cytol, vol II. Academic Press, London New York, pp 475-498

Hanford W, Arfin S (1977) J Biol Chem 252:6695-6699

Hannah A (1951) In: Caspari E (ed) Advances in genetics, vol IV. Academic Press, London New York, pp 87-125

Hannah A, Nelson O (1976) Biochem Genet 14:547-560

Harboe M, Osterland P, Mannik M, Kunkel H (1962) J Exp Med 116:719-738

Harding J, MacDonald R, Przybyla A, Chirgwin J, Pictet R, Rutter W (1977) J Biol Chem 252:7391-7397

Hardman M, Jack P (1977) Eur J Biochem 74:275-283

Harris H (1970) Cell fusion. Clarendon Press, Oxford, pp 1-196

Harris H (1973) Nucleus and cytoplasm. Mir, Moscow

Harris S, Means S, Mitchell W, O'Malley B (1973) Proc Nat Acad Sci USA 70:3776-3781

Harrison P, Birnic G, Hell A, Humphries S, Young B, Paul J (1974) J Mol Biol 84: 539-554

Harrison P, Affara N, McNab A, Paul J (1977) Exp Cell Res 109:237-246

Hartman F, Saskind S (1966) Gene action. Mir, Moscow

Havkin EE (1969) Induceable synthesis of enzymes during growth and morphogenesis of plants. Nauka, Moscow

Havkin EE (1974) Formation of metabolic systems in the growing plant cells. Autoreferate doctoral dissertation, L (Russ)

Hearing V (1973) Nature New Biol 245:81-82

Hebb DO (1949) The organization of behaviour. Wiley, New York

Hechter O, Halkerston I (1964) Perspect Biol Med 7:183-193

Hei E (1969) Regeneration. Mir, Moscow (Russ)

Heitz E (1931) Planta 12:775-844

Hemmingsen A, Krarup N (1937) K Dan Vidensk Selsk Biol Medd 13:1-61

Henderson N (1965) J Exp Zool 158:263-273

Henikoff S, Meselson M (1977) Cell 12:441-451

Hennig W, Hennig I, Stein H (1970) Chromosoma 32:31-63

Heron WT (1941) J Genet Psychol 59:41

Herzenberg L (1970) J Cell Physiol 76:303-310

Herzenberg L, Herzenberg L (1966) In: Genetic variations in somatic cells. Academiae, Prague, pp 227-232

Herzenberg L, McDevitt H, Herzenberg L (1968) In: Roman H (ed) Annu Rev Genet, vol II. Annu Rev Inc, Palo Alto, pp 209-241

Hexly J, De Beer G (1936) Basis of experimental embryology. Biomedgis, Moscow (Russ)

Heywood S, Kennedy D, Bester A (1974) Proc Natl Acad Sci USA 71:2428-2431

Hill R, Watt F (1977) Chromosoma 63:57-78

Hillman N (1975) In: Early Dev Mamals. Univ Press, Cambridge, pp 189-206

Hillman N, Hillman R (1975) J Embryol Exp Morphol 33:685-695

Hillman N, Tasca R (1973) J Reprod Fertil 33:501-506

Hillman R (1973) Genet Res 22:37-50

Hillman R, Shafer S, Sang J (1973) Genet Res 21:229-238

Hinuma G, Grace J (1967) Proc Soc Exp Biol Med 124:106-111

Hitzeroth H, Klose J, Ohno S, Wolf U (1968) Biochem Genetics 1:287-300

Hochman B (1971) Genetics 67:235-252

Hochman B (1973) Genetics 74:2, pt 2, 116-117

Holliday R, Pugh J (1975) Science 187:226-232

Holmgren P, Rasmuson B, Johansson T, Sundqvist G (1976) Hereditas 84:243

Holmquist G (1972) Chromosoma 36:413-452

Holtfreter J (1948) In: Symp Soc Exp Biol, vol II. Univ Press, Cambridge, pp 17-49

Holtzer H (1970) In: Schnjeide O, Vellis J (eds) Cell differentiation. Van Nostrand Reinhold, New York, pp 476-496

Holtzer H, Weintraub G, Mayne R, Mochan B (1972) In: Current topics in developmental biology, vol 7. Academic Press, London New York, pp 229-256

Honjo T, Kataoka T (1978) Proc Natl Acad Sci USA 75:2140-2144

Honjo T, Reeder R (1973) J Mol Biol 80:217-228

Hook E, Brustman L (1971) Nature (London) 232:349-350

Hoppe P, Whitten W (1972) Nature (London) 239:520

Hotta Y, Benzer S (1972) Nature (London) 240:527-534

Hough B, Davidson E (1972) J Mol Biol 70:491-511

Hough B, Smith M, Britten R, Davidson E (1975) Cell 5:291-299

Howells A (1972) Biochem Genet 6:217-230

Hozumi N, Tonegawa S (1976) Proc Natl Acad Sci USA 73:3628:362

Hsu C, Van Dyke J (1948) Anat Res 100:745

Hsu W (1952) Q J Microsc Sci 93:191-205

Hsu W, Weiss S (1969) Proc Natl Acad Sci USA 64:345-351

Huang R, Bonner J (1962) Proc Natl Acad Sci USA 48:1216-1222

Huez G, Marbaix G, Hubert E, Leclercq M, Nudel U, Soreq H, Solomon R, Lebleu B, Revel M, Littauer U (1974) Proc Natl Acad Sci USA 71:3143-3146

Huez G, Marbaix G, Hubert E, Cleuter Y, Leclercq M, Chantrenne H, Devos R, Soreq H, Nudel U, Littauer U (1975) Eur J Biochem 59:589-592

Huez G, Marbaix G, Weinberg E, Gallwitz D, Hubert E, Cleuter Y (1977) Biochem Soc Trans 5:936-937

Hughes A (1968) Aspects of neural ontogeny. Logos Press, London

Hughes M, Lucchesi J (1977) Science 196:1114-1115

Hummel K, Coleman D, Lane P (1972) Biochem Genet 7:1-13

Humphries S, Windass J, Williamson R (1976) Cell 7:267-277

Hunt J (1974) Biochem J 138:487-498

Hunt T, Hunter T, Munro A (1969) J Mol Biol 43:123-133

Hunter R, Markert C (1957) Science 125: 1294-1295

Hutton J (1971) Biochem Genet 5:315-325

Hyden H (1959) Nature (London) 184:433-435

Hyden H (1960) In: The Cell, vol 4. Academic Press, London New York, pp 215-323

Hyden H (1968) In: Genetic apparatus of cell and some problems of ontogenesis. Nauka, Moscow, pp 116-139

Hyden H (1974) Proc Natl Acad Sci USA 71:2965-2968

Hyden H, Lange P (1969) J Evol Biochem Phys (Russ) 5:145-157

Hyden H, McEwen B (1966) Proc Natl Acad Sci USA 55:354-358
Hyden H, Lange PW, Sejfried C (1973) Brain Res 61:446-451
Ignatijeva GM (1967) In: Advances of science. Series biology. Moscow, pp 5-62
Ignatijeva GM (1971) In: Advances of science. Series biology. Moscow, pp 5-81
Ikejima T, Takeuchi T (1974) Jpn J Genet 49:37-43
Ilan J, Ilan J (1971) Dev Biol 25:280-292
Ilan J, Ilan J (1972) In: Insect juvenile hormones. Academic Press, London New York, pp
    43-68
Ilan J, Ilan J (1975) Dev Biol 42:64-74
Illmensee K (1972) Roux Arch 170:267-298
Illmensee K, Mahovald A (1974) Proc Natl Acad Sci USA 71:1016-1020
Ingram V (1963) The hemoglobins in genetics and evolution. Academic Press, London New
    York
Ingram V (1972) Nature (London) 235:338-339
Ingram V, Chan L, Hagopian H, Lippke J, Wu L (1974) Dev Biol 36:411-427
Inoue A, Fujimoto D (1969) Biochem Biophys Res Commun 36:146-150
Iordansky AB, Pavulsone SA (1970) II Symposium on structure and function of chromo-
    somes. Inst Cytol Genet (Russ) Novosibirsk p 10
Ish-Horowitz D, Holden J, Gehring W (1976) 5th Eur Drosophila Res Conf Louvain-la-
    Neuve, p 49
Izquierdo I (1972) Behav Biol 7:669-698
Izquierdo I, Orsingher OA, Ogura A (1972) Behav Biol 7:699-707
Jacob F, Monod J (1961) J Mol Biol 3:318-356
Jacob F, Monod J (1964) In: Regulatory mechanisms of cell. Mir, Moscow, pp 278-304
Jacob F, Wollman E (1962) Sex and genetics of bacteria. IL, Moscow
Jacobson K (1971) Nature New Biol 231:17-19
Jacobson K, Calvino J, Murphy J, Warner C (1975) J Mol Biol 93:89-97
Jakovleva VI (1968) Adv Mod Chem (Russ) 9:55-94
Jamrich M, Greenleaf A, Bautz E (1977) Proc Natl Acad Sci USA 74:2079-2083
Janning W (1973) Drosophila Inf Serv 50:151-152
Janning W (1974) Roux Arch 174:313-332
Jansing R, Park W, Stein J, Stein G, Ross J (1977) In Vitro 13:196-197
Jarry B, Falk B (1974) Mol Gen Genet 135:113-120
Jazdowska-Zagrodzinska B (1966) J Embryol Exp Morphol 16:391-399
Jelesnjak M (1970) In: Ostin K, Perri J (eds) Factors affecting fecundity. Medicina, Moscow,
    pp 263-276
Jensen E, DeSombre E (1972) In: Royer P (ed) Annu Rev Biochem, vol 41. Annu Rev Inc,
    Palo Alto, pp 203-230
Jobbagy A, Aversa N, Denoya C (1975) Biochem Genet 13:813-831
Joffe J (1969) Prenatal determinants of behaviour. Pergamon Press, London New York
John B, Lewis K (1968) In: Protoplasmatologia. Wien 6/A
Johns EW (1972) In: Cellular nucleus. Mir, Moscow, pp 61-74
Johnson J, Leone C (1955) J Exp Zool 130:515-554
Johnson K, Chapman V (1971a) J Exp Zool 178:313-318
Johnson K, Chapman V (1971b) J Exp Zool 178:319-324
Johnson R, Grossman A (1974) Biochem Biophys Res Commun 59:520-526
Johnson R, Harris H (1969) J Cell Sci 5:625-640
Jones G, Marcuson E, Roitt I (1970) Nature (London) 227:1051-1052
Jones K, Elsdale I (1963) J Embryol Exp Morphol 11:135-154
Jones K, Robertson F (1970) Chromosoma 31:331-345
Jost A (1961) Harvey Lect Ser. 55, Academic Press, London New York, pp 201-226
Judd B, Young N (1974) Cold Spring Harbor Symp Quant Biol 24:573-579
Kabat D (1972) Science 175:134-140
Kabat D, Chappell M (1977) J Biol Chem 252:2684-2690
Kafatos F (1972) In: Current topics in developmental biology, vol 7. Academic Press, Lon-
    don New York, pp 125-191
Kafiani KA, Timofeeva MJa (1965) In: Cell differentiation and induction mechanisms.
    Nauka, Moscow, pp 61-68

Kafiani KA, Timofeeva MJa, Neyfakch AA, Rachkus JuA, Melnikova NL (1966) Biochimia (Russ) 31:365-371
Kahan B, DeMars R (1975) Proc Natl Acad Sci USA 72:1510-1514
Kaighn M, Prince A (1971) Proc Natl Acad Sci USA 68:2396-2400
Kaneko A, Ikeda I, Onoe T (1970) Biochim Biophys Acta 222:218-221
Kaneko A, Dempo K, Onoe T (1972) Biochim Biophys Acta 284:128-135
Kaplan N, Everse J (1972) In: Weber G (ed) Advances in enzyme regulation, vol 10. Pergamon Press, London New York, pp 323-336
Kaplan N, White Sh (1963) Ann NY Acad Sci 103:835-848
Kaplan R, Plaut W (1968) J Cell Biol 39:71a
Karakin EI, Lerner TJa, Kiknadze II, Korochkin LI, Sviridov SM (1977) Cytologia (Russ) 19:111-119
Karasik GI, Sukojan MA, Korochkin LI, Beljaev DK, Golubitsa AN, Maletskaya EI, Maximovsky LF (1971) J High Nerve Act (Russ) 21:184-190
Karasik GI, Korochkin LI, Maximovsky LF (1976) J High Nerv Act (Russ) 26:1066-1073
Karasik G, Eugenev M, Kulguskin V, Maximovsky L, Korochkin L (1978) XIV Int Cong Genet Abstr, part I, Sect 13-20, p 365
Karl T, Chapman V (1974) Biochem Genet 11:367-372
Karlsson L, Palmer P (1971) Comp Biochem Physiol 38B:299-308
Kazazian H, Young W, Childs B (1965) Science 150:1601-1602
Kauffman S (1967) J Theoret Biol 17:483-497
Kauffman S (1973) Science 181:310-318
Kauffman S, Shymko R, Trabert K (1978) Science 199:259-270
Kauffman T, Shannon M, Shen M, Judd B (1975) Genetics 79:265-282
Keep E (1962) Can J Genet Cytol 4:206-211
Kennedy D, Bester A, Heywood S (1974) Biochem Biophys Res Commun 61:365-373
Kennett R, Sueoka N (1971) J Mol Biol 60:31-44
Keppy D, Welshons W (1977) Genetics 85:497-506
Khesin RB (1972) Adv Mod Biol (Russ) 74:171-197
Khesin RB, Leibovitch B (1974) Chromosoma 46:161-172
Khesin RB, Leibovitch B (1976) Mol Biol (Russ) 10:3-34
Kiknadze II (1972) Functional organization of chromosomes. Nauka L
Kiknadze II, Belyaeva ES (1967) Genetica (Russ) 7:149-161
Killewich L, Feigelson P (1977) Proc Natl Acad Sci USA 74:5392-5396
Kindas-Mügge I, Lane C, Kreil G (1974) J Mol Biol 87:451-462
King J (1967) In: Behaviour genetic analysis. Academic Press, London New York, pp 130-157
Kischer C, Gurley L, Shepherd G (1966) Nature (London) 212:304-306
Kiselev LL, Nikiforov VG, Astaurova OB, Gottich BP, Kraevsky AA (1971) Molecular basis of the protein biosynthesis. Nauka, Moscow
Klebe R, Tchaw-ren Chen, Ruddle F (1970) Proc Natl Acad Sci USA 66:1220-1227
Klein J, Raska K (1968) Proc 12th Int Congr Genet, vol I. Veno Park, Tokyo, p 149
Klinger H, Schwarzacher H (1962) Cytogenetics 1:266-290
Klose J, Wolf U (1970) Biochem Genet 4:87-92
Klose J, Hitzeroth H, Ritter H, Schmidt E, Wolf U (1969) Biochem Genet 3:91-97
Klukas C (1977) Nature 265:297
Knöchel W, Kohnert-Stavenhagen E (1977) Hoppe Seylers Z Physiol Chem 358:835-842
Koch E, Smith P, King R (1967) J Morphol 121:55-67
Kofman-Alfaro S, Chandley A (1970) Chromosoma 31:404-420
Kolesnikov NN, Zhimulev IF (1974) Ontogenes (Russ) 6:177-182
Kolombet L (1977) Ontogenes (Russ) 8:269-274
Komma D (1966) Genetics 54:497-504
Konyukhov BV (1969) Biological modeling of the inherited diseases in humans. Medicina, Moscow
Konyukhov BV (1973) Adv Mod Biol (Russ) 16:171-188
Konyukhov BV, Sazhina MV (1966) Folia Biol 12:116-123
Konyukhov BV, Sazhina MV (1970) In: Genetic information and individual development. Inst Cyt Gen (Russ), Novosibirsk, pp 25-30

Korge G (1970a) Nature (London) 225:386-388
Korge G (1970b) Chromosoma 30:430-464
Korge G (1975) Proc Natl Acad Sci USA 72:4550-4554
Korge G (1977a) Chromosoma (Berl) 62:155-174
Korge G (1977b) Dev Biol 58:339-355
Kornberg R (1974) Science 184:868-871
Kornberg R (1977) Annu Rev Biochem 46:931-954
Korochkin LI (1965) Differentiation and aging of vegetative neurone. Nauka, Moscow
Korochkin LI (1970) In: Recent advances in anatomical researches in the USSR. Mir,
    Moscow, pp 53-65
Korochkin LI (1972a) Isozyme Bull 5:38
Korochkin LI (1972b) Cytologia (Russ) 15:670-673
Korochkin LI (1972c) In: Problems of developmental genetics. Nauka, Moscow, pp 107-118
Korochkin LI (1973) In: Problems of theoretical and applied genetics. Inst Cytol Genet
    (Russ), Novosibirsk, pp 30-43
Korochkin LI (1975) In: Markert C (ed) Isozymes, vol III. Academic Press, London New
    York, pp 99-117
Korochkin LI (1976) J Gen Biol (Russ) 37:184-191
Korochkin LI (1977) Priroda (Russ) 7:68-79
Korochkin LI (1978) In: Danielli J (ed) Int Rev Cytol, vol 49. Academic Press, London New
    York, pp 171-228
Korochkin LI (1979) Isozyme Bull 12:45-46
Korochkin LI, Belyaeva ES (1972) Ontogenes (Russ) 3:11-26
Korochkin LI, Korochkina LS (1971) J Hirnforsch 13:97-104
Korochkin LI, Matveeva N (1974) Biochem Genet 12:1-7
Korochkin LI, Olenev SN (1966) Adv Mod Biol (Russ) 62:77-96
Korochkin LI, Shumskaja IA (1976) In: Mechanisms of memory modulation. Nauka L, pp
    192-198
Korochkin LI, Sviridov SM, Ivanov VN, Maletskaja EI, Bachtina TK (1972a) Dokl Acad
    Sci USSR (Russ) 204:468-470
Korochkin LI, Korochkina LS, Serov OL (1972b) Folia Histochem Cytochem 10:287-292
Korochkin LI, Mertvetsov N, Matveeva N, Serov OL (1972c) FEBS Lett 22:213-216
Korochkin LI, Maximovsky LF, Karasik GI, Onitshenko AM, Kusin BA (1973a) In: Struc-
    ture and function of cellular nucleus. Nauka, Novosibirsk, pp 42-43
Korochkin LI, Matveeva N, Evgeniev M, Golubovsky M (1973b) Biochem Genet 10:363-
    393
Korochkin L, Matveeva NM, Kerkis AYu (1973c) DIS 50:130-131
Korochkin L, Aronstam A, Matveeva N. (1974) Biochem Genet 12:9-24
Korochkina LS, Korochkin LI, Grusdev AD, Kostomacha AN (1975) Cytologia (Russ)
    7:576-579
Korochkin LI, Belyaeva E, Matveeva N, Kuzin B, Serov O (1976a) Biochem Genet 14:161-
    182
Korochkin L, Kuzin B, Jakovleff V, Matveeva N (1976b) Isozyme Bull 9:41
Korochkin LI, Matveeva NM, Kuzin BA, Karasik GI, Maximovsky LF (1978a) Genetica
    (Russ) 14:632-643
Korochkin LI, Matveeva NM, Kuzin BA, Karasik GI, Maximovsky LF (1978b) Biochem
    Genet 16:709-726
Korolev MB, Neyfach AA (1965) J Gen Biol (Russ) 26:352-357
Koslov JuV, Georgiev GP (1971) Mol Biol (Russ) 5:789-796
Klukas C (1977) Nature (London) 265:297
Kress H (1972) Das Puffmuster der Riesenchromosomen in den Larvalen. Speicheldrüsen
    von D.virilis.-Bayer Akad Wiss Math Naturwiss Kl Ser 8:131-149
Krider H, Plaut W (1972) J Cell Sci 11:675-687, 689-697
Kritsman MG, Konikova AS (1968) Enzyme function in norma and pathology. Medicina,
    Moscow
Kröger H (1963) Nature (London) 200:1234-1235
Kröger H (1964) Chromosoma 15:36-70

Krushinsky LV, Molodkina LN, Romanova LG (1968) In: Genetics and pathology. Medicina, Moscow, pp 186-199

Kubli E (1970) Z Vergl Physiol 70:175-195

Kubli E, Weideli H, Chen P (1971) 2nd Eur Drosophila Res Conf Abstr, Zürich, p 15

Kugelberg E (1976) J Neurol Sci 27:269-289

Kuhn D, Cunningham G (1977) Science 196:875-877

Kulichkov VA, Belyaeva ES (1975) Dokl Acad Sci USSR (Russ) 221:463-466

Kumar S, Calef E, Szybalski W (1970) Cold Spring Harb Symp Quant Biol 35:331-339

Kuroda Y (1974/75) Annu Rep Natl Inst Genet Jpn 25:23-24

Kusin BA, Aronstam AM (1974) In: V Conference of embryologists. Nauka, Moscow, p 106

Kuwano M, Ono M, Endo H, Horik K, Nakamura K, Hirota Y, Ohnishi Y (1977) Mol Gen Genet 154:279-285

Labarca C, Paigen K (1977) Proc Natl Acad Sci USA 74:4462-4465

Lai E, Woo S, Dugaiczyk A, Catterall J, O'Malley B (1978) Proc Natl Acad Sci USA 75:2205-2209

Laird C (1973) In: Roman H (ed) Annu Rev Genet, vol 7. Annu Rev Inc, Palo Alto, Calif, pp 177-237

Laird C, McCarthy B (1969) Genetics 63:865-882

Lajtha L, Schofield R (1974) Differentiation 2:313-320

Lajtha L, Gilbert C, Guzman F (1971) Br J Haematol 20:343-356

Lakhotia S (1970) Genet Res 15:301-307

Lakhotia S, Mukherjee A (1969) Genet Res 14:137-150

Lakhotia S, Mukherjee A (1970) J Cell Biol 47:18-33

Lalley P, Shows T (1974) Science 185:442-444

LaMantia G, Graziani F (1977) Atti Assoc Genet Ital 22:187-189

Lambert B (1972) Repeated nucleotide sequences in polytene chromosomes. Civiltryck, Stockholm, AB

Landucci-Tosi S, Tosi R, Mage R (1975) Immunochemistry 12:865-872

Laskey R, Gurdon J (1970) Nature (London) 228:1332-1334

Lawrence P, Morata G (1977) Dev Biol 56:40-56

Lawton A, Kinkade P, Cooper M (1975) Fed Proc 34:33-39

Lea A (1972) Gen Comp Endocrinol 18:602-608

Lebherz H (1974) Experientia 30:655-658

Lebherz H (1975) In: Markert C (ed) Isozymes, vol III. Academic Press, London New York, pp 253-279

Lee C, Thomas C (1973) J Mol Biol 77:25-42

Lee G (1968) Genet Res 11:115-118

Lee J, Ingram V (1967) Science 158:1330-1332

Lee-Huang S, Sierra J, Naranjo R, Filipowicz W, Ochoa S (1977) Arch Biochem Biophys 180:276-287

Lefevre G (1974) In: Roman H (ed) Annu Rev Genet, vol 8. Annu Rev Inc, Palo Alto, pp 51-62

Lehman H (1957) In: Beginning of embryonic development. Am Assoc Adv Sci Wash, pp 201-230

Leibenguth F (1973) Biochem Genet 10:231-242

Leibovitch BA, Khesin RB (1974) Mol Biol (Russ) 8:467-474

Leibovitch BA, Belyaeva ES, Zhimulev IF, Khesin RB (1974) Ontogenes (Russ) 5:544-556

LeMeur M, Gerlinger P, Ebel J (1976) Eur J Biochem 67:519-526

Lengyel J, Penman S (1977) Dev Biol 57:243-253

Leonova SN (1974) In: V Conference of embryologists. Nauka, Moscow, p 115

Leskowitz S (1963) J Immunol 90:98-103

Less A, Waddington C (1942) Proc R Soc (London) Ser B 131:87-110

Levere R, Granick S (1965) Proc Natl Acad Sci USA 54:134-137

Levere R, Lichtman H (1963) Blood 22:334-341

Levi-Montalcini R, Angeletti P (1965) In: Organogenesis. Academic Press, London New York, pp 187-198

Levina SE (1974) Assays on the sex development during early ontogenesis of the higher ver-
    tebrates. Nauka, Moscow
Levisohn S, Thompson E (1972) Nature New Biol 235:102-104
Levy B, McCarthy B (1975) Biochemistry 14:2440-2446
Lewin B (1975) Cell 4:77-93
Lewis E (1950) In: Caspari E (ed) Advances in genetics, vol III. Academic Press, London
    New York, pp 73-115
Lewis E (1951) Cold Spring Harbor Symp Quant Biol 16:159-174
Lewis E (1963) Am Zool 3:33-56
Lewis E (1964) In: Locke M (ed) The role of chromosomes in development. Academic Press,
    London New York, pp 231-252
Lewis H, Lewis H (1963) Ann N Y Acad Sci 100:827-839
Lewis J (1973) J Theoret Biol 39:47-54
Lewis J, Summerbell D, Wolpert L (1973) Nature (London) 239:276-279
Lewis M, Helmsing P, Ashburner M (1975) Proc Natl Acad Sci USA 72:3604-3608
Liberman LL (1966) Adv Mod Biol (Russ) 61:260-271
Lima-de-Faria A (1969) In: Handbook of molecular cytology. North Holland Publ Co, Am-
    sterdam, pp 278-309
Lima-de-Faria A, Mozes M (1966) J Cell Biol 30:177-192
Lima-de-Faria A, Reitalu J, Bergman S (1961) Hereditas 47:695-717
Lima-de-Faria A, Nilsson B, Cave D, Jaworska H (1969) Chromosoma 25:1-20
Lindsley D, Grell E (1967) Genetic variations of *Drosophila melanogaster*. Carnegie Inst
    Washington Publ, Washington
Lis J, Hogness D (1977) 9th Miami Winter Symp Mol Clon Recomb DNA Genetic Manip.
    Affects Cancer Problems. Miami, pp 14-15
Litt M, Kabat D (1972) J Biol Chem 247:6659-6664
Lizardi P, Brown D (1975) Cell 4:199-205
Lodish H (1969) Nature (London) 224:867-870
Loor F, Kelus A (1977) Basel Inst Immunol Annu Rep, Basel, pp 41-42
Lopashov G (1935) Biol Zentralbl 55:606-615
Lopashov GV (1937) Dokl Acad Sci USSR (Russ) 15:135-137
Lopashov GV (1965) In: Cell differentiation and induction mechanisms. Nauka, Moscow,
    pp 242-270
Lopashov G (1977) Differentiation 9:131-137
Love W, Bennett E, Rosenzweig MR (1968) Physiol Behav 3:819-825
Lowenstein J, Smith S (1962) Biochim Biophys Acta 56:385-387
Lubimova T (1978) XIV Int Cong Genet Abstz, part I, Sect 13-20, p 594
Lubimova TI, Korochkin LI (1979) Ontogenes (Russ) 9:622-625
Lucchesi J (1973) In: Roman H (ed) Annu Rev Genetics, vol 7. Annu Rev Inc, Palo Alto,
    pp 225-238
Lucchesi J (1977) Am Zool 17:685-693
Lucchesi J, Rawls J (1973) Biochem Genet 9:41-51
Lucchesi J, Rawls J, Maroni G (1974) Nature (London) 248:564-567
Lucchesi J, Belote J, Maroni G (1977) Chromosoma (Berl) 65:1-7
Lummus Z, Cebra J, Mage R (1967) J Immunol 90:737-743
Lundin L, Allison A (1966) Biochim Biophys Acta 127:527-531
Lyon M (1961) Nature (London) 190:372-374
Lyon M (1968) In: Roman H (ed) Annu Rev Genet, vol II. Annu Rev Inc, Palo Alto, pp
    31-52
Lyon M (1974) Proc R Soc London Ser B 187:306-243-268
Lyon M, Searle A, Ford C, Ohno S (1964) Cytogenetics 3:306-323
MacIntyre R (1974) Isozyme Bull 7:23
MacIntyre R, O'Brien S (1976) Ann Rev Genet 10:281-318
Mage R (1967) Cold Spring Harbor Sympos Quant Biol 32:290-295
Mage R (1975) Fed Proc 34:40-46
Mage R, Dray S (1965) J Immunol 95:525-535
Mage R, Young G, Dray S (1967) J Immunol 98:502-511

Mage R, Young-Cooper G, Rejnek I, Ansari A, Alexander C, Appella E, Carta-Sorcini M, Landucci-Tosi S, Tosi R (1977) Cold Spring Harbor Symp Quant Biol 47 part 2 pp 677-686
Mahovald A (1971) J Exp Zool 176:345-352
Mahovald A (1977) Am Zool 17:551-563
Maletsky SI (1972) Cytol Genet (Russ) 4:354-361
Malup T, Sviridov S (1978) Brain Res 142:97-103
Malva C, Graziani F, Boncinelli E, Polito L, Ritossa F (1972) Nature New Biol 239:135-137
Mandel P, Ayad B, Hermetet J, Ebel A (1974) Brain Res 72:65-70
Maniatis G, Ingram V (1971) J Cell Biol 49:372-400
Maniatis G, Steiner L, Ingram V (1969) Science 165:67-69
Maniatis G, Ramirez F, Nudel U, Rifkind R, Marks P, Bank A (1978) FEBS Lett 85:43-46
Manner H (1971) Current Mod Biol 3:332-348
Manning J, Schmid C, Davidson N (1975) Cell 4:141-155
Marbaix G, Huez G, Burny A, Cleuter Y, Hubert E, Leclercq M, Chantrenne H, Soreq H, Nudel U, Littauer U (1975) Proc Natl Acad Sci USA 72:3065-3067
Marbaix G, Huez G, Burny A, Hubert E, Leclercq M, Cleuter Y, Chantrenne H, Soreq H, Nudel U, Littauer U (1977) Acta Biol Med Ger 36:319-321
Marchok A, Wolff J (1968) Biochim Biophys Acta 155:378-393
Mariano E, Schram-Doumont A (1965) Biochem Biophys Res Commun 103:610-622
Markert C (1963) In: Cytodifferentiation and macromolecular synthesis. Academic Press, London New York, pp 65-84
Markert C (1964) Harvey lectures, Ser 59, Academic Press, London New York, pp 187-218
Markert C (1968) Ann N Y Acad Sci 151:14-30
Markert C, Appella (1963) Ann N Y Acad Sci 103:915-929
Markert C, Faulhaber I (1965) J Exp Zool 159:319-332
Markert C, Hunter R (1959) J Histochem Cytochem 7:42-49
Markert C, Ursprung H (1962) Dev Biol 5:363-381
Markert C, Ursprung H (1963) Dev Biol 7:560-577
Markert C, Ursprung H (1973) Genetics of development. Mir, Moscow
Markert G, Whitt G (1968) Experientia 24:977-991
Maroni G, Plaut W (1973a) Chromosoma 40:361-377
Maroni G, Plaut W (1973b) Genetics 74:331-342
Marrakechi M, Prud'homme N (1971) Biochem Biophys Res Commun 43:273-280
Martensson L (1963) Lancet 1:946-948
Martensson L, Kunkel G (1965) J Exp Med 122:799-811
Martin D, Tomkins G, Granner D (1969) Proc Natl Acad Sci USA 62:248-260
Martin G, Epstein C, Travis B, Tucker G, Yatziv, S, Martin D, Clift S, Cohen S (1978) Nature (London) 271:329-333
Marx J (1978) Science 199:517-518
Marx P, Kovach J (1966) In: Current topics in developmental biology. Academic Press, London New York, pp 213-252
Marzluf G (1965) Genetics 52:503-512
Massaro E, Markert C (1968) J Exp Zool 168:223-238
Masters C, Hinks M (1966) Biochim Biophys Acta 113:611-613
Masters C, Holmes R (1972) Biol Rev 47:309-361
Masui I (1972) Ontogenesis (Russ) 3:574-587
Masui I, Markert C (1971) J Exp Zool 177:129-146
Matsuzawa T, Hamilton J (1973) Proc Soc Exp Biol Med 142:232-236
Matveeva NM, Korochkin LI (1974) Ontogenesis (Russ) 5:92-95
Maurer G (1971) Disk-electrophoresis. Mir, Moscow
Maximovsky LF (1970) Cytologia (Russ) 12:887-893
Maximovsky LF, Korochkin LI (1973) In: Problems of theoretical and applied genetics. Inst Cytol Genet, Novosibirsk, pp 120-128
McCarl R, Shaler R (1969) J Cell Biol 40:850-854
McCarthy B, Hoyer B (1964) Proc Natl Acad Sci USA 52:915-920
McCartney P, Penwick J, Munday M (1977) Heredity 38:37-45

McClearn G, Rogers D (1959) Q J Stud Alcohol 20:691-695
McClintock B (1934) Z Zellforsch 21:294-328
McClintock B (1950) Proc Natl Acad Sci USA 37:344-355
McClintock B (1956) Cold Spring Harbor Symp Quant Biol 21:197-216
McClintock B (1961) Am Nat 95:265-276
McClintock P, Papaconstantinou J (1974) Proc Natl Acad Sci USA 71:4551-4555
McCulloch E, Simonovitch L, Till J (1964) Science 144:844-846
McDowell EC (1924) Science 59:302-304
McGregor H (1968) J Cell Sci 3:437-444
McKenzie S, Henicoff S, Meselson M (1975) Proc Natl Acad Sci USA 72:1117-1121
McLaren A (1972) Nature (London) 239:274-276
McLaren A (1973) Ontogenes (Russ) 4:227-239
McLaren A, Bowman P (1969) Nature (London) 224: 238-240
McLaren A, Gauld I, Bowman P (1973) Nature (London) 241:180-183
McLean N, Jurd R (1971) J Cell Sci 9:509-528
McMillan P (1967) J Histochem Cytochem 15:21-31
McNeil M (1957) Heredity 11:261-264
Mechelke F (1961) Naturwissenschaften 1:29
Medvedev JA (1963) Biosynthesis of proteins and problems of ontogenesis. M. Medgis
    (Russ)
Medvedev JA (1968) Molecular-genetical mechanisms of development. Medicina, Moscow
Medvedev JA (1972a) In: 2nd Meeting VOGIS name Vavilov NI. Nauka, Moscow
Medvedev JA (1972b) Adv Mod Biol (Russ) 74:385-390
Meerson FS (1963) Interrelation between physiological function and genetic apparatus of
    cell. Nauka, Moscow
Meisler M, Paigen K (1972) Science 177:894-896
Mellors R, Korngold L (1963) J Exp Med 188:387-396
Mertvezov NP, Saprikin VA, Tchesnokov VN, Salganik RI (1974) Biochimia (Russ) 39:3-8
Mglinez VA (1973) Adv Mod Biol (Russ) 76:189-198
Michinoma M, Kaji S (1970) Jpn J Genet 48:307-310
Migeon B (1972) Nature (London) 239:87-89
Migeon B, Der Kaloustian V, Nyhan W, Young W, Childs B (1968) Science 160:425-427
Miller L, Knowland J (1972) Biochem Genet 6:65-73
Miller O, Beatty B (1969) Science 164:955-957
Milstone L, Zelenka P, Piatigorsky J (1976) Dev Biol 48:197-204
Minna J, Glaser D, Nirenberg M (1972) Nature (London) 235:225-231
Mintz B (1964a) J Exp Zool 157:85-100
Mintz B (1964b) J Exp Zool 157:273-283
Mintz B (1965) Science 148:1232-1233
Mintz B (1971) Fed Proc 30:935-943
Mintz B (1972) In: Harris R, Allin P, Viza D (eds) Cell differentiation. Scand Univ Book,
    Munksgaard Copenhagen, pp 176-181
Mintz B (1974) In: Roman H (ed) Annu Rev Genet, vol 8. Annu Rev Inc, Palo Alto, pp
    411-470
Mintz B, Sanyal S (1970) Genetics 64:43-44
Mischke D, Kloetzel P, Schwochau M (1975) Nature (London) 255:79-80
Mitashov VI, Korochkin LI (1974) Ontogenes (Russ) 3:513-517
Mitchell G, Williamson R (1977) Nucl Acids Res 4:3557-3562
Mitskevich MS (1966) In: Establishment of endocrinal functions in embryonic development.
    Nauka, Moscow, pp 7-25
Moalic J, Padieu P (1977) J Mol Cell Cardiol 9:31-32
Mohan J, Ritossa F (1970) Dev Biol 22:495-512
Molinaro M, Cusimano-Carollo I (1966) Experientia 22:246-247
Molinaro M, Mozzi R (1969) Exp Cell Res 56:163-166
Monesi V, Salfi V (1967) Exp Cell Res 46:632-633
Monk M, Ansell J (1976) J Embryol Exp Morphol 36:653-662
Monk M, Petzoldt V (1977) Nature (London) 256:338-339

Monroy A (1970) In: Genetic information and individual development (Russ). Inst Cytol Genet, Novosibirsk, pp 32-34

Monroy A, Maggio R, Rinaldi A (1965) Proc Natl Acad Sci USA 54: 107-111

Moog F (1959) In: Gorbman A (ed) Comparative endocrinology. Academic Press, London New York, pp 624-638

Moore B (1973) Proc Natl Acad Sci USA 50:1018-1026

Moore G, Lintern-Moore S, Peters H, Faber M (1974) J Cell Biol 60:416-422

Moore J (1963) J Cell Comp Physiol 60:19-34

Moore W, Mintz B (1972) Dev Biol 27:55-70

Morata G, Lawrence P (1977) Nature (London) 265:211-216

Morgan I, Bridges C (1919) Carnegie Inst Wash Publ 278:1-22

Morgan E, Kaplan S (1976) Biochem Biophys Res Commun 68:969-976

Morgan TH (1924) Structural basis of heredity. Gosisdat, Moscow

Morgan TH (1937) Development and heredity. Biomedgis, Moscow

Moscovkin GN (1975) Adv Mod Biol (Russ) 79:79-84

Mukherjee A (1966) Nucleus 9:83-96

Müller H (1950) Evidence of the precision of genetic adaptation. Harvey Lect Ser 43:165-229

Müller H, League B, Offermann C (1931) Anat Rec 51:110

Muto M (1977) Microbiol Immunol 21:451-468

Nace G (1963) Ann N Y Acad Sci 103:980-988

Nace G, Suyama T, Smith N (1960) In: Symp Germ Cells Dev. Inst Int Embryol, Baselli, pp 564-603

Nadal-Ginard B (1976) Proc Natl Acad Sci USA 73:3618-3622

Nadijcka M, Hillman N (1975) J Embryol Exp Morphol 33: 697-713

Naeser P (1974) Acta endocrinol 77:23

Nagamine M (1974) Clin Chim Acta 50:173-179

Nakatsu S, Masek M, Landrun S, Frenster J (1974) Nature (London) 248:334-335

Navashin M (1934) Cytologia 5:169-203

Necheles T, Allen D, Finkel M (1969) Clinical disorders of hemoglobin structure and synthesis. Academic Press, London New York

Nelson O (1968) Genetics 60:507-524

Nemchinskaya VA, Ganelina LM, Braun AD (1968) Cytologia (Russ) 10:322-328

Nemer M, Surrey M (1975/76) Inst Cancer Res 21st Sci Rep Philadelphia Pa, s a 172

Nemer M, Dubroff L, Graham M (1975) Cell 6:171-178

Nesbitt M (1971) Dev Biol 26:252-263

Nesbitt M (1974) Dev Biol 38:202-207

Nevers P, Saedler M (1977) Nature (London) 268:109-115

Newlon C, Gussin G, Lewin B (1975) Cell 5:213-225

Newmark P (1976) Nature (London) 261:544-545

Neyfach AA (1962) Problem of interaction of nucleus and cytoplasm during development. M Inst Morphol (Russ)

Neyfach AA (1963) J All Union Chem Soc MD Mendeleev (Russ) 4:403-412

Neyfach AA (1964) Nature (London) 201: 880-884

Neyfach AA (1965) In: Cell differentiation and inductional mechanisms. Nauka, Moscow, pp 38-59

Neyfach AA, Timofeeva MJa (1978) Problems of regulation in molecular biology of development. Nauka, Moscow

Niewisch H, Vogel H, Matioli G (1967) Proc Natl Acad Sci USA 58:2261-2267

Niewisch H, Hajdik I, Sultanian I, Vogel H, Matioli G (1970) J Cell Physiol 76:107-116

Nikitina LA (1964) Dokl Acad Sci USSR (Russ) 156:1468-1471

Nilsson L (1967) Drosophila Inf Serv 42:60

Nix C (1973) Biochem Genet 10:1-12

Noll M (1977) Nature (London) 251:249-251

Nomenclature of Multiple Forms of Enzymes. Recommendations, 1971. Eur J Biochem 24:1-3

Norby S (1973) Hereditas 73:11-16

Nöthiger R (1972) In: Ursprung H, Nöthiger R (eds) Results and problems in cell differenti-
    ation. Springer, Berlin Heidelberg New York, pp 1-34
Novak V (1960) Insektenhormone. Verlag Tschechosl Akad Wiss, Prague
Nover L, Luckner M, Parthier B (1978) Zelldifferenzierung. VEB Gustav Fischer Verlag,
    Jena
Nur U (1967) Genetics 56:375-389
Nur U (1973) Genetics 74:199
Nyhan W, Bakay B, Conner J, Marx J, Keele D (1970) Proc Natl Acad Sci USA 65:
    214-218
Oakesholt J (1976) Aust J Biol Sci 29:365-376
O'Brien S, Gethman R (1973) Genetics 75:155-167
O'Brien S, MacIntyre R (1972a) Biochem Genet 7:141-161
O'Brien S, MacIntyre R (1972b) Genetics 71: 127-138
Offerman C (1936) J Genet 32:103-116
Oganji G (1971) J Exp Zool 178:513-522
Ohno S (1967) Sex chromosomes and sex linked genes. Springer, Berlin Heidelberg New
    York
Ohno S (1971) Nature (London) 234:134-138
Ohno S, Hauschka T (1960) Cancer Res 20:541-551
Ohno S, Lyon M (1970) Clin Genet 1:121-127
Ohno S, Kaplan W, Kinosita R (1959) Exp Cell Res 18:415-418
Ohno S, Makino S, Kaplan W, Kinosita R (1961) Exp Cell Res 24:106-110
Ohno S, Stenius C, Christian L, Harris C (1968) Biochem Genet 2:197-204
Ohno S, Christian L, Stenius C, Castro-Sierra E, Muzamoto J (1969) Biochem Genet 2:361-
    370
Ohno S, Stenius C, Christian L (1970) Clin Genet 2:128-140
Ohtaku J, Spiegelman S (1963) Science 142:493-495
Okada M, Kleinman I, Schneiderman H (1974a) Dev Biol 37:43-54
Okada M, Kleinman I, Schneiderman H (1974b) Dev Biol 39:286-294
Okazaki K, Holtzer H (1966) Proc Natl Acad Sci USA 56:1484-1490
Olenev SN, Korochkin LI, Demin DV (1968) Adv Mod Biol (Russ) 65:246-266
Olenov Ju/M (1967) Cell inheritance, cell differentiation and cancerogenesis as problems of
    evolutionary genetics. Nauka, L
Olenov M/Ju (1972) Ontogenes (Russ) 3:566-573
Oliverio A (1974) Prog neurobiol 3:191-215
O'Malley B, Woo S, Harris S, Rosen J, Means A (1975) J Cell Physiol 85:343-356
Osterman LA (1971) Adv Mod Biol (Russ) 71:353-373
Oudin J (1966) J Cell Physiol 67:77-108
Ouweneel W (1976) Adv Genet 18:179-248
Paigen K (1961) Exp Cell Res 25:286-301
Paigen K, Ganschow R (1955) In: Genetic control of differentiation. Academic Press, Lon-
    don New York, pp 99-114
Paigen K, Swank R, Tomino S, Ganschow R (1975) J Cell Physiol 85:379-392
Paigen K, Meisler M, Felton J, Chapman V (1976) Cell 9:533-539
Painter L, Taylor A (1942) Proc Natl Acad Sci USA 28:311-317
Palmer C, Funderburk S (1965) Cytogenetics 4:261-276
Palmer L, Kjellberg B (1967) Experientia 23:800-801
Palmiter R, Wrenn J (1971) J Cell Biol 50:598-615
Palotta D, Berlowitz L, Rodriguez L (1970) Exp Cell Res 60:474-477
Papaconstantinou J (1967) Science 156:338-346
Papermaster B (1967) Cold Spring Harbor Symp Quant Biol 32:447-460
Pardue M (1969) J Cell Biol 43:101a
Park W (1957) J Anat 91:369-373
Park W, Jansing R, Stein J, Stein G (1977) Biochemistry 16:3713-3721
Parker G (1962) Am Nat 66:147-158
Pastan I (1972) Sci Am 227:97-105
Paterson B, Bishop J (1977) Cell 12:751-765

Patten B (1959) Human embryology. M. Medgis, Moscow
Paul G (1972) Nature (London) 238:444-446
Paul G, Gilmour R (1968) J Mol Biol 34:305-316
Paul J, Fottrell P (1961) Biochem J 78:418-424
Pavan C, Perondini A (1967) Exp Cell Res 48:202-205
Pavlov IP (1937) Lectures about the activity of the large cerebral hemispheres. Biomedgis, Moscow
Payne P, Gordon M, Dobrzanska M, Parker M, Barlow P (1977) Colloq Int CNRS 261:487-499
Pearson M, Fowler W, Wright S (1963) Proc Natl Acad Sci USA 50:24-31
Pederson T (1974) J Mol Biol 83:163-183
Peeples E (1966) Diss Abstr 27:w10
Peeples E, Ireland P (1974) Biochem Genet 12:367-373
Peeples E, Geisler A, Whitcraft C, Oliver C (1969) Biochem Genet 3:563-570
Pellegrini M, Manning J, Davidson N (1977) Cell 10:213-224
Pelling C (1966) Proc R Soc London Ser B 164:279-291
Pemberton R, Housman D, Lodish H, Baglioni C (1972) Nature New Biol 235:99-102
Penhoet E, Rajkumar T, Rutter W (1966) Proc Natl Acad Sci USA 56:1275-1282
Perez-Davila Y, Baker W (1967) Dev Biol 16:18-35
Perlman S, Phillips C, Bishop J (1976) Cell 8:33-42
Perlman S, Ford P, Rosbach M (1977) Proc Natl Acad Sci USA 74:3835-3839
Pernis B (1966) In: Genetic variation in somatic cells. Academicae, Prague, pp 219-226
Pernis B (1967) Cold Spring Harbor Symp Quant Biol 32:333-341
Pernis B, Chiappino G, Kelus A, Gell Ph (1965) J Exp Med 122:853-865
Pernis B, Forni L, Amante L (1970) J Exp Med 132:1001-1005
Pero J, Nelson J, Fox T (1975) Proc Natl Acad Sci USA 72:1589-1593
Perondini A, Dessen E (1969) Genetics 61:250-260
Perry R, Kelly D (1970) J Cell Physiol 76:127-139
Peterson R (1970) Genetica 41:33-56
Petras M, Biddle F (1967) J Genet Cytol 9:704-710
Pevsner LZ (1963) Ukr Biochem (Russ) 35:3, 448-477
Pevsner LZ (1972) Functional biochemistry of neuroglia. Nauka, Moscow
Pevsner R (1974) Concepts in bioenergetics. Prentice-Hall. Englewood Cliffs, New Jersey
Phillips J, Forrest H, Kulkarni A (1973) Genetics 73:45-56
Pipkin S, Bremner I (1969) Genetics 61:48
Pipkin S, Hewitt N (1972) J Hered 63:267-271
Plagens V, Greenleaf A, Bautz E (1976) Chromosoma 59:157-165
Pokrovsky AA, Korovikov KA (1969) Probl Med Chem (Russ) 15:382-385
Poluektova EV (1970a) Genetica (Russ) 6:68-76
Poluektova EV (1970b) Genetica (Russ) 6:110-117
Popp R (1965) J Hered 56:107-108
Portin P (1975) Genetics 81:121-133
Portin P (1977) Hereditas 87:77-84
Pospelov VA (1973) Cytologia (Russ) 15:1327-1337
Pospelov VA, Pupishev AB (1973) Mol Biol (Russ) 7:67-72
Postlethwait J, Schneiderman H (1971) Dev Biol 24:477-519
Poulson D (1940) J Exp Zool 83:271-236
Poulson D (1968) Proc 12th Int Congr Genet vol I. Ueno Park, Tokyo, p 143
Poznakhirkina N, Serov O, Korochkin L (1975) Biochem Genet 13:65-72
Prader A, Anders G, Habich H (1962) Helv Paediatr Acta 17:271
Prives J, Paterson B (1974) Proc Natl Acad Sci USA 71:3208-3211
Prokofieva-Belgovskaja AA (1946) Dokl Acad Sci USSR (Russ) 54:169-172
Prokofieva-Belgovskaja AA (1960) In: Problems of cytology and general physiology. Akademdat, Moscow, pp 215-253
Prokofieva-Belgovskaja AA (1963) Cytologia (Russ) 5:5-23
Prokofieva-Belgovskaja AA, Gindilis VM (1965) Proc Acad Sci USSR (Russ) 2:188-200

Puckett L, Snyder L (1974) Biochem Genet 11:249-260
Quail P, Scandalios J (1971) Proc Natl Acad Sci USA 68:1402-1406
Quevedo M (1973) J Invest Dermatol 60:407:417
Rabbits T (1978) Nature (London) 275:291-296
Rabbits T, Millstein C (1977) Contemp Top Mol Immunol 6:117-143
Rachkus JuA, Timofeeva MJa, Kuprijanova NS, Kafiani KA (1969a) Mol Biol (Russ)
    3:428-440
Rachkus JuA, Kuprijanova NS, Timofeeva MJa, Kafiani KA (1969b) Mol Biol (Russ)
    3:617-626
Rao M, Blackstone M, Busch H (1977) Biochemistry 16:2756-2762
Rapacz J, Haster J (1968) Proc XIth Eur Conf Blood Groups Protein Polymorphism in Ani-
    mals, Warsaw, pp 101-108
Rapp J (1965) Endocrinology 76:486-490
Rapp J (1969) Am J Physiol 212:1135-1146
Rasch M, Lewis A (1968) J Histochem Cytochem 16:508-511
Ratner VA (1970) Ontogenesis (Russ) 1:166-175
Ratner VA (1972) Principles of organization and mechanisms of the molecular-genetical
    processes. Nauka, Novosibirsk
Ratner VA (1974a) Molecular-genetical system of regulation. Autoreferate doctoral disser-
    tation, Novosibirsk (Russ)
Ratner VA (1974b) Adv Mod Biol (Russ) 77:3-17
Ratner VA (1975) Molecular-genetical system of regulation. Nauka, Novosibirsk
Ratner VA, Tchuraev RN (1971) Genetica (Russ) 7:175-179
Ratner VA, Furman DP, Nikoro ZS (1969) Genetica (Russ) 5:72-82
Rauch N (1969) J Exp Zool 172:363-368
Raushenbach IJ, Korochkin LI (1972) Ontogenesis (Russ) 3:53-62
Raven Ch (1964) Oogenesis. Mir, Moscow
Rawls J, Lucchesi J (1974a) Genet Res 24:59-72
Rawls J, Lucchesi J (1974b) Genet Res 24:73-80
Rayle RE (1967) Genetics 56:583
Rechcigl M, Heston W (1967) Biochem Biophys Commun 27:119-124
Rechsteiner M (1970a) J Insect Physiol 16:957-977
Rechsteiner M (1970b) J Insect Physiol 16:1179-1192
Rechsteiner M, Parsons B (1976) J Cell Physiol 88:167-179
Reeder R, Bell E (1965) Science 150:71-72
Reeves R (1977) Eur J Biochem 75:545-560
Reeves R, Jones A (1976) Nature (London) 260:495
Reich E (1964) Science 143:684-689
Renart J, Sebastian J (1976) Cell Diff 5:97-107
Revel M, Aviv H, Groner Y, Pollack Y (1970) Fed Eur Biochem Soc Lett 9:213-217
Revel M, Groner Y, Pollack Y, Cnaani D, Zeller H, Nudel V (1973) In: Karolinska Symp
    Res Methods Reprod Endocrinol 6th Symp. Stockholm, pp 54-74
Reyer R, Coulombre A, Yamada T, Papaconstantinou J (1966) Science 154:1682-1687
Ribbert D (1972) In: Beerman W (ed) Results and problems in cell differentiation, vol IV.
    Academic Press, London New York, pp 153-179
Richards B, Pardon J (1970) Exp Cell Res 62:184-196
Rickwood D, Threlfall G, Mac Gillavary A, Paul J, Rickes P (1972) Biochem J 129:50-51
Riggs A (1975) Cytogenet Cell Genet 14:9-25
Riley M, Solomon L, Zipkas D (1978) XIVth Int Cong Genet Abstr P I Sect 1-12. Nauka,
    Moscow, p 148
Rimoin D, Schimke R (1971) Genetic disorders of the endocrine glands. C Mosbi Comp,
    St Louis
Ringertz N (1974) In: Davidson R, de la Cruz F (eds) Somatic cell hybridization. Raven
    Press, New York, pp 239-264
Ringertz N, Carlsson S, Ege T, Bolund L (1972) Proc Natl Acad Sci USA 68:3228-3237
Ringertz N, Carlsson S, Savage R (1972) In: Advances in biosciences. Pergamon Press, Lon-
    don New York, pp 1-18

Ris H, Kubai D (1970) Annu Rev Genet 4:263-294
Rischkov VL (1965) Proc Acad Sci USSR (Russ) 4:533-543
Ritossa F (1968a) Proc Natl Acad Sci USA 59:1124-1131
Ritossa F (1968b) Proc Natl Acad Sci USA 60:509-516
Ritossa F (1972) Nature New Biol 240:109-111
Ritossa F, Atwood V, Spiegelman S (1966) Genetics 54:663-676
Ritossa F, Malva C, Bonicinelli E, Graciani F, Polito L (1971) Proc Natl Acad Sci USA
    68:1580-1583
Rittenhouse E (1968) Dev Biol 25:351-365, 366-381
Roberts D (1971) Nature (London) 233:394-397
Roberts E (1962) J Am Med Soc 93:277-279
Roberts P (1972) Genetics 72:607-614
Roberts R, Baker W (1973) Am Nat 107:709-726
Rodman T (1964) J Morphol 115:419-445
Roehrdanz R, Kitchens J, Lucchesi J (1977) Genetics 85:489-496
Rogers M, Shearn A (1977) Cell 12:915-921
Rokitsky PF (1929) J Exp Biol (Russ) 5:182-214
Rollins J, Flickinger R (1972) Science 178:1204-1205
Ronichevskaya GM (1975) Spontaneous and induced somatic mutability in mammals. Au-
    toreferate doctoral dissertation, Novosibirsk (Russ)
Roos P, Martin J, Westman-Naeser S, Hellerström C (1974) Horm Metab Res 6:125-128
Rose R, Hillman R (1969) Biochem Biophys Res Commun 35:197-200
Rose R, Hillman R (1972) Genet Res 21:239-245
Rosenberg M (1970) Proc Natl Acad Sci USA 67:32-36
Rosenberg M (1972) Nature (London) 239:520-522
Rosenzweig M, Krech D, Bennett E (1958) In: Harlow R, Woolsey C (eds) Biological and
    biochemical bases of behavior. Univ Wisconsin Press, Madison, pp 367-400
Ross L, Goldsmith E (1955) Proc Soc Exp Biol Med 90:50-55
Rubinstein N, Pepe F, Holtzer H (1977) Proc Natl Acad Sci USA 74:4524-4527
Ruddle F (1971) Fed Proc 30:921-925
Ruddle F, Rapola J (1970) Exp Cell Res 59:399-412
Ruddle F, Roderick T (1965) Genetics 51:445-454
Ruddle F, Chapman V, Chen T, Klebe R (1970) Nature (London) 227:251-257
Ruderman J, Pardue M (1977) Dev Biol 60:48-68
Rudkin G (1964) In: Bonner J, Tso P (eds) The nucleohistones. Holden Day Inc, San Fran-
    cisco, pp 184-192
Ruffini A, Compere A, Baglioni C (1970) J Immunol 104:1511-1519
Rundles R, Falls H (1964) Am J Med Sci 211:641-658
Russell L (1964) In: The role of chromosomes in development. Academic Press, London
    New York, pp 153-181
Rutter W, Kemp J, Bradshaw W, Clark R, Sanders F (1968) J Cell Physiol 72:1-18
Rutter W, Morris P, Goldberg M, Paule M, Morris R (1973) Biochem Gene Express Higher
    Organisms. Proc Symp Sydney, Sydney, pp 89-104
Ryabov SI, Shostka GD (1973) Molecular-genetical problems of erythropoiesis. Medici-
    na L
Sadovnikova-Koltzova MP (1925) J Exp Biol Med (Russ) 1:40-48
Saedler H, Starlinger P (1967) Mol Gen Genet 100:178-189
Saiduddin S, Bray A, York R, Swerdloff R (1973) Endocrinology 93:1251-1256
Sakai R, Tung D, Scandalios J (1969) Mol Gen Genet 105:24-29
Salganik RI (1968) News Acad Med Sci USSR (Russ) 23:3-10
Sallei J, Zuckerkandl E (1974) Biochimie 56:547-553
Salpeter M (1965) J Morphol 117:201-212
Sandler L, Hiraizumi Y (1961) Can J Genet Cytol 3:34-96
Sanno Y, Holzer M, Schimke R (1970) J Biol Chem 245:5668-5676
Sato K, Sato T, Morris H, Weinhouse S (1975) In: Markert C (ed) Isozymes, vol III. Aca-
    demic Press, London New York, pp 951-968
Saunders J, Cairns J, Gaseling M (1957) J Morphol 101:57-87

Saxen L (1973) Arch Oral Biol 18:1469-1479
Saxen L, Toivonen S (1963) Primary embryonic induction. IL, Moscow
Scalenghe R, Ritossa F (1976) 5th Eur Drosophila Res Conf Convain-le-Neuve, p 109
Scherbakov ES (1968) Genetica (Russ) 4:60-69
Scherrer K (1973) Acta Endocrinol 74:95-129
Scherrer K, Marcaud L (1968) J Cell Physiol 72:181-182
Schimke R (1973) In: Meister A (ed) Advances in enzymology. J Wiley and sons, New York,
    pp 135-187
Schmid C, Manning J, Davidson N (1975) Cell 5:159-172
Schmidtke J, Kuhl P, Engel W (1976) Nature (London) 260:319-320
Schneiderman H, Young W, Childs B (1966) Science 151:361-363
Schochterman C, Perry R (1972) J Mol Biol 63:591-596
Schrader F, Hughes-Schrader S (1931) Q Rev Biol 6:411-438
Schultz G (1974) Exp Cell Res 86:190-192
Schultz G, Manes C, Hahn W (1973) Biochem Genet 9:247-259
Schultz J (1956) Cold Spring Harbor Symp Quant Biol 21:307-318
Schwartz D (1971) Genetics 67:411-425
Schwartz D (1973a) MGC News Lett 47:53-55
Schwartz D (1973b) Genetics 75:639-641
Schwartz M, Sofer W (1976) Genetics 83:125-136
Searle A (1968) Comparative genetics of coat colour in mammals. Academic Press, London
    New York
Sederoff R, Clynes R, Poncz M, Hachtel S (1973) J Cell Biol 57:538-550
Seecoff R, Kaplan W, Futch D (1969) Proc Natl Acad Sci USA 62:528-535
Sell S, Gell P (1965a) J Exp Med 122:423-430
Sell S, Gell P (1965b) J Exp Med 122:923-932
Sellers L, Granner D (1974) J Cell Biol 60:337-345
Sena E (1966) Developmental variation of alkaline phosphatase in *D. melanogaster*. M S
    Thesis. Cornell Univ Ithaca, New York
Serebrovsky AS (1938) Dokl Acad Sci USSR (Russ) 19:77-78
Serebrovsky AS, Dubinin NP (1929) Adv Exp Biol (Russ) 4:235-247
Serfling E, Panitz R, Wobus U (1969) Chromosoma 28:107-119
Serov OL (1968) Genetica (Russ) 4:135-145
Serov OL, Chlebodarova JN (1973) Genetica (Russ) 9:45-48
Serov OL, Korochkin LI (1971) Ontogenes (Russ) 2:471-478
Serov OL, Zakijan SM (1974) In: Vth Meeting of embryologists. Nauka, Moscow, pp 160-
    161
Serov OL, Zakijan SM, Kulichkov VA, Korochkin LI, Vladimirov AV (1976) Genetica
    (Russ) 12:44-60
Shapiro NI (1966) In: Actual problem of modern genetics. MGV (Russ), pp 266-280
Shaw Ch (1965) Science 149:936-943
Shaw Ch (1969) In: Bourne G, Danielli J (eds) Int Rev Cytol, vol 25. Academic Press, Lon-
    don New York, pp 297-332
Shaw Ch, Barto E (1963) Proc Natl Acad Sci USA 50:211-214
Shearn A, Garen A (1974) Proc Natl Acad Sci USA 71:1392-1397
Shellenbarger D, Mohler J (1975) Genetics 81:143-162
Sherbet G (1968) Histones and transport of genetic information. Mir, Moscow, pp 105-119
Shin T, Bonner J (1970) J Mol Biol 50:333-344
Shiokawa K, Yamana K (1975) Dev Biol 47:303-309
Shiokawa K, Nada O, Yamana K (1967) Nature (London) 213:1027-1028
Shire J (1974) Endocrinology 62:173-207
Sholl AD (1956) The organization of the cerebral cortex. Academic Press, London New
    York
Shows T, Ruddle F (1968) Proc Natl Acad Sci USA 61:574-581
Shumskaya IA, Marchenko N, Korochkin LI (1975) Genetica (Russ) 9:74-80
Shumskaya IA, Belyaev AI, Korochkin LI (1980) Genetica (Russ) in press
Silverman W, Shapiro F, Heron WT (1940) J Comp Psychol 30:279-282

Silvers W, Russell E (1955) J Exp Zool 130:199-220
Siracusa G (1972) Exp Cell Res 78:460-462
Skaer R (1977) J Cell Sci 26:251-266
Skoblina MN (1974) Ontogenesis (Russ) 5:334-340
Slegers H, Mettrie R, Kondo M (1977) FEBS Lett 80:390-394
Slizynski B (1964) Cytologia 29:330-336
Sluyser M (1977) Trends Biochem Sci 2:202-204
Smalgausen II (1938) Organism as a unity in individual and historical development. Acad
    Sci USSR Publ, Moscow
Smirnov VG, Vatti KP (1971) Genet Res (Russ) 4:76-110
Smith C (1962) Science 138:889-890
Smith C, Kissano J (1963) Dev Biol 8:151-164
Smith D, Blizzard R, Wilkins L (1957) Pediatrics 19:1011-1012
Smith D, Meltzer V, McNamara M (1974) Biochim Biophys Acta 349:366-375
Smith I (1977) In: Weissbach H, Pestka S (eds) Molecular mechanisms of protein biosynthe-
    sis. Academic Press, London New York, pp 627-700
Smith L (1956) J Exp Zool 132:1-83
Smith P, Lucchesi J (1969) Genetics 61:607-618
Smith P, Koenig P, Lucchesi J (1968) Nature (London) 217:1286-1288
Smithies O (1964) Cold Spring Harbor Symp Quant Biol 29:309-319
Smithies O, Connell G, Dixon G (1968) J Mol Biol 21:213-224
Snatkin A (1976) PAABS Rev 5:425-433
Soga K, Takahashi Y (1975) Nature (London) 256:233-234
Sokolov NN (1959) Interaction of nucleus and cytoplasm in the interspecific hybridization
    of animals. Acad Sci USSR Publ, Moscow
Solovijeva IA, Timofeeva MJ, Sosinskaja IE (1973) Ontogenesis (Russ) 4:549-556
Solter D, Schachner M (1976) Dev Biol 52:98-104
Sonnenschein C, Richardson V, Tashian A (1971) Exp Cell Res 69:336-344
Sorenson J, Scandalios J (1974) MGC News Lett 48:152
Sorenson J, Scandalios J (1976) Plant Physiol 57:351-352
Sorsa V (1978) Abstr XIV Int Cong Genet Part I, Sect 1-12. Nauka, Moscow, p 56
Sorsa V, Green M, Beermann W (1973) Nature New Biol 245:34-37
Sorsby A, Franceschetti A, Joseph R, Davey J (1952) J Ophtalmol 36:547-581
Speiser Ch (1974) Theoret Appl Genet 44:97-99
Spelsberg T, Tankersley S, Hnilica L (1969) Proc Natl Acad Sci USA 62:1218-1255
Spemann H (1938) Embryonic development and induction. Yale University Press, New Ha-
    ven
Sperry R (1960) In: Experimental psychology. IL Moscow, pp 319-404
Spickett S, Shire J, Stewart J (1967) In: Spickett S (ed) Endocrine genetics. Univ Press,
    Campbridge, pp 271-288
Spiegelman M, Bennett D (1974) J Embryol Exp Morphol 32:723-738
Spirin AS, Belitsina NV, Aitchojin MA (1965) In: Cell differentiation and inductional mech-
    anisms. Nauka, Moscow, pp 18-37
Spradling A, Penman S, Pardue M (1975) Cell 4:395-404
Spradling A, Pardue M, Penman S (1977) J Mol Biol 109:559-587
Sprey T (1977) Roux Arch 183:1-15
Stambaugh R, Bucley J (1967) J Biol Chem 242:4053-4059
Starlinger P, Saedler H, Rak B, Tillmen E, Venkow P, Waltschewa L (1973) Mol Gen Genet
    122:279-286
Stedman E, Stedman E (1950) Nature (London) 166:780-781
Stein K, Rudin I (1953) J Hered 44:59-69
Steitz J (1969) Nature (London) 224:957-964
Steitz J, Jake K (1975) Proc Natl Acad Sci USA 72:4734-478
Stent G (1974) Molecular Genetics Mir, Moscow
Stern C (1929) Biol Zentralbl 49:261-290
Stern C (1936) Genetics 21:625-730
Stern C (1954) Caryologia 6:355-369

Stern C (1968) Genetic mosaics and other essays. Harvard Univ Press, Cambridge
Stern C, Sekiguti K (1931) Biol Zentralbl 51:194-199
Stern P, Martin G, Evans M (1975) Cell 6:455-465
Stevens R, Williamson A (1973) Proc Natl Acad Sci USA 70:1127-1131
Stewart A, Lloyd M, Arnstein H (1977) Eur J Biochem 80:453-459
Stewart B, Merriam J (1975) Genetics 79:635-647
Stewart J, Papacostantinou J (1966) Biochim Biophys Acta 121:69-78
Stewart J, Papacostantinou J (1967) Proc Natl Acad Sci USA 58:95-102
Stewart W (1973) New Sci 57:527-529
Stiles C, Lee Kai-Lin, Kenney F (1976) Proc Natl Acad Sci USA 73:2634-2638
Stockard C (1921) Am J Anat 28:115-227
Stockdale F, Holtzer H (1961) Exp Cell Res 24:508-519
Strehler B, Hirsch G, Gusseck D, Johnson R, Bick M (1971) J Theoret Biol 33:429-474
Strom C, Dorfman A (1976) Proc Natl Acad Sci USA 73:3428-3432
Sturtevant A (1925) Genetics 10:117-147
Subtelny S, Wright D (1969) J Cell Biol 43:141a
Sullivan D, Palacios R, Stavnezer J, Taylor J, Faras A, Kiely M, Summers N, Bishop J,
    Schimke R (1973) J Biol Chem 248:7530-7539
Sutherland E (1972) Science 177:401-408
Suzuki Y, Brown D (1972) J Mol Biol 63:409-429
Suzuki Y, Gage L, Brown D (1972) J Mol Biol 70:637-649
Svetlov PG (1960) In: Problems of cytology and general physiology. Nauka, Moscow, pp
    265-285
Sviridov SM, Poliakova EV, Korochkin LI (1970) Ontogenesis (Russ) 1:463-472
Sviridov S, Korochkin L, Poliakova E, Matveeva N (1971) Biochem Genetics 5:379-396
Sviridov S, Korochkin L, Ivanov V, Maletskaya E, Baktina T (1972) J Neurochem 19:713-
    718
Swanson K, Mertz T, Young W (1967) Cytogenetics. Prentice Hall, Englewood Cliffs, New
    Jersey
Syr-Young Lin, Riggs A (1975) Cell 4:107-111
Szeinberg A, Mor A, Vernia H, Rescher S (1966) Life Sci 5:1233-1238
Szybalski W (1972) In: Ledoux L (ed) Uptake of informative molecules by living cells. North
    Holland Publ Comp, Amsterdam, pp 61-79
Szybalski W, Bovre K, Fiandt M, Hayes S, Hradecna Z, Kumar S, Lozeron H, Nifkamp
    II, Stevens W (1972) Cold Spring Harbor Symp Quant Biol 35:341-353
Takagi N (1976) Human Genet 24:207-211
Takagi N, Sasaki M (1975) Nature (London) 256:640-642
Talwar G, Jailkhani B, Narayanan P, Narasimhan C (1973) Acta Endocrinol 6:341-356
Tartof K (1969) Genetics 62:781-790
Tartof K (1971) Science 171:294-297
Tartof K (1973) Genetics 73:57-71
Tartof K (1974a) Proc Natl Acad Sci USA 71:1272-1276
Tartof K (1974b) Cold Spring Harbor Symp Quant Biol 38:491-500
Tartof K, Perry R (1970) J Mol Biol 51:171-183
Tashiro Y, Morimoto T, Matsuura S, Nagata S (1968) J Cell Biol 38:574-588
Tata J (1971) In: Control mechanisms of growth and differentiations. Univ Press, Cam-
    bridge, pp 163-181
Teng C, Teng V, Allfrey V (1970) Biochim Biophys Res Comm 41:690-698
Terada M, Banks J, Marks P (1971) J Mol Biol 62:347-360
Tettenborn U, Dofuku R, Ohno S (1971) Nature New Biol 234:37-39
Thaler M, Villee C (1967) Proc Natl Acad Sci USA 58:2055-2062
Thomas C, Zimm B, Dancis B (1973) J Mol Biol 77:85-100
Thomson J (1969) Currents in Modern Biology 2:333-338
Thompson L, McCarthy B (1968) Biochem Biophys Res Commun 30:166-172
Thompson W, Sontag L (1956) J Comp Physiol Psychol 49:454-456
Tidwell T, Allfrey V, Mirsky A (1968) J Biol Chem 243:707-715
Tiedemann H (1966) In: Current topics in development biology, vol I. Academic Press, Lon-
    don New York, pp 85-112

Tiedemann H (1967) In: Biochemistry of animal development, vol II. Academic Press, London New York, pp 3-55

Tiedemann H (1970) In: Genetic information and individual development. Inst Cytol Genet, Novosibirsk, pp 45-47

Tiedemann H (1976) J Embryol Exp Morphol 35:437-444

Tiepolo L, Diaz G, Landani U (1974) Chromosoma 45:81-89

Tilghman S, Tiemeier D, Leder P, Curtis P, Weissmann C (1978) Proc Natl Acad Sci USA 75:1309-1310

Till E, McCulloch E, Siminovitch L (1964) Proc Natl Acad Sci USA 51:29-39

Timofeeva MJa, Kafiani KA (1964) Biochemia (Russ) 29:110-115

Timofeev-Ressovsky NV, Ivanov VI (1966) In: Actual problems of modern genetics. MGU, Moscow, pp 114-130

Tissieres A, Mitchell H, Tracy V (1974) J Mol Biol 84:389-398

Tjian R, Pero J, Losick R, Fox T (1976) In: Nierlich D, Rutter W, Fox I (eds) Molecular mechanisms in the control of gene expression. Academic Press, London New York, pp 89-99

Tobler J, Bouman J, Simmons J (1971) Biochem Genet 5:111-117

Tocchini-Valentini G, Mandavi V, Brown R, Crippa M (1974) Cold Spring Harbor Symp 38:551-558

Toivonen S, Tarin D, Saxen L (1976) Differentiation 5:49-55

Tolman E (1924) J Comp Psychol 4:1-18

Tomkins G, Martin D (1970) In: Roman H (ed) Annu Rev Genet. Academic Press, London New York, pp 91-106

Tomkins G, Gelehrter T, Granner D, Martin D, Samuels H, Thompson E (1969) Science 166:1474-1480

Tompkins R (1970) Dev Biol 22:59-83

Tonoue T, Eaton J, Frieden E (1969) Biochem Biophys Res Commun 37:81-88

Travers A (1976) Nature (London) 263:641-646

Treiman D, Fulker D, Levine S (1970) Dev Psychobiol 3(2):131-140

Tseng Mi-pai (1960) Acta Biol Exp Sin 7:59-66

Tunnicliff G, Wimer C, Wimer R (1973) Brain Res 61:428-434

Turner S, Laird Ch (1973) Biochem Genet 10:263-274

Twardzik D, Grell E, Jacobson K (1971) J Mol Biol 57:231-245

Tyler B, DeLeo A, Magasanik B (1974) Proc Natl Acad Sci USA 71:225-230

Ugolev AM (1972) Membranic digestion. Nauka, Moscow

Uphouse L, Bonner J (1975) Dev Psychol 3:171-185

Ursprung H (1964) Fed Proc 23:990-993

Ursprung H (1965) Naturwissenschaften 52:375-379

Ursprung H, Huang R (1967) Prog Biophys Mol Biol 17:151-177

Ursprung H, Smith K (1965) In: Genetic control of differentiation. Academic Press, London New York, pp 1-13

Ursprung H, Smith K, Sofer W, Sullivan D (1968) Science 160:1075-1081

Ursprung H, Sofer W, Burrough N (1970) Roux Arch 164:201-208

Vachtin JB (1972) Variability and selection in the tumor cell populations. Autoreferat Doctor Dissertation.

Vachtin JB (1974) Genetics of somatic cells. Nauka, Moscow

Vachtin JB, Borchsenius TV (1969) Cytologia (Russ) 11:1313-1322

Van Blerkom J, Manes C (1974) Dev Biol 40:40-51

Van Blerkom J, Barton S, Johnson M (1976) Nature (London) 259:319-321

Varshavsky A (1976) Adv Mol Biol 81:209-224 (Russ)

Van de Berg J, Cooper D, Sharman G, Poole W (1977) Aust J Biol Sci 30:115-125

Verdonk N (1968) J Embryol Exp Morphol 19:33-42

Verne J, Hebert S (1949) Ann Endocrinol 10:456-460

Vesell E, Philip J, Bearn A (1962) J Exp Med 116:273-306

Vice J, Hunt W, Dray S (1969a) J Immunol 103:629-638

Vice J, Hunt W, Dray S (1969b) Proc Soc Exp Biol Med 130:730-733

Vice J, Hunt W, Dray S (1970) J Immunol 104:38-48

Vlasova IE, Kiknadze II (1975) Cytologia (Russ) 17:518-523

Vogel H, Niewisch H, Matioli G (1969) J Theoret Biol 22:249-271
Vogel H, Hajdik I, Niewisch H, Sultanian I, Matioli G (1970) J Cell Physiol 76:117-126
Volobuev VE, Radjabli SI (1974) Genetica (Russ) 9:77-82
Vorontzova MA, Liosner LD (1955) Physiology of regeneration. Nauka, Moscow
Vyasov OE (1962) Immunology of embryogenesis. Medgis, Moscow
Vyasovaya EA, Malup TK, Sviridov SM (1975) Dokl Acad Sci USSR 225:1194-1197
Waardenburg P, Van den Bosch J (1956) Ann Human Genet 21:101-122
Wachsmuth E, Pfleiderer G, Wieband T (1964) Biochem Z 340:80-90
Waddington C (1964) Morphogenesis and genetics. Mir, Moscow
Waddington C, Perkowska E (1965) Nature (London) 297:1244-1246
Wagner R, Mitchell G (1958) Genetics and metabolism. IL, Moscow
Wagner R, Radman M (1975) Proc Natl Acad Sci USA 72:3622-3629
Wagner R, Selander R (1974) Annu Rev Entomol 19:117-138
Wainwright S (1971a) In: Regulation of erythropoiesis and haemoglobin synthesis. Proc Int
    Symp Erythropoiesis. Praha, pp 304-309
Wainwright S (1971b) Cancer Res 31:694-696
Wainwright S, Wainwright L (1972) Can J Biochem 50:1165-1173
Wake N, Takagi N, Sasaki M (1976) Nature (London) 262:580-581
Wales R (1975) Biol Reprod 12:66-81
Walker A, Walsh M, Pennica D, Cohen P, Ennis M (1976) Proc Natl Acad Sci USA
    73:1126-1130
Wall D, Blackler A (1974) Dev Biol 36:379-390
Wallace R (1963) Biochim Biophys Acta 74:505-515
Wallis B, Fox A (1968) Biochem Genetics 2:141-158
Ward K, Marzuki S, Haslam J (1973) In: Pollak J, Wilson J (eds) The biochemistry of gene
    expression in higher organisms. Reidel Publ Co, Dordrecht Boston, pp 105-116
Ward R (1975) Genet Res 26:81-93
Ward R, Hebert P (1972) Nature New Biol 236:243-244
Warner C, Hearn T (1977) J Reprod Fertil 50:315-317
Warner C, Versteegh L (1974) Nature (London) 248:678-680
Warner ZH (1932) J Genet Psychol 41:57
Watkins W (1966) Science 152:172-176
Weber R (1967) In: Biochemistry of animal development, vol II. Academic Press, London
    New York, pp 227-301
Weideli H (1971) Mol Gen Genet 112:167-196
Weideli H, Kubli E, Chen P (1969) Rev Suisse Zool 76:788-797
Weiler E (1965) Proc Natl Acad Sci USA 54:1765-1772
Weiler E, Melletz E, Breuninger-Peck E (1965) Proc Natl Acad Sci USA 54:1310-1317
Weinman R (1972) Genetics 72:267-276
Weintraub H, Grondine M (1976) Science 193:848-856
Weiss M (1974) In: Davidson R, de la Cruz F (eds) Somatic cell hybridization. Raven Press,
    New York, pp 151-158
Weitlauf H (1974) J Exp Zool 189:197-202
Wellauer P, Reeder R (1975) J Mol Biol 94:151-161
Wells R, Royer H, Hollenberg C (1976) Mol Gen Genet 147:45-51
Welshons W (1965) Science 150:1122-1129
Wessels N, Rutter W (1969) Sci Am 220:36-44
Wessels N, Wilt F (1965) J Mol Biol 13:767-779
West J, Frels W, Chapman V, Papaioannu V (1977) Cell 12:873-882
Whilmart C, Urbain J, Givol D (1977) Proc Natl Acad Sci USA 74:2526-2530
White B, Tener G, Holden J, Suzuki D (1973) J Mol Biol 74:635-651
White R, Hogness D (1977) Cell 10:177-192
Whiteley A, McCarthy P, Whiteley H (1966) Proc Natl Acad Sci USA 55:519-525
Whitmore D, Whitmore E, Gilbert L (1972) Proc Natl Acad Sci USA 69:1592-1595
Whitmore D, Gilbert L, Ittycheriah P (1974) Mol Cell Endocrinol 1:37-54
Whitt G (1969) Science 166:1156-1158
Whitt G (1970) J Exp Zool 175:1-36

Whitt G, Cho P, Childers W (1972) J Exp Zool 179:271-282
Whitt G, Childers W, Cho P (1973a) J Hered 64:54-61
Whitt G, Miller E, Shaklee J (1973b) In: Schroder J (ed) Genetics and mutagenesis of fish. Springer, Berlin Heidelberg New York, pp 243-276
Wilde Ch (1961) Colloq Int CNRS 101:183-198
Wiley L, Calarco P (1975) Dev Biol 47:407-418
Wilkinson J (1965) Isozymes. E and F Spon Ltd, London
Will B (1977) Behav Biol 19:143-171
Williams G, Fritz P (1973) Fed Proc 32:621
Williams R (1960) Biochemical individuality. IL, Moscow
Williamson J, Procunier J, Church R (1973) Nature New Biol 143:190-191
Willis M, Baseman J, Amos H (1974) Cell 3:179-184
Wilson E (1940) Cell, vol II. Acad Sci USSR, Moscow
Wilt F (1967) In: Krause G, Sander K (eds) Advances in morphogenesis. Academic Press, London New York, pp 89-119
Wimer C, Roderick T, Wimer R (1969) Psychol Rep 29:363-368
Wingaarden J (1970) Biochem Genet 4:105-125
Witschi E (1967) In: The biochemistry of animal development, vol II. Academic Press, London New York, pp 193-225
Wobus U, Serfling E, Panitz R (1971) Exp Cell Res 65:240-245
Wobus U, Popp S, Serfling E, Panitz R (1972) Mol Gen Genet 116:309-322
Wolf U, Engel W (1972) Humangenetik 15:99-118
Woodland A, Graham C (1969) Nature (London) 221:327-332
Woodland H, Gurdon J, Lingrel J (1974) Dev Biol 39:134-140
Wosnick H, White B (1977) Nucl Acid Res 4:3919-3930
Wright D, Moyer F (1966) J Exp Zool 163:215-229
Wright D, Moyer F (1968) J Exp Zool 167:197-206
Wright D, Shaw Ch (1969) Biochem Genet 3:343-354
Wright D, Shaw Ch (1970) Biochem Genet 4:385-394
Wright D, Subtelny F (1971) Dev Biol 24:119-140
Wright D, Subtelny F (1972) J Cell Biol 43:160a-161a
Wright T (1970) In: Caspari E (ed) Advances in genetics, vol 15. Academic Press, London New York, pp 261-395
Wu J, Hurn J, Bonner J (1972) J Mol Biol 64:211-219
Yablonka Z, Yaffe D (1977) Differentiation 8:133-143
Yagie G, Feldman M (1969) Exp Cell Res 54:29-36
Yamana K, Shiokawa K (1966) Exp Cell Res 44:283-293
Yamana K, Shiokawa K (1975) Dev Biol 47:461-463
Yamamoto M, Strychardz W, Nomura M (1976) Cell 8:129-144
Yang S, Comb D (1968) J Mol Biol 31:139-142
Yazaki I (1960) Annot Zool Jpn 33:217-225
Young R (1977) Biochem Biophys Res Commun 76:32-39
Young W (1966) J Hered 57:58-60
Zacharov AF (1968) Adv Mod Biol (Russ) 65:83-106
Zavadovsky MM (1922) Sex and development of its characters. Gosisdat, Moscow
Zemaitis M, Hill R, Greene F (1974) Biochem Genet 12:295-308
Zhimulev IF (1974) Cytological aspects of transcriptional activity of polytenic chromosomes in D. melanogaster. Autoreferate candidate dissertation. Novosibirsk (Russ)
Zhimulev IF, Litshev VA (1970) Ontogenesis (Russ) 1:318-324
Zhimulev IF, Litshev VA (1972) Ontogenesis (Russ) 3:199-201
Zhimulev IF, Belyaeva ES (1975a) Genetica (Russ) 9:175-181
Zhimulev IF, Belyaeva ES (1975b) Chromosoma (Berl) 49:219-231
Zibina EV (1964) Cytologia (Russ) 6:583-586
Zibina EV, Tichomirova MM (1965) Cytologia (Russ) 7:585-601
Ziegler R, Emmerich H (1973) Drosophila Inf Serv 50:18-182
Zinkham W, Blanco A, Kupchyk L (1964) Science 144:1353-1354
Zuckerkandl E (1964) J Mol Biol 8:128-147

Zuckerkandl E (1976) J Mol Evol 9:73-104
Zuckerkandl E, Pauling L (1962) In: Kasha M, Pullman B (eds) Horizonts in biochemistry. Academic Press, London New York, pp 189-195
Zwaan J (1968) J Cell Physiol 72:47-72
Zwilling E (1968) In: Emergence order development system. Academic Press, London New York, pp 184-207

Subject Index

Monographs on Theoretical and Applied Genetics

Editors: R. Frankel (coordinating editor)
G. A. E. Gall, M. Grossmann, H. F. Linskens,
D. de Zeeuw

Volume 1:
J. Sybenga

Meiotic Configurations

A Source of Information for Estimating Genetic
Parameters

1975. 65 figures, 64 tables. X, 251 pages.
ISBN 3-540-07347-7

„The present book treats meiotic chromosome configurations from a quantitative genetic point of view.
Its pupose is not primarily to increase understanding
of chromosome behaviour, but rather to construct
generally applicable systems for estimating genetic
parameters related to recombination.
There exist few previous books covering this
important field of cytogenetics. Dr. Sybenga's book
is therefore especially welcome...
Several tables and diagrams aid in the understanding of meiotic principles. Fortunately, the book
contains simple figures usually explaining in a
perfect manner the various phenomena...
in conclusion the book is highly recommended for
anyone interested in cytogenetics."

Norwegian Journal of Botany

Volume 2:
R. Frankel, E. Galun

Pollination Mechanisms, Reproductions and Plant Breeding

1977. 77 figures, 39 tables. XI, 281 pages.
ISBN 3-540-07934-3

"This book, which is based on an advanced course
in plant bredding... sets out to give a comprehensive
account of the botanical, genetic and breeding
aspects of the reproductive biology of sperma-
tophytes, mainly with respect to angiosperms..."

The line drawings are of a high quality, the photo-
graphs clear and the schematic representations of
various phenomena described in the text and of
breeding procedures well laid out. This ably written reference book will be of use to botanists,
geneticists and plant breeders alike, bringing
together as it does so many aspects of pollination and
reproduction in plants"

Plant Breeding Abstracts

Volume 3:
D. de Nettancourt

Incompatibility in Angiosperms

1977. 45 figures, 18 tables. XIII, 230 pages.
ISBN 3-540-08112-7

"Rarely, in a new subject, is a book to be found that
is both well balanced and up to date. *Incompatibility
in Angiosperms* is just that. From its beginnings in
'classical' genetics the study of incompatibility
mechanisms in plant breeding systems has emerged
as a subject of considerable scientific and commer-
cial interest, and rapid progress has been made from
genetical, physiological and structural points of
view. This book, written by an acknowledged expert
in the field, manages to do justice to all these aspects
of the work, and to preface them with first class intro-
ductory material... The illustrations that accompany the text are in general of exceptional clarity..."

Phytochemistry

Springer-Verlag
Berlin
Heidelberg
New York

Advanced Series in Agricultural Sciences

Co-ordinating Editor: B. Yaron

Editors: D. F. R. Bommer, G. W. Thomas,
B. R. Sabey, Y. Vaadia, L. D. van Vleck

Distribution rights for India: Allied Publishers
Private Ltd., New Delhi

A Selection

Volume 2
H. Wheeler

Plant Pathogenesis

1975. 19 figures, 5 tables. X, 106 pages
ISBN 3-540-07358-2

"...This concise and very clearly written book
presents an excellent account of the process of
plant pathogenesis. ...The book is very stimulating
in that it presents not merely facts (in a very systematic way), but also the author's interpretation
thereof. The author is to be congratulated on his
use of examples to illustrate the phenomena described and to make his points clear. ...The excellent
integration of biochemical, ultrastructural and
genetic information is an important aspect of this
book."
Netherlands Journal of Plant Pathology

Volume 3
R. A. Robinson

Plant Pathosystems

1976. 15 figures, 2 tables. X, 184 pages
ISBN 3-540-07712-X

"...because *Plant Pathosystems* is already a very good
book and unique in many ways. The text is based
on long and varied experience of field plant pathology in Africa and essentially is a series of percep-
tive analyses of interactions between populations of
host plants and populations of pathogens, ...The
book is well and clearly written but must be taken
slowly because each paragraph on each page contains one or more substantial points which must be
understood if the main themes which are pro-
pounded are to have their full impact. But the effort
called for is very well rewarded because few books
published on plant pathology during the past three
decades are so stimulating, challenging, and have
been written with such worthy objectives. *Plant
Pathosystems* will undoubedly cause quite s stir
among plant pathologists and plant breeders, for
most of whom it must be regarded as compulsory
reading."
Nature

Volume 6
J. E. Vanderplank

Genetic and Molecular Basis of Plant Pathogenesis

1978. 3 figures, 36 tables. XI, 167 pages
ISBN 3-540-08788-5

"This book is an extended essay in which a
distinguished plant pathologist puts forward a new
molecular hypothesis for disease resistance in
plants...
The book is well written throughout. Each
chapter is divided into a series of connected sections
which is just as well since it makes 'concentrated'
reading. The effort is well worth while and I would
like to recommend it to all plant biochemists
working on aspects of plant resistance to disease."
Phytochemistry

"...Vanderplank takes a fresh look at a great many
facets of the host-pathogen relationship and analyses them with deep understanding, clear logic and
precise words. He sets us all thinking and has made a
notable cintribution to the development of ideas on
pathogenicity."
Exp. Agriculture

Springer-Verlag
Berlin
Heidelberg
New York